Schooling in Modern European Society

A Report of the Academia Europaea

Academia Europaea

The *Academia Europaea* was created in 1988 as an independent organization of leading scholars in Europe. It now has over a thousand members from across the whole of Europe, East and West, and covers the whole range of scholarly activities including humanities, social sciences, life and physical sciences, medicine and engineering. It organizes conferences particularly of an interdisciplinary nature and has recently discussed issues such as the responsibility of scientists, space research, human origin and future, problems of old age, the future of museums, and the future of academic institutions.

This book is the end product of the work in the Academia's first study group. Other groups are investigating the problems of youth and old age and conduct research on the human genome and risk assessment. The work of a study group looking into the problems and prospects of university education in Europe has just started. It will be a follow-on of this study of school education.

Schooling in Modern European Society

A Report of the Academia Europaea

Edited by

T. Husén, A. Tuijnman and W. D. Halls

PERGAMON PRESS
OXFORD · NEW YORK · SEOUL · TOKYO

U.K.	Pergamon Press plc, Headington Hill Hall, Oxford OX3 0BW, England
U.S.A.	Pergamon Press, Inc., 395 Saw Mill River Road, Elmsford, New York 10523, U.S.A.
KOREA	Pergamon Press Korea, KPO Box 315, Seoul 110-603, Korea
JAPAN	Pergamon Press Japan, Tsunashima Building Annex, 3-20-12 Yushima, Bunkyo-ku, Tokyo 113, Japan

Copyright © 1992 Pergamon Press plc

All Rights Reserved. No part of this publication may be reproduced, stored in a retrieval system or transmitted in any form or by any means: electronic, electrostatic, magnetic tape, mechanical, photocopying, recording or otherwise, without permission in writing from the publishers.

First edition 1992

Library of Congress Cataloging-in-Publication Data
Schooling in modern European society: a report of the Academia Europaea/edited by Torsten Húsen, Albert Tuijnman and W. D. Halls. 1st ed.
p. cm.
Study commissioned by the Academia Europaea.
Includes bibliographical references and index.
1. Education – Europe - Aims and objectives. 2. Education – Social aspects - Europe.
I. Húsen, Torsten, 1916- . II. Tuijnman, Albert. III. Halls, W. D. IV. Academia Europaea.
LA622.S34 1991 370'.94 – dc20   91-41534

British Library Cataloguing in Publication Data

A catalogue record for this book is available from the British Library

ISBN 0-08-041393-5

Printed and bound in Great Britain by BPCC Wheatons Ltd, Exeter

Schooling in Modern European Society

The following team accepted the invitation of the *Academia Europaea* to carry out the work presented here.

TORSTEN HUSÉN (Chairman)
Stockholm University, Sweden

ALBERT TUIJNMAN (Scientific Secretary)
University of Twente, The Netherlands

BILL HALLS (Member)
Formerly of the University of Oxford, United Kingdom

MICHEL DEBEAUVAIS (Member)
Université de Paris VIII, France

HARTMUT VON HENTIG (Member)
University of Bielefeld, Germany

GEORGE MARX (Member)
Roland Eötvös University, Hungary

Foreword

Scenarios for the future development of Europe abound, but essential to all of them is a steadily increasing cooperation and closer understanding between countries and regions. Europe can be regarded as a great arena within which much of the culture and technology of the modern world evolved. Ideas developed in individual regions over a long period; some of these were so persuasive that they spread and became pan-European, but many others remained local and essentially private to a single region. Part of the current trend in Europe involves a more vigorous comparison of cultural patterns, partly motivated by internal dissatisfaction and the thought that other peoples' experiences may be enlightening.

We live in a period of rapid change when many of the time honoured ways of doing things are questioned, but we are not sure what should replace them. Nowhere is this questioning more pressing and more relevant than in education, which constitutes the preparation of young people for the 'New Europe' and for the world. Everyone accepts that education should be provided for all, that it needs to be extended, and indeed that education for life is a phrase of real significance. How is it best provided, have we gone too far down the path of thinking that "for all" means an elimination of all differentiation?

Can we afford the improvements of education we seek? Are there economies available which permit improvements in quality without an overall increase in budget? The newer view of Europe presupposes greater mobility and, since there is no European *lingua franca*, the acquisition of other languages has become a pressing requirement. Some educational systems have dealt with this very successfully, others less so. Educationalists recognise that the home and school play complementary rôles, yet the family has recently undergone a remarkable change. Formerly most of our population were rural and the family large and multi-generational, now the absence of siblings and of one parent is a commonplace. The balance of school and home is inevitably distorted.

Yet another problem is the changed status of teachers. Once they constituted the main educated group in society together with ministers of religion. Mass education has changed that and the teachers have become just one group in a competitive labour market.

This book explores these and many other questions about school education in depth and in a mainly European dimension. It makes no assumptions about a single educational system for Europe and indeed the purpose of the study is investigative and not prescriptive. Its aim is to provide the information from which educators and policymakers can draw their own conclusions about the merit of various approaches to problems.

This study was commissioned by the *Academia Europaea* in the conviction that education is the foundation of all our activities and that it needs to adapt and develop in order to prepare our children for the great changes in society already in train, and even more to make them adaptable to the greater changes to be expected in the next century, whose nature we can barely guess at.

The *Academia Europaea* is most grateful to the members of the study group, Bill Halls, Michel Debeauvais, George Marx and Hartmut von Hentig, but especially to Torsten Husén, its chairman and guiding spirit and to Albert Tuijnman who has not only contributed chapters but also has so ably nourished the group with information and taken overall responsibility for the preparation of the report. Among others who have made a contribution our thanks go to John Lewis who wrote a major part of the essay on challenges for science education.

We are also most grateful to the Tercentenary Fund of the Bank of Sweden that has taken so sympathetic an interest in the study and has funded it most generously.

June 1991

Sir Arnold Burgen
President
The Academia Europaea

Chairman's Preface

The *Academia Europaea* Council, in February 1990, decided to sponsor a study group on education in modern society. The study was to be conducted with a European perspective. At that time the political and cultural map underwent dramatic changes.

The *Academia* Vice-President, David Magnusson, had prior to the Council decision approached me with the request to take responsibility for setting up a study group with the task of analysing problems facing basic education in modern Europe. As a starting point I prepared a memorandum for the Academy with a list of what I regarded to be pervasive problems. The project was envisaged as a one-year exercise with a study group, consisting of five to six members, meeting at least twice during the work period in order to discuss the problems we wanted to concentrate on. The study group had to be assisted by a secretary with expert competence. A proposal was submitted to the Bank of Sweden Tercentenary Fund, which generously decided to finance the project.

It may well be asked, incidentally, whether there is any point in setting up a study group and to charge it with the task of identifying and spelling out the ramifications of the problems basic schooling in Europe is faced with today and in the foreseeable future? Europe is highly diversified, economically, socially and culturally. This diversification is reflected in a broad array of national systems of education shaped by different traditions and different languages. All this being recognized, I would submit that there are three important reasons for analysing the problems in basic education in Europe.

First, and most obvious, in a Europe in the process of being integrated politically and with regard to labour and trade markets where temporary or permanent movements across borders become increasingly frequent, where mass media bring to the attention of the individuals what is happening in the other countries, there are ample reasons to ask ourselves what the implications of this are for our schools. In our report we have dealt with four problem areas which easily come to mind: school quality, examinations, the teaching of foreign languages and science education.

Secondly, less obvious but nevertheless of utmost importance is the challenge to analyse the rôle of the school as an institution in an increasingly complex technological and meritocratic society. This is not a specific European problem; many of our nations share it with the United States of America, for example.

We want to draw attention to the proper rôle of the school at a time when the family structure has drastically changed, particularly in the advanced industrial societies in Western Europe. What has been referred to as the ecology of education, the interplay between various educative agents, has changed dramatically over the few decades since the end of the Second World War. An increasing number of young people are staying in school an increasing number of years. Social status and life career in modern society are more and more dependent upon the amount and quality of formal education a person has been able to acquire. Post-initial or "lifelong" education is increasingly becoming a prerequisite for keeping one's position in the workforce or for upward social mobility.

The third, and by no means the least important reason for setting up a study group, has been to spell out the rôle of basic education in establishing a European identity and consciousness. Elusive as it may seem, its concrete importance in the decades to come cannot be denied or neglected. As the Europeans are searching for a new identity in response to and in anticipation of change in the political economy, geography and cultures of the region, it is only natural that careful thought is devoted to the question of what European school education is like.

Our work began in earnest in April 1990. The first session of the study group was held in Stockholm, at the Royal Swedish Academy of Sciences. We then decided what issues we should focus on, planned the collection of information and agreed on the "homework" each member of the group was expected to complete before the next meeting, which took place in late September 1990 in Budapest at the Hungarian Academy of Sciences. At that session we agreed on the outline of the report for which already first drafts of certain chapters were available. The third session again took place at the Royal Swedish Academy of Sciences in March 1991, when the report was available in penultimate draft. The *Academia* then sent it out to a number of expert referees for review. We have tried to consider their comments in preparing the final manuscript.

A study of the kind we are now reporting, given the scope of our tasks and the time-frame, must necessarily have certain limitations.

We are keenly aware of the fact that in a region as heterogeneous as Europe the priority among issues is open for discussion, since they obviously vary from country to country. We are also aware that, in spite of our ambition to confine ourselves to the identification of problems and their ramification, we have sometimes, at least by implication, not entirely succeeded in avoiding recommendations.

We have regarded it as important to try to confine ourselves to identify what we regard as problems and issues in basic education in Europe today, problems which are particularly prominent in urbanized and highly industrialized countries. We also wanted to spell out the ramifications of these problems and issues. It is our conviction that school education has to be viewed in a much broader perspective than that of the single schools or classrooms. In order to provide a background for a fruitful discussion on educational problems among policymakers, practitioners and the general public, this broader perspective has to be kept open, something we have tried to do in our report.

In elucidating the problems we have analysed we have drawn upon demographic and educational statistics, compiled by the Council of Europe, UNESCO and OECD, such as information on educational expenditures, enrolment and demand and supply of teachers.[1] We have also made use of data collected for studies conducted by the International Association for the Evaluation of Educational Achievement (IEA). We are grateful to persons in the secretariats of these bodies for the help they have given us. The same applies, of course, to the secretariat of the *Academia Europaea* in London. I have had the privilege of discussing our project a couple of times with *Academia* President, Sir Arnold Burgen.

Given the time-frame the preparation of the report has been a heavy task for the authors. I would in particular like to acknowledge the competent work of our project coordinator, Dr. Albert Tuijnman, who, in spite of other commitments, has been able to gather the information we needed, and also to conduct a major portion of the writing, revising and editing of the report.

We are particularly grateful to our colleague, Dr. Bill Halls, who not only prepared the drafts of three chapters but also, being the only member of the group whose mother-tongue is English, carefully scrutinized the drafts the other group members had produced.

I would like to thank the Royal Swedish Academy of Sciences for providing the venue for two of our meetings. The *Academia* Vice-President, David Magnusson, and Professor Kjell Härnqvist (both *Academia* members), the latter in his capacity of Chairman of the Governing Board of the Bank of Sweden Tercentenary Fund, have shown a great interest in the project, for which we are grateful.

Last, but not least, thanks go to Sandra Schele at the Institute for Educational Research, University of Twente, Enschede, the Netherlands, for the typing and the lay-out of the manuscript.

Again, we hope that the present report could serve as a background for a debate on basic education in a Europe which, in so many respects, is in the process of change.

Torsten Husén
November 1991 Stockholm

---

[1] Most of the statistics were collected before the unification of the two Germanies. It must be noted that the data presented in this report refer to the situation in former *West Germany*, unless indicated otherwise.

Contents

Foreword	vii
Chairman's Preface	ix
Table of Contents	xiii
List of Tables	xviii
List of Figures	xxi
Abbreviations	xxii

CHAPTER 1. The Rise of the New Europe: The Repercussions on School Systems

Introduction	1
The European Identity	2
Demographic and Economic Factors Shaping School Policy	5
The Question of Pan-European Cooperation	7
The Single European Act: Some Implications Relevant to Education	10
National School Policy in a European Context	13
Questions of Diversity and Harmonization in School Policy	16
Knock-on Effects of Economic Integration	22
Concluding Remarks	27

CHAPTER 2. Trends and Basic Concerns in School Education: A Retrospective View

Introduction	31
Major Themes of Educational Policy from 1960 to 1985	32
Perceptions of the Role of the School in the 1960s	33
"World Crisis in Education"	37
School Education in Question	39
A Summary of Criticisms of the School	41
Constraints on Secondary Education	43
The Idea of the Comprehensive School	45

Main Trends in Secondary School Reforms 47
Equality of Opportunity and the Meritocratic
Tendencies 50
Social Demand and Problems of Access 54
Gifted Pupils and Specialist Schools 58
Access to Continuing Education 61
Professionalism and Vocationalism 62
Decentralization and Parental Choice 63
School Education in "Nations at Risk" 65
The Quality of European Schooling 67
Concluding Observations 74

CHAPTER 3. Emerging Issues in School Education

Introduction 77
The Role of the School 78
Goal Conflicts in Basic Education 81
The Governance and Management of School Education 85
Grading and Assessment of Students 86
National Assessment Programmes 88
Criticisms and Dilemmas of National Assessment in
Education 92
The Quality and Coherence of Schooling 95
School as a Part of Lifelong Education 100
Conclusion 104

CHAPTER 4. Home and School Relationships

Introduction 107
Statistics on Changing Household Structures 108
The Changing Role and Structure of the Family 111
The Changing Role of the School 119
Discrepancies and Discontinuities Between Home and
School 121
Consequences for School Education 122
The School and "Social Capital" 124
The "Old" and The "New" Educational Underclass 126
The School Intervention Idea and Intervention
Programmes 130
Some Crucial Problems 132
A Concluding Note 134

CHAPTER 5. New Challenges for Science Education in Europe

Introduction	137
The Beginning of Science Education	138
The Beginning of the Change	139
International Cooperation in the Development of Science Education	143
Primary Science	144
Science for All and a Balance of the Sciences	146
Science and Society	148
The Changing Role of Mathematics Education	152
Information Culture and Information Technology	152
Design and Technology	154
How to Teach Science	155
Enrolment in Science Education	157
Conclusion	160

CHAPTER 6. The Teaching of Languages in Europe

Introduction	163
Language and Literature as Expressions of Culture	163
Developments in the Teaching of Foreign Languages	168
The "Small" Languages of Europe	172
Approaches and Methods in Language Teaching	175
Language Teaching and the Syllabuses	177
Language Policy and Practice in the New Europe	180
Conclusion	183

CHAPTER 7. The Education of Linguistic, Cultural, and Migrant Minorities in Europe

Introduction	185
National Minorities in Europe	186
Immigration: The Overall Picture	189
Educational Problems and General Policies	192
Policies in Action: Germany and France	198
Immigrants and the Majority School System	200
The Teaching of Languages in Intercultural Society	201
The Teaching of Culture	206
Religions in European Society and Schools	210
Culture Gaps and School Policy	211
Immigrants and School Performance	213

Schooling and Employment ... 216
A Note on Teacher Education ... 217
Concluding Remarks ... 218
Annex: Recommendations of the Dutch Advisory Council for Education with respect to the education of minorities. ... 219

CHAPTER 8. School Leaving Examinations and the Role of the School in Preparing for Higher Education

Introduction ... 221
Goals of Upper Secondary Schooling ... 222
The Uses of School Leaving Examinations ... 224
National Differences in Admission Procedures to Higher Education ... 228
Examinations in Europe: The Case of Sweden ... 231
Insufficiencies of the School Leaving Examination ... 234
Leaving Examinations and the Syllabuses ... 236
Pedagogical Problems and Examining Methods ... 238
Coping with the Rising Demand for Higher Education ... 240
The Upsurge in Technical and Vocational Education ... 242
Links between Secondary School and Higher Education ... 243
Efforts towards Cross-national Convergence ... 244
Concluding Observations ... 247

CHAPTER 9. The Recruitment and Training of Teachers

Introduction ... 249
Shortages and Surpluses: Introduction ... 251
Demographic Factors Influencing Recruitment ... 252
Problems of Supply and Demand ... 253
The Impact of Vocational Education on Recruitment ... 254
Competition from Other Occupations ... 255
The Present Position: Oversupply? ... 256
Enlarging the Pool of Recruitment ... 258
The Limitations of Educational Planning ... 260
Quality Recruitment and Teachers' Salaries ... 261
Factors Influencing the Decision to Become a Teacher ... 267
Elements of a European Model for Teacher Training ... 275

Contents

The Trainers of Teachers	278
In-service Training of Teachers	279
Teacher Training Programmes	282
Specialization	285
The Educational Career of the Would-be Recruit	286
Summary and Concluding Remarks	286

CHAPTER 10. The Economic and Financial Aspects of Education: Prospects for the 1990s.

Introduction	289
Formal Education and the Economy	290
The Supply and Demand of Basic Education	293
Financial Issues	296
Educational Expenditures: What are the Limits?	296
The Costs of European School Education	303
The Rising Costs of Schooling	304
The Financing of European School Education	308
Expenditures, Costs and Financing: Challenges and Prospects	310
Conclusion	314

CHAPTER 11. Challenges and Opportunities for School Education

The Task	315
The European School Setting Today	316
Present and Future Challenges in School Education	319
Education in a North-South Perspective	327
Some Concluding Observations	329

Statistical Supplement	333
References	361
Index	385

List of Tables

Table 2.1	Enrolment Rates in Pre-primary Education in European Countries	57
Table 2.2	Science Achievement at the Lower Secondary Level in Different Systems	71
Table 3.1	Stages in the National Assessment of Student Performance in France	89
Table 3.2	Estimates and Projections of Illiteracy in European Countries	96
Table 4.1	Marriage Rate, Divorce Rate, Fertility Rate, Modal Age of Mother at Birth of First Child, and Life Expectancy in European Countries	109
Table 4.2	Percentage of Net Absolute Increase in the Number of Single-parent Families	111
Table 4.3	Female Labour Force Participation Rates, 1967-1988, European Countries	112
Table 4.4	Number of Places in Regulated Child Care for Pre-school Children	113
Table 4.5	Indicators of Household Conditions, Sweden, 1900-1985	114
Table 4.6	Effects of Various Measures of Social Capital Provided by the Parent-child Relation on Estimated Percentage of Students Dropping Out	125
Table 4.7	Poverty Incidence in EC Countries in 1980 and 1985	128
Table 5.1	Ratios of Male to Female Students in Science Education and Indices Indicating Sex Differences in Science Achievement at the Terminal Grade of Secondary School	147
Table 5.2	Teaching of Environmental Education Themes by Subject in German Secondary Schools	150
Table 5.3	Introduction of Computers and Provision of Computer Education in Secondary Schools	154
Table 5.4	Participation in Science Courses at the Terminal Grade Level of Secondary Education	158
Table 6.1	Percentages of Countries Teaching Major International Languages as the First Foreign Language in the Secondary School	170
Table 6.2	Linguistic Minorities in EC Countries	174
Table 6.3	Early Teaching of Modern Language	178
Table 7.1	Refugees and Displaced Persons in Western European Countries	190
Table 7.2	Percentage of Foreigners in European Countries in 1989	192

List of Tables

Table 7.3	School Enrolments of Foreigners in Selected European Countries	194
Table 7.4	Percentage Increase or Decrease in Number of Nationals and Foreigners in School Education	195
Table 7.5	Foreign School Enrolment in Selected Countries	195
Table 7.6	Foreigners Working in Germany	199
Table 7.7	Immigrants Studying their Mother Tongue in the Land of Hesse	204
Table 7.8	Turkish Children in German Schools	215
Table 8.1	Proportion of an Age Group Obtaining a Second-level Qualification Allowing Access to Higher Education, and Proportion Entering Higher Education	226
Table 8.2	Enrolment Rates for Ages 16-24, Full-time and Part-time Students	234
Table 8.3	Students in Vocational Education at the Secondary and Post-secondary Level	243
Table 8.4	Higher Education in the EC Countries	245
Table 9.1	Population Estimates, 5 to 19 Year-olds	252
Table 9.2	Population Size and Change, 15 to 19 Year-olds	253
Table 9.3	Enrolments at Secondary Level by Type of Education	255
Table 9.4	Comparative Income Position by Teacher Category	263
Table 9.5	Student Achievement in Mathematics in Relation to Instruction Time and Teachers' Salaries	264
Table 9.6	Years of Teaching Service Required to Reach Maximum Seniority	265
Table 9.7	Indexed Comparison of Teachers' Salaries in the EC Countries	266
Table 9.8	Females in Teaching by Level of Education	270
Table 9.9	Number of Teacher Contact Hours per Year	272
Table 9.10	Provision of Staff Development for Teachers Using Computers in Upper Secondary Schools, and Agencies Giving Support with Teacher Training	281
Table 10.1	Public Educational Expenditure in the Netherlands, 1900-1988	299
Table 10.2	Indices of Public Educational Expenditure in the Netherlands by Type of Provision	301
Table 10.3	Child Care in the Nordic Countries	307
Table 10.4	Expenditure on Adult Education and Job Training in EFTA Countries	313
Table 10.5	Sources of Finance for Job Training in EC Countries	314
Table 11.1	Out-of-school Youth	328

Supplementary Tables

Table S.1	Demographic and Economic Indicators of European Countries	334
Table S.2	Eastern Europe, Basic Economic Indicators	335
Table S.3	Government Spending on Education, European Countries	336
Table S.4	Length of Compulsory and Secondary Education in Europe	337
Table S.5	Selected Education Indicators by European Country	338
Table S.6	Number of Children Less than 12 Years of Age in European Countries	339
Table S.7	Development of the Cohort Aged 16 to 18 Years, 1968-2000	340
Table S.8	Total Unemployment Rate, Youth Unemployment Rate, and Female Labour Force Participation Rate in European Countries	341
Table S.9	Total and Direct Effects of Four Predictor Variables on Life Career Outcomes from Age 25 to 56	342
Table S.10	Mean Scores of Student Populations in European Countries and Japan on the IEA Scale "Like School"	343
Table S.11	Attitudes Towards Cooperation with the European Community	344
Table S.12	Country Scores, International Physics Olympiad, 1989	345
Table S.13	Percentage of Mathematics and Physics Lessons Compared to all School Subjects in Some European Countries	346
Table S.14	"L'Enseignement des langues vivantes"	347
Table S.15	Science Achievement of 10 and 14 Year-olds	352
Table S.16	Linguistic Relations with Other European Countries	353
Table S.17	Foreign Students in Eleven EC Countries by Country of Origin, 1986	354
Table S.18	Trends and Projections of School Enrolment in Europe	356

List of Figures

Figure 1.1 Opinions on Decision-making on a National Level or by the European Community — 15

Figure 1.2 Participation Rates in Full-time Education for 16 Year-olds in 10 EC Member States — 20

Figure 2.1 Major Themes and Principles of Educational Policy in the United Kingdom — 33

Figure 2.2 Mathematics Achievement at the Terminal Level of Upper Secondary Education in Different School Systems — 69

Figure 2.3 Percentage of Primary Schools in Various Countries with Lower Average Scores in Science than the Lowest Scoring School in Japan — 70

Figure 2.4 Mean Scale Scores on the Science Achievement Scale, 14 Year-olds — 73

Supplementary Figures

Figure S.1 Physics Achievement of 10 Year-olds on 40 Test Items in Eight Countries — 357

Figure S.2 Physics Achievement of 14 Year-olds on the Core Test in Ten Countries — 358

Figure S.3 Physics Achievement of 18 Year-olds on Reduced Number of Test Items (as taken by United States' students) in Nine Countries — 359

Abbreviations

ASEAN	Association of South East Asian Nations
CEDEFOP	European Centre for the Development of Vocational Training
CERI	Centre for Educational Research and Innovation
CMEA	Council for Mutual Economic Assistance
COMETT	Community in Education and Training for Technology
CPSU	Communist Party of the Soviet Union
EC	European Community
ECTS	European Community Course Credit Transfer System
EFTA	European Free Trade Association
EMS	European Monetary System
EMU	Economic and Monetary Union
EPC	European Political Cooperation
ERASMUS	European Community Action Scheme for the Mobility of University Students
ESN	Educationally Subnormal School
EUREKA	European Framework for Research Cooperation
EUROSTAT	Statistical Office of the European Communities
FORCE	Action Programme for the Development of Continuing Vocational Training in Europe
GATT	General Agreement on Tariffs and Trade
GDP	Gross Domestic Product
GNP	Gross National Product
IEA	International Association for the Evaluation of Educational Achievement
ISIP	International School Improvement Project
LINGUA	Community Programme for the Promotion of Teaching and Learning of Foreign Languages
NAEP	National Assessment of Educational Progress
NARIC	European Community Network of National Academic Recognition Information Centres
NATO	North Atlantic Treaty Organization
OECD	Organization for Economic Cooperation and Development
PETRA	Community Action Programme on Youth Training
PSSC	Physical Science Study Committee
SEA	Single European Act
SEM	Single European Market
TEMPUS	Trans-European Mobility Programme for University Students
UNESCO	United Nations Educational and Scientific Organization
USSR	Union of Soviet Socialist Republics
YES	Youth for Europe Scheme

Chapter 1

The Rise of the New Europe: The Repercussions on School Systems

1. Introduction

Spectacular changes have recently taken place in European political geography. Popular mass movements and uprisings have supported demands for grassroot influence and human rights vis-à-vis the strong nation-state and its apparatus of government. Free communication and mobility across national borders have been facilitated in Central and Eastern Europe. In the West, the process of trade and labour market integration can be expected to be reinforced, at least formally, within the next few years. The Iron Curtain has been removed both as a symbol and a reality. Europe is no longer militarily divided into two competing blocs.

In this unprecedented situation, the Europeans are groping for a new identity which, paradoxically, may well be less Eurocentric than the former one, dominated as it was by military blocs and the carefully guarded provincialism of nation-states. The challenges the Europeans are now facing derive from changes both within and without the region. The growing competition on the world trade market, for example, is a clear signal that European supremacy can no longer be taken as self-evident. European economic and cultural cooperation has to be established in order to compete with two other important blocs, the American and the East Asian.

In Europe the traditional nation-state has come to play a less important role. The flow of global forces and regional influences can just as well be handled by local bodies as by the central state. Industrial processes and their ecological and social consequences give rise to coordination and management issues above the national level. Central government often finds itself in a situation where it has to adapt itself to the process of internationalization rather than to govern independently. The state was an important unit at the early stage of industrialization. A national market developed around national industries. National labour and employer organizations developed as well as a national welfare policy guided by the state. We can now note how national power centres have to yield to international ones. The single state has to adapt itself to international power centres.

What is true of politics and economics applies equally well to school education. Indeed, the question of what an appropriate and high quality schooling should be like involves, by definition, the making of a political choice about the shaping of future society. Moreover, it is also recognized that school education constitutes an economic asset, a factor of importance in determining the economic and cultural wellbeing of nations and regions. As the Europeans are searching for a new identity in response to and in anticipation of even more drastic changes in the political economy and geography of the region, it is only natural, and indeed inevitable, that careful thought is given to the question of what school education in and for the new Europe ought to be like.

There has been a tendency to apply a narrow perspective and take a piecemeal approach to problems besetting school education in today's highly industrialized society. In the same way as national economies are no longer the unique framework for economic decisions to be made, and governments now have to balance domestic policy in a broad international setting, an over-emphasized nationalist and ethnocentric perspective on the goals and content of basic education can no longer be maintained. Indeed, the time has finally arrived when decision-makers, teachers, parents, and others concerned with the scope and quality of the education our schools provide, are required by the circumstances to consider school policy in its European, and even global, context. The approach taken should not merely be one of pursuing an intellectual discussion devoid of practical meaning, however interesting such an exercise could be to those taking part.

2. The European Identity

The map of Europe is being redrawn in the last decade of the 20th century. Borders, not only custom boundaries but political walls as well, are disappearing. Beyond the signature of agreements about abolishing cumbersome visas and customs, a huge task is emerging for education, i.e. for helping the young to take a broad view of the larger unit in which they will live and work, take responsibility and, hopefully, maintain peace.

Opening up for the 'larger unit Europe' may be spelled out as overcoming a narrow view of one's own country, and understanding the opportunities which the current situation provides. In relating to this broad view of Europe, schools could show the contribution of each nation to a common cultural heritage.

Some teaching of European history and culture has been part of basic school education in all countries; the geographical, economic and political outline of Europe is in the minds of practically all of its inhabitants, put there and reinforced if not by the school then by the daily news in the mass media; an ever growing number of people, among them mainly the young, are travelling through Europe; the desire to do so is strong with those whose possibilities to travel have, so far, been restricted.[1]

Yet all this is neither a sufficient nor a safe introduction into the Europe that may emerge during the last decade of this millennium. The views one has of the other nations are shaped by former experience, biased by one's own nationalistic outlook, limited by one's own interests and determined by one's own economic and social conditions. In many countries the subjects taught in school are still dominated by the ideas of the period in which the nations ascended to statehood and independence and thus beset with antagonism to other nations. Their military victories, their art and literature, their scientific discoveries and technological inventions are being accentuated. But they will miss their chance in the wider European context if they cannot define their new role: as givers and takers, as communicators of their problems and their solutions, as cooperators in common endeavours and mediators in conflicts, as interpreters or ambassadors of the common culture and philosophy to other Continents.

Nationalism is not dead. After the collapse of the totalitarian regimes with their imposed ideologies and with the subsequent erosion of the centres which held together vast power blocs, the idea of the nation has gained a new vigour. The end of this century will bring neither the end of the nation nor even the end of the nation-state. But their roles will be different. The individual nation may well come to serve within a wider scope set by a common market, a common currency, a common system of laws, and common cultural institutions. The coming about of Europe is not so much the result of the 'disappearance' of the nations and nationalism, but of the growing awareness of what brings the nations together − a result of understanding what the word "common" means in the new European context.

---

[1] For example, the very day when they were no longer required to have a visa, over 100,000 Polish tourists crossed the border to Germany.

In a sense 'common ground' already exists in the *cultures* containing Europe. For example, the Dutch child is brought up on Pinocchio, as the Italian is on Andersen's fairy-tales. A German Gymnasium student may read Homer and Plato as well as Schiller, and a British schoolgirl may study Racine, Chaucer and Shakespeare. The concert halls in Europe resound to Tchaikovsky and Bartók; Galileo and Faraday, Marx and Freud provide 'universal' keys to understanding nature, society, mankind – and they are considered to be part of our 'common' European heritage.[2]

Of course, the Americans and the Australians, the New Zealanders and the Chinese or Indians or Africans claim to partake in this culture, and many contributions have come from them to what now might be called modern world civilization. Yet if one takes the ideas which the names mentioned above stand for and puts them together, the result will be a rich impression of a cohesive whole rightly identified with Europe, with the Continent of their origin: the history of ideas rooted in Antiquity, Christianity, Humanism, and the Enlightenment.

European cultures rely on change, crisis, controversy, even revolution for their continuous selfrenewal, and though not all Europeans believe in 'progress' they might still agree that Socratic examination of the *condition humaine* and the resulting consciousness can help in improving their lives. Since the cultural heritage is, at least in part, transmitted by the school this clearly poses a challenge to national curricula. What do schools today do to foster the perception of a peculiar European identity? When and where do students realize both the importance and the relativity of national differences? How are they introduced to the cultures of this world and thereby to the difficult role to which Europe is committed on account of its tradition? By what experience or knowledge do they become aware of the responsibility which the European ideas quoted above impose on them?

Elaborated in such broad terms, the agenda for the schools appears to be monumental. It is moreover evident that the idea of "Eurocentricity" that was prevalent in Western Europe in the 1960s is no longer educationally relevant. European ideas, after all, extend far beyond the boundaries of one Continent. The educational systems of the old Continent may well have to be outward-looking, reflecting the view of a *Europa über die Grenzen hinaus*. This indeed chimes well with the sentiments of young people. Travelling around Europe during the summer, they already take their Continent – now including the onetime "Soviet Bloc" – for granted. This is one factor that educationists cannot neglect.

---

[2] Of course, this list of "great Europeans" is incomplete. One could, for instance, also mention Mozart, Cervantes and Comenius, Goethe and Corneille, Beethoven and Chopin, Hugo and Tolstoy, Voltaire, and many others.

The recent removal of the Iron Curtain that descended from the Baltic to the Mediterranean over forty years ago, dividing most of the world into power blocs, and in particular, separating Western Europe from Central and Eastern Europe, cannot fail to have repercussions on education. Education is concerned with the preservation, transmission and enlargement of culture, with the role of schools being largely concerned with passing on the heritage of a civilization, preparing the young for life and for earning a living, – there is a distinction to be drawn between the two tasks – and, if possible, for adding to the cultural patrimony. That there is a peculiarly pan-European culture to be passed on is indisputable. But it has now also, with its changing population, to assimilate elements from widely differing religious, political and economic ideas.

Where and how do the young Europeans learn to live with such vast differences in ideas and perspectives – not only to tolerate them but to understand, to welcome and to protect those that deviate from the norms defined by a majority? What justifications other than charity or abstract reason can be given to young people for this demand? Free movement of goods and persons between European countries will exact an unprecedented effort of concertation and mutual understanding; governments and elected representatives may not be able to persuade their own national public opinion to accept the compromises that will be implicit in the process of economic and political harmonization and the construction of the 'New Europe', unless there exists a core of common convictions.

3. Demographic and Economic Factors Shaping School Policy

As hinted at above, the withering of the old Europe and the rise of the new poses numerous challenges for educational policy and practice. It is also evident that repercussions on school education of developments occurring in rapid succession in European countries, and particularly those put in motion in the 12 member states of the European Community (EC) as a consequence of the Single European Act [3] and the decision to

---

[3] The momentum of the process of European consultation in the mid-1980s can be inferred from the speed with which important agreements were negotiated. The main event in this connection was the Draft Treaty Establishing the European Union, which was approved by the European Parliament in February 1984 (for a discussion see Taylor, 1989). This initiative was followed up in the summer of the following year when the European Heads of Government endorsed the Commission's so-called "White Paper" on completing the internal market (Commission of the EC, 1985). In turn, this led to the Single European Act, which came into force on July 1, 1987. Decisions on the Economic and Monetary System (EMS) and the Political Union will be taken in 1991 (Eyskens, 1991).

create, by December 31, 1992, a Single European Market, [4] are difficult to identify, substantiate and analyse.

Besides the reasons for a common Europe to be found in its diverse and rich cultural heritage there are reasons relating to economic conditions. As Europe moves towards some form of economic integration, the education dispensed will not only have to reflect its common cultural heritage, but also the requirements of the labour market.

There is a realization at present that certain competencies are in short supply. It is argued – more forcibly than 25 years ago when the electronic revolution had only just begun – that the demand for high-order skills has increased in Europe as a result of technological innovation in key areas, such as telecommunications, systems engineering and biotechnology, advanced materials and product design. The information revolution has transformed many occupations and is rendering much previous education and training obsolete. Skills shortages are growing also as a consequence of demographic developments. The number of young persons in Europe is declining. This is accompanied by a considerable ageing of the labour force. Because school leavers are the main source of new abilities and skills brought into the labour market, a critical requirement is to improve the quality of school education.

As mentioned in Section 1, an enlarged Common Market has to face the trade competition from across the Atlantic and from the Far East. Schools are affected by the stiffening of international trade competition to no small extent. Indeed, in many European countries the current concern with school effectiveness derives from what is perceived as a challenge presented by competing industrial countries, not least on the Pacific Rim. [5] There may well be a causal relationship between the strong commitment to school education and the economic strength of these countries. One could mention that 95 per cent of young Japanese are in full-time education up to the age of 18; in Europe the figure is often less than 50 per cent. [6] In 1988 the EC had a foreign trade deficit

---

[4] The deadline of December 31, 1992, has been incorporated as an amendment into the Treaty of Rome, although not in the form of a legally binding agreement. Thus, even though 1992 is commonly used in texts concerned with aspects of European integration, the Single European Market is not scheduled to become a reality until January 1, 1993.

[5] Asian countries on the Pacific Rim are generally regarded as the main competitors of European industry. Some of these 'Asian Tigers' have been highly successful. The balance of gains and losses in export shares among 10 EC countries, the United States and Japan over the period from 1979 to 1985 shows that both the European countries and the United States have lost significant market shares to the latter (Source: OECD, *Main Economic Indicators*, Paris, 1989, p. 24; see also T. Inoguchi: "Japan and Europe: Wary Partners", *European Affairs*, Vol. 5, pp. 54-58).

[6] Source of data: *COMETT Bulletin*, No. 10, February 1991.

of some $948 million, as compared with the Japanese surplus of $6,459 million. [7]

Comparative studies conducted by IEA (Chapter 2) show that Japan, Hong Kong and Korea also have a high yield in science education. If the high economic performance of the Asian 'Dragons' may be explained, at least in part, as an outcome of an effective school system, [8] then underinvestment in skill formation and low school performance in particular, may be factors of importance in an explanation of the unfavourable economic performance of *some* Western European countries compared to Japan. The upshot is that, in industrial manufacturing generally, the countries of the Pacific Rim as a whole may eventually outstrip Europe in competitiveness.[9] This is a material challenge that education in Europe may have to meet.

The question of what Europe is at present, and what it may well become, is central not only to decisions about monetary policy, industrial and labour market policy. It has, of course, practical implications for the challenges and problems that will confront our schools, not least in terms of what they should teach. As the Europe of the future is taking shape at present, it is imperative that the question of the European dimension of school policy be put on the agenda now.

The ramifications of some of the overriding issues that will have to be taken into account in a discussion focused on the likely consequences of the new European context for school education are outlined below. The aim is to highlight some major cultural, ethical and economic issues which those responsible for education will have to consider, and to draw out some implications of current developments for educational policy. However, before the focus is narrowed to education, some major aspects of change in political economy will have to be discussed, since these are at the heart of school policy in future Europe.

4. The Question of Pan-European Cooperation

The first cautious steps on a course of political action that eventually would create the current, unprecedented situation of non-confrontation and normalization in the relations between, firstly, the Soviet Union and the United States, and secondly, the different countries associated with

---

[7] Source: OECD, *op. cit.*
[8] This case is made by Lawrence P. Grayson: *Japan's Intellectual Challenge*, 1983.
[9] The data presented above (cf. [5]) have been used to argue that European industry is declining and that the economy is becoming service- rather than production-oriented. The position of European industry and the implications for education are discussed in a report to the European Round Table of Industrialists, edited by Kari Kairamo: *Education for Life: A European Strategy,* Brussels, 1989.

the European Community (EC), the COMECON and the, now ineffective, Council for Mutual Economic Assistance (CMEA), were taken in Helsinki in 1975, when countries recognized the importance of protecting human rights and the need for extending their trade relations and other forms of economic cooperation.

Rapprochement became a key concept in the gradual process of normalizing external relations between socialist regimes and pluralist democratic systems in Europe.[10] The political will to deal with the challenge of extending European cooperation was reflected in the adoption, in Luxemburg in June 1988, of the Joint Declaration on the establishment of official relations between the EC and the CMEA. [11]

That Europe had really entered into a new phase of *rapprochement* was confirmed in the summer of 1989, when Bush and Gorbachev met at the Malta "Summit". The significance of this meeting became clear after the fall of the Berlin Wall on November 9, 1989. Its removal was the clearest sign of the dawn of a new era for all-European cooperation. [12] The sheer speed with which the one-party political system supporting the German Democratic Republic ceased to exist surprised many observers. [13]

The present political situation in the countries of Central and Eastern Europe is very different compared with a few years ago, when changes began with the rise of the Solidarity Movement and the imposition of martial law in Poland and the partial economic liberalization of the Hungarian economy. The present position is one of precarious instability – and of hope. There are conflicting views regarding the pace of reform. Many citizens of, for instance, Czechoslovakia, Poland and Romania consider processes of change to be proceeding too slowly and are failing to improve the quality of life. Others, especially people formerly protected by being government employees and those displaced by the restructuring of agriculture and industry, argue that rapid radical change results in chaos and hence that *perestroika* should be carried through

---

[10] The turning point in the histories of Eastern European countries may well be said to have begun at the April Plenary of the CPSU Central Committee in April 1985, when the essential ideas and objectives of a course for *perestroika* were outlined (Gorbachev, 1987; Bykov, 1989).

[11] The Joint Declaration on the establishment of official relations between the European Economic Community and the Council for Mutual Economic Assistance. (Official Journal of the EC, Luxemburg 157/35, June 25, 1988). Reprinted in Maresceau (1989), pp. 319-320.

[12] The removal of the Berlin Wall came shortly after Hungary had opened up its border to Austria, thus making it possible for East Germans to travel freely to the West.

[13] There were scholars who anticipated revolutionary change in the relations between the two Germanys years before it actually happened. One example is Tsakaloyannis (1987, pp. 152-153).

more gradually – if at all. This can be seen as a sign of the difficulties encountered in the change to a market system.

However, there is an atmosphere of expectation, if not optimism, with respect to the possibilities resulting from a future re-entry of Central and Eastern European nations into a pan-European market economy system. [14] But these expectations, which are accompanied by an increasing awareness of the common problems besetting European countries in an industrial age, such as unemployment and the preservation of the natural environment, also give rise to apprehensions: Will economic, technological, and cultural developments weaken or even do away with national identity, historical variety, the special aggregate of political experience which has accrued in the two halves of Europe during the past fifty years? What will this Europe be other than one expressed in economic terms? What can work for a common consciousness between Bulgarian and Dutch, Portuguese and Finnish, German and Irish people? What will distinguish them from people in Asia and the Americas?

Change in Central and Eastern European countries raises questions concerning the architecture of Europe as a whole. Today, as the centre of gravity of Europe shifts to the East, the bargaining position of the countries cooperating within the framework of the European Free Trade Association (EFTA) vis-à-vis the Community has been weakened, since new agreements for strengthening cooperation in Europe will have to take account of the former socialist countries. [15]

Taken together, the developments referred to above undoubtedly give the impression of a Europe in flux. The overall impression that the situation conveys to the detached observer seems to be one of optimism about the new possibilities for Europe that will open up during this last decade before the year 2000. A cautious note may be struck here, however, as not all the changes at present occurring are necessarily for the better in the long-term perspective. Many latent, potentially dangerous, conflicts accompany the present state of economic instability

---

[14] A poll conducted in Poland in 1990 (reported in *European Affairs*, Vol. 5, February/March, pp. 65-68, 1991) showed 55 per cent of the electorate "in favour of joining the European Community now", and another 17 per cent "in favour of joining in five years' time". Only two per cent said they were either somewhat or strongly opposed. Similar results were also obtained in Hungary, whereas in Czechoslovakia the attitude to the EC was, in the main, also positive.

[15] That the former socialist countries and EFTA will be placed on exactly equal footing in trade with the EC countries seems unlikely in the short-term, however, as the value of EC-EFTA trade exceeds that of the Community trade with, first, the United States and, secondly, Japan. On June 14, 1991, former Prime Minister Ingvar Carlsson announced before the *Riksdag* that Sweden, which was an important foreign investor in the EC countries in 1989 and 1990, would submit its application for EC membership on July 1, 1991.

and political upheaval. Severe problems of regional economic imbalance, poverty and mass unemployment, and even armed struggle, continue to beset certain areas in Europe.

Compared with the position a few years ago, when the Cecchini study and the T. Schioppa Report [16] added to already high expectations about the benefits of the Common Market, the governments of most EC countries have recently become somewhat less optimistic about their short-term economic prospects. Doubts as to whether economies of scale are going to make a fundamental difference are also voiced. Some fear that a large hierarchical organization, such as the Commission might become, will weaken rather than increase flexibility. As a result of a slow-down in economic growth, inflation, and a levelling off in the growth rate of new employment opportunities in the EC countries, "Europhoria" [Cf. A. Smith, 1980] is turning into "Eurealism".

The impact of economic hardship is increasingly being felt in the former socialist countries. Difficulties are encountered also in the Soviet Union, where deep-rooted political controversies over ethnic problems and growing nationalist tensions add to the ills of a stagnating economy. A sure sign of the difficulties faced in the Soviet Union is that, contrary to most other countries, male life expectancy has actually decreased. [17]

Thus all European countries at present find themselves in a state of change. As the outcome of economic and political developments is uncertain, it may be too early to draw out clear implications for educational policy of changes in the institutional fabric of European nations, particularly because educational systems interact with political, social and economic conditions. Difficulties arise because a discussion of central problems besetting the functioning of schools in Europe necessarily involves an anticipation of the likely effects of changes in the politics and economics of Europe on the provision of school education. Hence in a discussion of the problems and opportunities the integration process brings to school policy, some speculation cannot be avoided.

5. The Single European Act: Some Implications Relevant to Education

The possible repercussions of the international integration process on school systems in Europe provide part of the rationale for this study. Integration may be deemed to refer to the adoption and implementation

---

[16] This optimism is reflected in the T. Padoa-Schioppa *et al.* Report (Commission of the EC, 1987) and the P. Cecchini *et al.* Report (1988), which summarizes the findings of studies on the benefits of the internal market.

[17] Cf. Zbigniew Brzezinski, "To Strasburg or Sarajevo?", *European Affairs,* Vol. 5 (1), p. 22.

of collective goals by the countries concerned, such as the revitalization of the economy, the growth in employment opportunity, and the development of education, in addition to the goals of each individual country. [18] In this perspective, economic and political integration processes in Europe also have policy consequences in education.

Political *rapprochement* is an essential precondition for establishing successful cooperation in the economic sphere. However, if the scope is widened to encompass the aim of international economic integration, then political cooperation needs also to be strengthened. [19] This understanding has been formalized in the Single European Act, with the governments of member states committing themselves to attempt "jointly to formulate and implement a European foreign policy". [20] This commitment can be regarded as an expression of the wish on the part of some political groups in member states to achieve a foreign policy based on common grounds. [21]

Yet the strengthening of the Commission's mandate to decide on important legal issues and aspects of foreign policy, which is envisaged in the Single European Act, may eventually reduce the decision-making role of national governments and narrow the scope of national sovereignty, [22] because some degree of political integration runs parallel to economic integration. [23] As has become increasingly clear over recent years, European political cooperation has made advances as a result of the internal dynamic of developing practical cooperation but has also, to some extent, been forced upon certain EC member states by world events. [24] Even though EC member governments can still delay or block

---

[18] This explanation is presented and elaborated by Axelrod (1984) and Guerrieri and Padoan (1989).

[19] Schwartze (1987) and Roney (1990) discuss the "balance of power" and the, sometimes conflicting, interests of the Commission, the EPC, the European Parliament and the individual member states, with regard to the formulation and implementation of a European foreign policy.

[20] This quotation of a passage in the Single European Act is taken from Roney (1990, p. 137).

[21] The Gulf War illustrated the difficulty of achieving practical cooperation among the EC countries. Subsidized agricultural overproduction provides another example.

[22] The French President, François Mitterand, acknowledged this possibility when he employed the term *finalité fédérale* in his address to the European Summit meeting in Dublin in June 1990. Mitterand seemed to agree that an extension of the EPC, following the establishment of the European Monetary System, should be accepted as the logical next step in what Cecchini *et al. op. cit.* refer to as "Project 1992".

[23] Political and economic integration in Europe are regarded by some politicians as a *sine qua non* for reducing the influence of the superpowers on the European Continent. It can also be seen, at least in part, as an answer to what in the past has been called "the German question" (*die deutsche Frage*.)

[24] The Schengen Accord of 1985 provides another example of the difficulty of establishing a common approach to European foreign policy. A description of the Accord and a discussion of implementation problems can be found in studies by the Commission

decisions, they are nevertheless becoming increasingly constrained in their freedom of action and decision-making power by commitments and policy decisions previously introduced. [25]

The creation of the Single European Market guaranteeing free movement of goods, workers, services and capital requires that national barriers to the circulation of all the factors of production, including technology and labour, be abolished. [26] The internal market, because of its size and also because of the possibility it offers for reducing production costs in industry, creating new employment opportunities and enhancing scientific, technical and commercial cooperation, may provide the economic context for a competitive European industry in both goods and services. Some even go so far as to claim that the common market will give a permanent boost to the prosperity of the Europeans – and indeed of the world. [27]

Implicit in the idea that institutions of collective government should be strengthened is acceptance of the need to establish policy instruments to strike a balance between the rights of citizens, regions and territories. This has led the EC Commission to recommend priority to the "Social Charter" with the overarching goal of developing a "People's Europe". [28] The implication may well be that some of the already existing "action programmes", in which education is a central element, will be enlarged in the 1990s, even though gradually. [29]

The objective of achieving a Single European Market comprising the EC countries, and even of establishing a Free Economic Space in the whole of Europe, has set in motion an internal dynamic of reform in different policy domains in the various countries. Change in one area evidently calls for adjustment in others. The effects can already be observed clearly in areas traditionally controlled by governments, such as

---

of the EC (1984), Bulmer and Wessels (1987) and Lodge (1989). Problems with respect to immigration and drugs policy proved difficult to resolve. This led the American weekly *Time* to state, in 1991, that "Schengenland is not so much a country as a state of mind".

[25] One could mention that the EC has passed directives that in effect limit the decision-making power of the German *Länder*, something the *Bundestag* could not have done under current German law.

[26] In contrast to the notion of European Union, the Single European Act, because of its binding character, may well have important consequences for educational policy. The European Union is an expression without concise meaning and significance (Nicol and Salmon, 1990, p. 42).

[27] Lord Cockfield, Preface, in P. Cecchini, *op. cit.*, 1988.

[28] The Commission is seeking to bring about the aims set out in the European Social Charter. The improvement of living and working conditions, the right of freedom of movement and the right to vocational training are among the concerns included in the Social Charter.

[29] The following EC-sponsored programmes have a bearing on educational policy: DELTA, COMETT, YES, ERASMUS, LINGUA, FORCE, TEMPUS, and EUREKA.

monetary and agricultural policy, industrial policy, and legal affairs. [30] The precise characteristics of change may differ from one policy area to another, but one of the overriding goals is usually the same: harmonization.

How burning an issue is harmonization policy in the minds of the Europeans – for example with respect to the syllabuses of our schools? Many people seem so far to be fairly uninterested in the various aspects of integration. Recent polls show that, even in comparatively "supportive countries" such as Germany and the Netherlands, positive attitudes to EC integration are found among less than two-thirds of the electorate. [31] It may be noted here that, in 1990, as many as 30 per cent of the Europeans in the 12 member states had "never read or heard anything about the plan to create a single trading market for all EC member countries by 1992". [32] This public attitude is likely to change radically when matters such as the harmonization of value-added tax are brought to the agenda. This, of course, holds true also for educational policy, to which we now turn.

6. National School Policy in a European Context

Despite the common heritage set out in a previous section, Europe is far from a homogeneous cultural entity. The above discussion on economic and political integration processes may lead one to assume that Europe has already come a long way towards establishing one large educational unit. This, however, is far from the case. Indeed, there are many that question whether absolute unity is even worth achieving. They argue that cultural richness thrives on diversity, whilst acknowledging that some degree of educational integration is not only desirable but essential. It is on this premise that various European or partly-European intergovernmental organizations have concerned themselves with education.

Organizations such as the Council of Europe have tended to interpret school education as a means of promoting international understanding and solidarity. In the case of the EC, schools may well have the objective

---

[30] One of the consequences of the Single European Act is that national legal systems in the EC countries tend to develop towards a common European standard. An interesting discussion of fundamental legal issues arising from the adoption of the Single European Act can be found in Rasmussen (1986) and Wyatt and Dashwood (1987).

[31] Elections to the European Parliament typically attract only about half of the electorate. But in some countries this figure does not deviate sharply from elections at the communal and regional level and, in some cases, even national elections.

[32] In the same year 41 per cent of the U.S. electorate had "never read or heard anything about the European Community", whilst 67 per cent had "never read or heard anything about the plan to create a single trading market for all EC member countries by 1992" (Source: EUROBAROMETER, 1991, Vol. 2.)

and challenge of fostering intra-European understanding and of instilling in the new generation a tendency to see Europe as a unified whole. Schools provide a natural setting for realizing objectives, such as the teaching of human rights and environmental education, and the extension and improvement of foreign language learning. Education can thus contribute to the creation of a new ethic of responsibility for the quality of life in Europe.

Many decision-makers at the various levels of European governments are committed to the objective of seeking ways to maximize the relative advantages of a culturally rich and ethnically diversified region. It is clear that they tend to view education as an important instrument not only in realizing the possibilities for development offered in an integrating Europe, but also in dealing with the various problems that may arise in the change process.

Few doubt that nation-states will have to surrender a part of their sovereignty as a consequence of "Europeanization" and the envisaged strengthening of joint decision-making within the Community. A decade from now, European citizens may well find themselves living in a Community that has taken over many of the tasks and duties now carried out by national governments.

The question is, however, whether this will also be the case with respect to school policy. A representative sample of EC citizens was asked, in 1990, which areas of policy should be decided by (national) government, and which should be decided jointly within the EC. Figure 1.1 shows that a majority of European citizens consider that decisions on matters relating to scientific and technological research; the protection of the environment; cooperation with the Third World countries and foreign policy towards countries outside the EC fall mainly within EC jurisdiction. However, educational policy, as well as health and social welfare, are seen by a majority of Europeans as matters for which the national governments are responsible. This attitude may explain why school policy has, until now, not been a priority issue on the EC agenda. This does not negate the possibility that such a discussion could take place in future, once the common market has been fully implemented and initial problems of government have been cleared up, because the school curriculum determines what the new generation will be taught: European cultures, history and geography, economics and politics. Some, therefore, will maintain that 'European Society' may eventually require a 'European Basic Curriculum', for exactly the same reasons as were applied previously in nation building and the consolidation of sovereign states.

Education, by its very nature, tends to be provincial and ethnocentric. Each region or country has an educational system closely tied to its culture and traditions. Here one should note that safeguarding the cultural identity of a country has always been one of the principal goals legitimizing government policy. Primary education played an important and powerful role in the process of building the nation-state. Industrialization was a parallel development.

However, it has become apparent that the successful structural adjustment of the economic sectors of a post-industrial European society depends on coordination beyond the national level. The nation-state is playing a less powerful role both on the national and international scene as a result. The central governments of European countries find themselves in a situation where they are coerced by the circumstances of globalization to yield power to international decision-making centres. One likely effect of market integration in Western European countries is therefore that national power centres will need to consider domestic educational policy issues in a broad international context. A second effect is that a policy of promoting the European dimension in school education may become necessary. Hence the upshot may be not only that parochialism in educational policy-making may be increasingly challenged but also that the harmonization of school policy will be promoted. For much the same reasons as were used when universal primary schooling was once introduced as a means of nation-state building, a future European government, whether federal, confederal or unitary, is likely to emphasize the importance of providing a common basic education of high quality for all children in the region.

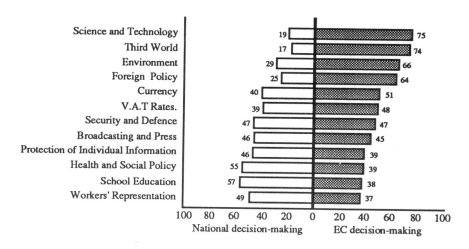

Figure 1.1 Opinions on Decision-making on a National Level or by the European Community.[33]

---

[33] Source: EUROBAROMETER, op. cit.

Yet, given the diversity of Europe's cultural heritage and its present political structure, one needs to be realistic about the limits of educational harmonization that can be achieved. It can be expected that major controversies in educational policy are likely to come to the fore, because objectives of national government policy, such as ensuring cultural diversity and safeguarding national interests in the area of education, may conflict with general economic and political objectives sought by countries drawn into the process of integration in Europe. There is a potential conflict between the interest of the independent nation-state in using education for its own purposes and the trend towards the increasing internationalization of learning.

Internationalization is promoted not only by developments in the EC but also by the opening up of educational institutions in Eastern European countries and the Soviet Union. New information technology makes internationalization not only a possibility, but also inevitable. The dialectic between two conflicting "laws of motion" creates pressures producing a partial convergence in educational policy. [34] Flows of new knowledge, scholars and students across systems and international convergence in the content of school curricula are reflections of the trend towards increased communication in education. There is a realization today that the quality and worth of school education may be enhanced and acquire "added value" if provisions were extended beyond the framework of national systems. [35]

7. Questions of Diversity and Harmonization in School Policy

It follows from the above that areas of public policy formerly deemed to be the exclusive prerogatives of national and local bodies, such as education, are now being brought into the international political and economic arena, not least by the EC Commission. A medium-term plan for educational policy in the European Community was launched in May 1987. [36] It spells out three broad aims to be pursued in education. The first objective is to seek ways of enhancing the contribution of education and training in order to realize the Single European Market. Secondly, the plan urges the Commission to develop innovative strategies for improving the match between the labour market and the educational

---

[34] The idea of conflicting laws of motion producing a pressure for convergence in educational policy is taken further by C. Kerr (1990).
[35] See Fothergill (1988), Teichler (1988) and Neave (1989) for a discussion of the benefits that eventually may derive, at least potentially, from the development of international education in Europe.
[36] Cf. Commission of the EC (1988).

system, both on a local and a European level. The third aim is to maximize the contribution that educational policies can make to the reduction of regional disparities and to the creation of social cohesion within the "People's Europe".

It also follows from the above that education will not fulfil its functions if it is looked at solely as a means better to attain these economic aims. From from various attempts made during the past three decades we may perceive that to instrumentalize education for social and economic development may lead to a deterioration of the educational climate in schools and thereby of the most important factor for successful learning. If maximizing the achievement and therefore competitiveness, accountability, standardization, and abstraction become dominant in school, as they dominate in economic and social life, they may very well turn out to be counterproductive – they may lead to frustration and alienation, to a lack of initiative and diminished self-responsibility, to blunted minds and crippled personalities. Some of the school experiments and field research have clearly revealed the dialectic nature of the relationship between school and life. [37] A good school from the point of view of the pedagogue will prepare better for life by counteracting its constraints rather than by complying with them. The result, not the ways and means of the school, must "match" the demands of life and the workplace.

***

In contrast to other international organizations, such as UNESCO and the Council of Europe, the Commission of the EC has, until recently, not taken school education as a major area of involvement. Organizations such as the OECD and the Council of Europe influence school policy and exert a pressure for harmonization. Because its organizational framework makes it possible to draw upon common resources and to coordinate the use of national means for international educational purposes, the EC may be in a position to go further than the Council of Europe or the OECD along the path to internationalization. For example, programmes run by the EC, such as ERASMUS, YES, LINGUA and TEMPUS, [38] can be

---

[37] Cf. Helmut Schultz and Annemarie von der Groeben, *Laborschulabsolventen '80. Zwei Interpretationen*, 1985; Silvia-Iris Lübke und Marianne Michael, *Absolventen '85 – Eine empirische Längsschnittstudie über die Bielefelder Laborschule*, 1989; and Karin Kleinespel, "Schule als biographische Erfahrung. Die Laborschule im Urteil ihrer Absolventen", *Studien zur Schulpädagogik und Didaktik, Bd. 3*, Weinheim und Basel, 1990 (Beltz). See also, i.e. Michael Rutter et al., *Fifteen Thousand Hours. Secondary Schools and their Effects on Children*, London, 1979 (Open Books); Edgar Z. Friedenberg, *Coming of Age in America: Growth and Acquiescence*, New York, 1963 (Random House).

[38] See the abbreviations for an explanation.

expected to produce a pressure for harmonization. Students and teachers who gain experience from education in other countries will become aware of discrepancies in the objectives of education, curricula and examinations as well as differences in applied methods and techniques. This will be reinforced if teacher training becomes a focal point for student exchange in an EC action programme. So far, however, the Commission has mainly approached educational policy indirectly, for example through social and industrial policy and the so-called structural programmes. This is in accordance with the agreement recognized in the original Treaty of Rome, in which it is implied that educational policy is not a matter for the Commission but falls squarely within the jurisdiction of the national governments. Hence the objective of harmonizing school policy has, until now, not been an important item on the agenda.

One might mention, however, that general guidelines for a programme in vocational training were adopted by the Council of European Ministers of Education in 1971. Thus improving the quality of vocational training has been a common concern for a long time. A modest European Community Action Programme in Education was launched in 1976. [39] This plan was also mainly concerned with encouraging the development of vocational training, which was useful in coping with the rising level of youth unemployment in Europe at that time. It was not until the late 1980s that the EC interested itself explicitly in general education. However, even the programmes then initiated were not intended primarily to harmonize school policy but rather to encourage cooperation in third-level education and to facilitate collaboration between institutions of higher education and industry. The Commission has long expressed a concern with the recognition and equal treatment of degrees and diplomas. This was the subject of a directive adopted by the Council of the EC, in 1988, stipulating that member states should mutually accept higher education qualifications.

In 1989 the Council of Ministers of the EC adopted guidelines for European cooperation in education. Five goals were formulated:

1. A multicultural Europe.
2. A mobile Europe.
3. An educational Europe for all.
4. A competent Europe.
5. A Europe open to the world.

Harmonization is not among these goals, although it may well be part of what may be termed the "hidden agenda". The official emphasis is

---

[39] This programme is discussed by G. Neave (1984), and S. Opper and U. Teichler (1989).

mainly on encouraging the exchange of knowledge and ideas about educational objectives, means and methods, and outcomes. Yet the purpose of the EC efforts in education is probably also to minimize regulations about the value of educational qualifications and differences that are incompatible with the goal of economic and labour market integration and the building of a "Social Europe". An effort to make education more alike in the EC countries may thus be called for, even though this is not stated as such.

Convergence and harmonization are different, at least in degree. Even if a partial convergence in education were to be achieved by drawing on the best practice from the rich diversity of the school systems within Europe, the goal of actively promoting convergence is bound to stir up strong emotions. The successful implementation of any major change in education requires the support of both critical decision-makers, influential interest groups and a large section of the general public. This is one of the main reasons why the Commission, although it is as a matter of principle involved in an effort to promote cooperation and harmonization in European education, has until now not transgressed its own agreement that it would not strive for the development of a unified European system of school education.

European school systems, needless to say, differ markedly in many respects. A number of these are dealt with in subsequent chapters. It may suffice for the moment to mention only one illustrative aspect of cross-national variation. Figure 1.2 shows the participation rates in full-time education for 16 year-olds in 10 EC member states in 1970 and 1983. There was significant variation between the countries in 1970. The participation rate was comparatively high in Belgium, France and the Netherlands, and low in Denmark, Germany and the United Kingdom. Even though more children stayed-on in full-time education in all of the 10 EC countries by the early 1980s, substantial variations were still found. The participation rate doubled in Denmark during this period, bringing the country on a level with the Netherlands. Yet the participation rate in Germany and the United Kingdom was still below the EC average in 1983-1984. [40] The important observation emerging from the data is not just that the retention rate has increased over the period, as would be expected, but that the spread among the countries

---

[40] Secondary education has continued to expand in most European countries during the 1980s. The enrolment rates for 16 year-olds were generally lower in 1983-1984 compared with 1986-1987. In that year, 70 per cent of all the 16 year-olds in the United Kingdom attended school. This figure can be compared with those for other countries (OECD: *Education in OECD Countries 1986-87: A Compendium of Statistical Information*, 1989, Table 4.2, p. 81): Belgium (93%), Denmark (90%), France (88%), Germany (100%), the Netherlands (93%), Norway (87%), Sweden (81%) and Switzerland (85%).

does not seem to have diminished. It may be noted, incidentally, that the retention rate is no safe criterion for a good educational system.

Equally important is the observation that the increase in the school retention rate was not always related to higher demands but to the fact that there were not enough openings – be it on the labour market, be it in apprenticeship – to absorb the growing number of school leavers due to demographic changes. In Germany the obligatory tenth year of the *Hauptschule* was introduced, during the second half of the 1970s, only when there was no other way of getting youth off the streets. On the other hand, what the Germans call their "dual systems" (*duales System*) – the carefully planned combination of apprenticeship with a regular attendance of the *Berufsschule* (vocational school) – produces a vocational training much envied and emulated by other countries, not least by the Netherlands' authorities and employers.

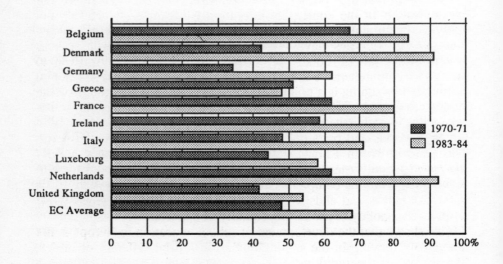

Figure 1.2: Participation Rates in Full-time Education for 16 Year-Olds in 10 EC Member States (in 1970-1971 and 1983-1984).
Source: EUROSTAT and OECD [41]

---

[41] Source of data: EUROSTAT and OECD.

Harmonization is pursued in connection with the structural adjustment of both economy and labour markets. However, some degree of harmonization in school education is, in due course, also likely to take place, regardless of whether or not it is pursued as an explicit objective. The supply-side "shock" of market integration to the European economic environment as a whole must be assumed to have implications not only for industrial and labour market policy but also for educational policy. If education is going to match the requirements of the workplace, then common definitions of major, long-term policy goals for educational systems will have to be sought by politicians and, not least, agreed by employers. Harmonization in the curricula for school education can be anticipated because of forces producing increased social, economic and political interdependence among nation-states in a highly competitive international environment. If the content of education is influenced by EC developments, then even examination policy may eventually be affected, as is argued in Chapter 8.

Even though the pressure for achieving a certain degree of harmonization in the content of education is mounting, this does not apply to the structure of the educational system. It is of interest to note that several countries have retained the parallel or tripartite school system while others, notably the Nordic countries, have adhered to models of comprehensive schooling. Switzerland, with its resistance to comprehensivization, is a good example of a country where the tripartite system has, with some adjustment, been kept in place. It seems reasonable to anticipate that this system will not be changed overnight in the Netherlands and parts of Germany. There are exceptions, however. The State of North Rhine Westphalia, for example, operates a fully integrated school system until the age of 19. One tendency seems to be to move away from the comprehensive model and towards a system that is given the means and ends to differentiate among students, albeit while adhering to a policy of promoting equality of opportunity.

This shows that the development of school education in Europe is not necessarily moving in the same direction in all the different countries. The inertia and the traditions of the school system are so great that even a degree of convergence in economic and political matters would not necessarily result in a real convergence in educational policy and in the organization of school education. The resistance of culture is perhaps too strong a force to let systems evolve in the same direction. The conclusion is that increased harmonization in the content of schooling does not necessarily call forth a demand for structural adjustment. A good example of this is in the Netherlands, where the long-debated integration of lower secondary education is mainly a matter of introducing a common core of curriculum elements and not the structural integration of schools catering for different student populations.

Exclusive attention to intra-Community developments in education is, in the context of the aims of this report, too narrow a perspective. This is so, for example, because the development of migration flows, which have consequences for educational policy, are pan-European and even global concerns. The differences in the ideological bases of educational policy in Eastern and Western European countries ought to be acknowledged. It should also be noted that some Eastern European countries have shown remarkable achievements in school education. For example, surveys of school performance conducted under the auspices of the International Association for the Evaluation of Educational Achievement (IEA) indicate that students in a country such as Hungary tend to outperform their peers in Western European countries in science. [42]. Even though this may in part be a consequence of a low participation rate in science (IEA figures in Chapter 5 indicate that only a very small portion of an age group is retained in Hungary in physics courses at the grade 12 level), the high level of student achievement in science is not in doubt. One could hypothesize that the schools in this country are not to blame for the partial collapse of the Hungarian economy at present, but that there are other factors that determine relatively poor economic performance. Recognition of high student achievement in Eastern European countries implies that there is no justification for Western Europe to lay claim to any educational hegemony. This raises the question as to what Eastern Europe can contribute to educational policy and practice in EC member states. Here one could consider examples such as the introduction of up-to-date science courses in upper secondary school and the education of highly gifted children.

8. Knock-on Effects of Economic Integration

Repercussions of the economic integration process on educational systems are most easily recognized at the post-compulsory level. [43] However, as is the case in any complex social system, changes in one sector call for policy responses in other parts of the system. Hence, the effects of European integration must be assumed to have consequences also for the teaching of children. What, then, could be the implications of the Single European Market for education?

---

[42] Studies comparing science achievement in Hungary with the performance of students in other countries are contributed by T.N. Postlethwaite and D.E. Wiley (1992) and J.P. Keeves (1992).

[43] L. Cerych (1989) and G. Neave (1990) discusses some implications of European integration for higher education in the member states of the European Community.

One outcome seems self-evident. The removal of physical, technical and financial barriers in Europe and the mutual recognition among countries of educational degrees and diplomas, including also professional qualifications and various certificates awarded by institutions specializing in vocational education, once fully implemented, will have implications for the mobility of teachers and students across countries and educational systems. Cross-border flows of teachers and students reinforce internationalization in policy and practice, scope and content. [44]

ERASMUS is based on bilateral agreements between educational institutions and not primarily among national governments. Yet the upshot of the promotion of exchanges through such formal programmes may be that a need is created for extended consultation among the educational authorities of nation-states, both inside and outside the EC. As countries are required to consider a broad diversity of views, demands and institutional concerns, internationalization may well have the unintended effect of adding to the complexity of decision-making, as the number of factors to be taken into account will be increased. If the present position in countries with pluralist forms of government is one of confusion about the goals of education, as is argued in a subsequent chapter, then consultations intended to encourage cohesion among the Community countries may at least initially further complicate matters. Yet the exchange programmes also provide incentives for undertaking experiments in school education, which may be seen as a positive development.

Another effect of the removal of barriers is that labour from Eastern European countries and the Mediterranean area will be available cheaply. Hence a drain of qualified manpower from these countries can be expected. Examples of this can already be seen in countries such as Hungary and Poland. It may be recalled that the Berlin Wall was built to prevent workers from leaving former East Germany for a future in the West. This illustrates that the brain drain will have negative repercussions on countries in transition to market capitalism. Even though much depends on political events in the Soviet Union and the economic situation in Eastern Europe, mass migration from these countries is likely to develop − even to such an extent that the EC countries would close their borders to immigrants. Whereas young, well-educated people looking for attractive jobs and increased prosperity will be among the first to move, there will also be others for whom the economic factor does not come first but who may flee from oppression or

---

[44] Among the most far-reaching programmes the EC has implemented in order to promote student exchanges and internationalization in Europe are ERASMUS, TEMPUS, LINGUA (see Chapter 6), PETRA and YES.

seek to enhance their cultural identity. Change in the demand for education will be the result.

A tremendous shift of populations may upset stability in the wealthy countries of the West: immigrants cause anxieties and antagonism as they are perceived as competing for jobs, aggravating the scarcity of housing, crowding the cities and slowing down the teaching processes in the classroom. There are open manifestations of xenophobia in many host countries, from Germany and the Netherlands to France and Sweden, Switzerland and the United Kingdom (See Chapter 7.) Although education will not be able to quench this fire, the school administration can help to avoid some of the most disastrous conditions, at least within schools by training and employing teachers who understand the language of the migrants, by providing for the right mix of 'guests' and 'hosts', by preparing curricula which give the children of the host country a good understanding of their guests and would-be compatriots.

The task set by the children of immigrants has been labelled "multicultural" or "intercultural education". That is a tall order! Yet public discussion of what this means has really gone beyond stating the need to strike a balance between respect for the cultures of origin and integration of the migrants into the host country. The underlying issue is a philosophical one: What makes a value a value?

Educators and school administrators are well advised not to avoid this question – and not to overestimate their role.

The fact that educational systems in the EC may be required to give credit for courses their students may take abroad implies that, by necessity, curricula must be compared and evaluated. It was hinted at in a previous section that this will most likely create a pressure for harmonization in the content of the school curriculum. This applies not only to subjects such as civics, history and geography, which may well have to become "European" or even global in orientation, but also to the science curriculum.

Most European governments have, not least during the years of austerity in the 1970s and early 1980s, expressed a concern with quality and excellence in education and are seeking ways of making schools more productive and effective, and more humane, by raising both the quality of teaching and standards for student learning, and by adapting school management. Countries are also putting a considerable effort into the development of national assessment programmes and indicators measuring aspects of local school performance and student outcomes. This new emphasis sets a major challenge for teachers, school administrators, and parents.

Another rather obvious implication of growth in the European harmonization process relates to language policy. Language may indeed come to be regarded as a key factor facilitating or inhibiting the internationalization of learning. An interim evaluation of the ERASMUS programme has clearly documented the decisive role that language competence plays in the degree to which students from different countries take part in international study exchange programmes. [45] The predominance of English and the linguistic imbalance vis-à-vis the "small language" countries tend to result in a lack of reciprocity in student exchanges. Socio-cultural differences may also be implicated. For example, many German students are interested in studying in France for some time, but few travel in the opposite direction. [46] The majority of students in Europe leave secondary school without having acquired an adequate knowledge of at least two foreign languages. Yet there can be no doubt that the basis of the language ability of adults is formed at school. It is for this reason that EC member states would have to assign priority to the development of language education in schools. It may even be necessary to provide instruction in languages other than those of the national majorities. Because the skill to communicate in a foreign language is becoming increasingly important in the new Europe, it can be anticipated that language policy will be among the first areas in school education where the EC Commission will formulate general directives, for example rules requiring all member states to implement national laws on language use in schools, so that equality of educational opportunity and treatment of children from indigenous and other cultures are guaranteed. There may also be attempts at early learning of the first foreign language (at the age of 7 or 8), or at teaching whole fields of subject-matter in a foreign language, or at providing all children with a minimal amount of instruction in their mother tongue regardless of where they live or plan to go – realizing the importance of "being at home" firmly in one language in order to be able to venture into others.

Yet another factor is the extension and intensification of technological cooperation. [47] The time factor plays a very important role here. Technology is developing rapidly, as are the economies of competitor countries. Thus an important challenge for the school is how to be flexible and adaptive to changing conditions. The EC is committed to a course that seeks to revitalize the European economy, for example by stimulating innovative industrial production and applying new technologies. Learning is seen as indispensable in this context, as success depends on the quality of the available human capital. This reinforces the need to update the curriculum for science and technology education.

---

[45] The evaluation report is discussed in G. Baumgratz-Gangl and N. Deyson (1990).
[46] This observation is elaborated in Teichler, U. et al.: *Auslandsstudienprogramm im Vergleich: Erfahrungen, Probleme, Erfolge,* Bonn, 1988.
[47] Some of the implications of technology for educational policy in the EC countries are discussed by Swann (1983) and Sharp and Shearman (1987). See also Cerych and Jallade (1986).

Updating in this context also sparks off a demand that elements of science subject-matter common to all Community countries be introduced. On the other hand, updating the curriculum should not only mean covering more subject-matter in a shorter time or modernizing the syllabus. Comprehension, independent thinking, the ability to apply, to transfer or to transform subject-matter should not be neglected in favour of quick acquisition and correct reproduction. In the future much of what has hither to been entrusted to the human memory will be left in the charge of computers.

Educational technology refers not only to technical products such as the 'traditional' audiovisual media and 'modern' media for storing and processing information, but also to a systematic or systems approach used in solving specific educational problems and designing instruction. [48] New information technology and educational technology as a problem-solving process may help in bringing about changes that improve learning and the school more exciting for students. Schools in Europe will have to come to grips with the staggering growth in scientific knowledge and its applications. Educational technology constitutes an urgent common concern for our Continent in this context. Has computer education, for example, found an adequate place in the curriculum and more so, in the minds of teachers yet?

Finally, from a different angle, developments in Soviet education warrant some consideration. In reflecting upon the situation at present, one ought not to forget that the Soviet Union was the first to introduce a 10-year (almost universal) common school for all children. Now the Soviet Union has embarked on a course of *glasnost* and *perestroika* aimed at obliterating the legacy of Stalinist totalitarianism and the rigidity of the pedagogical methods applied in the school. [49] Change has come not only in terms of a partial liberalization of economic and political life but also in the form of major social upheaval as a consequence of growing fragmentation, ethnic tensions and nationalistic tendency. These factors have had an impact on the provision of school education in the Soviet Union, even though it may be too early yet to appreciate in full the sort of changes that have occurred. One example indicating that perestroika is being extended to the area of education is that, in 1987, a semi-autonomous School Ad Hoc Research Group – Vremennyi Nauchno-Issledovatel'skii Kollektiv Shkola – was formed to reconsider the purposes of Soviet education, the assumptions guiding the

---

[48] A useful discussion of major definitions of educational technology is presented by D.P. Ely and Tj. Plomp in "The Promises of Educational Technology: A Reassessment", *International Review of Education*, Vol. 32 (2), 1986, pp. 231-249.

[49] The impact of social and political turmoil on Soviet education is the subject of a paper by S.T. Kerr in *Educational Researcher* (1990, October).

curriculum, and the way teachers and students interact in the school. [50] Led by the historian Eduard Dneprov, who became the Minister of Education of the Russian Republic following Boris Yeltsin's election as President, the study group presented its report in August 1988. [51] In some ways the conclusions signify a radical departure from "traditional" Soviet pedagogy. Educators are urged, for example, "to stop thinking of schools as 'government institutions' and to start seeing them as 'agents' of society". [52] There is also an emphasis on the spiritual and moral content of education. This may be a consequence not only of a loosening in the grip of the State over religious matters, but also of the break-down of school discipline in some parts of Central and Eastern Europe. There are proposals in some Soviet Republics calling for a reconsideration of the teaching of history and geography. Moreover, interdisciplinarity, the inclusion of different study electives, and the granting of freedom to teachers in interpreting curriculum elements and selecting appropriate teaching methods are encouraged.

These proposals have not met with complete approval from all sides, however. Teachers in particular have been highly critical. Another criticism is that too little attention may have been given to the important achievements of the Soviet school system, for example its high standard in the teaching of mathematics and science. Yet the report, and several others like it, may well be seen as landmarks on the brink of major reform, signalling the demise of the traditional, monolithic and centrally controlled Soviet educational system, and the inception of local school systems characterized by distinctive cultural traditions. Some republics have apparently already embarked on a programme aimed at the development of a more autonomous school system.

9. Concluding Remarks

Important challenges are confronting school education in the context of European integration, cultural change and recent economic development. Several large-scale cooperative projects in the field of education have been set up by the EC in response to and in anticipation of future developments. These projects can be seen as expressions of the policy ideas spelt out in the medium-term plan for education, namely that the relationship between education and the economy should be developed and that opportunities must be sought for promoting the "coherence" of educational policy in member states. It can be

---

[50] *Ibid.*
[51] "Kontseptsiia obshchego srednego obrazovaniia" [The concept of general secondary education], (1988, August 23), *Uchitel'skaia gazeta* [Teachers' Gazette].
[52] S.T. Kerr, *op. cit.*

hypothesized that one of the outcomes of the desire to develop close ties between schools and other educational institutions on the one hand, and industry and the public sector on the other, will be that the congruence between formal education and work is strengthened in the future, even more than is the case in European societies today. Furthermore, the responsibility of deploying and managing resources for the acquisition of specific vocational skills may well come to rest with the employment sector.

The creation of the new Europe may lead to difficulties and tensions between the Commission and the governments of EC member states. This is true also in education, because schools represent a wide range of diverse interests existing in a society and are deeply rooted in national history. The resistance of many educators to a further subjection of education to economic interests is another factor of importance. Realization of the breadth of this potential area of conflict has induced the Community institutions and governments of member states to proceed slowly and cautiously with initiatives that might impose a need for reform in school education on EC member states. Given the fact that integration started out as an economic experiment, and that education was not part and parcel of the programme from the beginning, the emphasis has eventually been put on education with a direct bearing on economics.

In view of the diverse interests described and demonstrated in national or international reports one tends to forget the original (and less measurable) pedagogical interest: to help the child to grow up in its respective surroundings – i.e. to be able to cope with life without losing his or her individuality. Many years ago, right after the Soviet Union launched the Sputnik, the United States engaged in a critical and thorough assessment of its educational policy. In one of the most influential papers, the Rockefeller Report, [53] it was said:

> "Paradoxical though it may seem, society as a whole must come to the aid of the individual – finding ways to identify him as a unique person, and to place him alongside his fellow men in ways which will not inhibit or destroy his individuality. By its educational system, its public and private institutional practices, and perhaps most importantly, by its attitude toward the creative person, a free society can actively insure its own constant invigoration".

It is generally agreed that full respect for cultural diversity and autonomy in education are key determinants of success in creating a new European social space. This is why the goal of increasing uniformity in the provision of school education is regarded with much wariness and

---

[53] Panel Report V of the Special Studies Project, Rockefeller Brothers Fund: *The Pursuit of Excellence, Education and the Future of America*, New York, 1958, p. ix.

suspicion by decision-makers at the various levels of government. Another reason is that diversity in school systems is regarded as a valuable asset. Accordingly, European cooperation in the area of school education has so far only implied voluntary intergovernmental consultation in a setting that guarantees complete national sovereignty over educational goals and policy. That encompassing reforms in school education will be brought about as a result of European economic and political integration seems therefore highly unlikely in the short term. One could note that the need for achieving internationalization and harmonization in the content and structure of school education is mainly driven by economic concerns and also by arguments about the necessity of stabilizing European political economy. But there is another factor, one that tends to receive less attention from politicians and the media: the imperative to encourage the development of a cultural and intellectual identity of Europeans.

In order to minimize the potentially disruptive effects arising from contradictions between Europeanization and strong national and regional interests in education, a framework of cooperation would have to be put in place. This would facilitate the solving of problems of interdependence and compatibility in education, while assisting governments in maintaining control over their national policy instruments. Thus a model might be found that would allow member states to retain their sovereignty in education whilst at the same time submitting themselves to an intergovernmental and consultative mechanism aimed at harmonizing – not standardizing – educational policy.

Chapter 2

Trends and Basic Concerns in School Education:
A Retrospective View

1. Introduction

Educational problems do not arise in a historical and social vacuum. Since the main purpose of this study is to identify and analyse those problems that relate to the present-day functioning of schools, one may appropriately reflect on their background, for example within the framework of the vast quantitative expansion of education that took place after the Second World War. The focus is on issues that concern many European countries, and which have usually appeared on the agenda of governments and at international education conferences.

As governments formulated and implemented policy, sometimes following, sometimes in advance of public opinion, certain principles and themes have emerged over the crucial quarter of a century since 1960 that for a while dominated the educational debate. These did not emerge necessarily in succession: some co-existed; others evolved strongly but seemed to die away quite swiftly. They may well not be identified for some while after they have begun to exert an influence. Yet they may be what by the year 2000 will be regarded as significant. What follows is therefore a tentative effort to identify such themes and principles. Some are common to all European countries; others are phenomena that only manifest themselves in some.

2. Major Themes of Educational Policy from 1960 to 1985

The chart in Figure 2.1 represents for one country, the United Kingdom, the currency of certain "slogans" or "watchwords" over the period from 1960 to 1985. By 1960 several national systems of education were embarking on vast projects of reform; after the mid-1980s it is difficult to perceive what new topics are emerging, or how far the old ones continue to be sustained. In any case, dates are only approximate and probably also disputable. Nevertheless, the chart shows phenomena that have tended to dominate the political and professional debate in the most "reformist" period ever in the history of State education since the nineteenth century. It may well be the chart also has some significance, *mutatis mutandis*, for other European countries.

In any case the themes are not usually mutually exclusive. Thus resources once poured into achieving "equality of opportunity" may now have been in part diverted to schooling in preparation for employment, and both will have been affected by "decrementalism" — the overall relative decline in resources available for education.

Moreover, the "slogans" cannot show the relative strength of each, which determines the effort and resources devoted to realizing them. Thus "equality of opportunity" in the United Kingdom is possibly less potent a factor now than it was in 1985, and less than it was in 1960, as well as being comparatively less prominent than, say, in Sweden at any time. Furthermore, it is likely that the way in which the terms are interpreted, and most certainly when they are translated into action, will differ considerably from one country to another. Local control, for example, will take on a different meaning in a hitherto highly centralized country such as France than it does in the United Kingdom, always far less centralized, where power and financial resources have, within certain constraints and in certain sectors of the system, devolved down to school level.

In Germany, early reading and modernization, curriculum revision and democratization were important themes; from the early 1960s to about 1970 in the case of the former and 1975 in the latter case, when other ideas took precedence, such as "conflict education" and school autonomy. The principles of lifelong learning and recurrent education also constitute interesting examples. They were strongly advocated during the late 1960s and early 1970s. After about a decade of quiet, during which very little happened and countries were unwilling to venture to any significant extent into this area of policy because of financial constraints, the two "catchwords" made a major come back in the 1980s.

It is perhaps interesting to note that in no other period of State education has it been possible to characterize in so many ways the direction of events. This is of course an inevitable concomitant of the politicization of the educational enterprise as a whole.

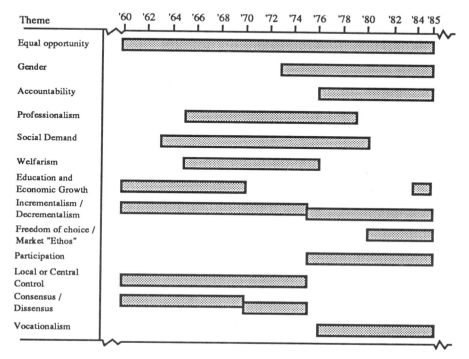

Figure 2.1: Major Themes and Principles of Educational Policy in the United Kingdom (1960-1985).

3. Perceptions of the Role of the School in the 1960s

The general background to education over the period from 1960 to about 1985 lies in the field of social policy. Until about 1960 war-torn Europe had priorities other than education: housing ("roofs over heads"), communications, and rebuilding the economy. In the latter case, it came to be realized that education had a significant part to play. Moreover, the "baby boom" of the post-War years meant that "roofs over heads" had as much an educational as a housing application.

It was in a climate of economic recovery and growth that educational change took place. As compared to before the War, families began to achieve a higher standard of living with the result that parents aspired to better education for their children. No longer was the weekly wage packet of the young worker so important in order for the family to live. Moreover, for perhaps the first time in European history, the aspirations of the individual and his parents corresponded to what at the time was perceived as the need for a well-educated and skilled workforce. The number of places at the top of the employment pyramid increased considerably, and it was argued that the way to fill these should be to award them to those that reached the highest level in education.

The launching of Sputnik in 1957 marked a turning point in the perception by policymakers and public of the role of school education in Western societies. The fact that the Soviets had managed to place a satellite into orbit earlier than the United States and Western Europe was regarded on both sides of the Atlantic as a reflection not only of the superior technology the Soviet Union possessed, but especially of superiority in the teaching of relevant school subjects, notably mathematics and physics. Within a year of Sputnik the U.S. Congress passed the National Defence Education Act, which was aimed at "upgrading" school education in the two subjects mentioned. The Act was inspired by a concern with military strength and not, as will be seen when discussing the background of educational change in the 1980s, with competitiveness in international trade. There was a feeling among influential interest groups that Western educational systems were not up to standard and were wasting the intellectual resources and reserves of talent of the nation.

Thus the 1960s was a decade when education and research were heralded as the main motor behind the spectacular economic growth that in the rich countries provided the resources for the vast expansion of the educational system. It was the era of growth, characterized by a firm confidence in what education might do to improve society and boost the welfare of the population.

As well as being the facilitator of life chances, by the 1960s education was in some countries seen as a prime instrument for social welfare. The doctrine of *welfarism* was advanced and widely acclaimed. The educational system was considered as a means of improving social conditions.

Moreover, the belief in what might be termed "education as a profitable investment" or "*economism*" was also held: as already mentioned, education would promote economic growth. A better educated labour force would bring about greater prosperity. It was emphasized, for example, that the system of vocational education and apprenticeship training in the Federal Republic of Germany known as the "dual system" was an important factor in the national "economic miracle". The same belief underlay the French "National Plans", in which education began to loom large in the 1960s, as it did in the abortive British imitation of the French example, the National Plan of 1965. Expansion of education was generally approved by European economists.

"Rescue" came from the intellectuals. Those like James B. Conant on the one hand, who gave a thorough, scientific look at the conditions, the weaknesses and the merits of the American educational system, and those who rather than concentrating on the academic aspects of their disciplines began to devise school curricula following the principles and ideas which the Woods Hole Conference (of leading scientists) had worked out in 1959. These ideas centred around the notion of the "structure of the discipline". The gist of the ensuing educational theory was formulated by Jerome Bruner: [1]

> "We begin with the hypothesis that any subject can be taught effectively in some intellectually honest form to any child at any stage of development".

That formal education played a special role in economic growth was an article of faith among most policymakers and research workers in the early 1960s. This was when the concept of "human capital" was making its way into the debate among economists and policymakers in education. The belief was inspired not least by the work of the economist and Nobel Laureate, Theodore Schultz. [2] Typical of the mood of the time is what President Johnson is reported to have said in 1965, when he signed the Elementary and Secondary Education Act: "We are going to educate them [the underprivileged] out of poverty".[3]

The two main concerns with regard to the development of school education in the 1960s were, first, the conception of education as a major agent of social and economic progress and, second, the "linear" expansion of enrolment, which raised many questions regarding the

---

[1] Jerome S. Bruner: *The Process of Education*, New York, 1960, p. 33. Significant expressions of the emerging optimism in the debate on the future of institutional schooling in the United States in the late 1950s can be found in *The Pursuit of Excellence: Education and the Future of America* (Panel Report V of the Special Studies Project, Rockefeller Brothers Fund, 1958) and the seminal investigation by J.B. Conant: *The American High School Today* (1959).

[2] Theodore W. Schultz (1961 and 1980). See also Gary Becker (1964) and Jacob Mincer (1962). Mincer published an excellent review of current developments in the economics of education in *Educational Researcher*, Vol. 18, 1989.

[3] Mentioned in T. Husén, *Education and the Global Concern*, Oxford, 1990. A recent Kappan Special Report states: "More than 25 years after America first declared war on poverty, the nation's children are worse off than ever" (*Phi Delta Kappan*, Vol. 72, p. 758/K1).

trade-off between, on the one hand, equality of opportunity and, on the other, quality of provision. The first issue was the subject of an OECD policy conference held in Washington in 1961. This meeting was important in launching the idea of education as a major agent of economic growth and social development. [4] Another major event in the same year was the OECD conference on ability and educational opportunity which took place in Kungälv, Sweden. Lionel Elvin, who was the conference Chairman, noted in his address:

> "There is a pressure on all European countries today to expand public education. This is demanded in part as a human right, in part as necessary for economic and social progress. But doubts have been expressed as to the wisdom of too rapid and too large an expansion. Will not the quality of our secondary and higher education suffer? No doubt access to secondary and higher education should be made less dependent on financial and social position, but do we have large reserves of ability in our populations which have been denied their opportunity through education?" [5]

The Kungälv conference was significant because, for the first time, the perspective on the formidable problems encountered in attempts to reform educational structures and adapt the system by making it responsive to the demands for equality and the needs of working life and society in general was widened beyond the provincial one. It was indeed characteristic of the time that OECD, as a newly instituted, innovative and facilitating organization of contemporary economic thought, should have brought together a conference on ability and opportunity in the educational systems of its member states. [6] The complexity of the relationship between education and the surrounding social structure was acknowledged. This was an important step forward because this relationship had at earlier times been viewed as relatively unproblematic. However, perhaps the most crucial contribution of the conference was that, probably for the first time in an international meeting of leading research workers in education, the academically selective and/or streamed junior secondary school for different reasons came under attack from different quarters. The debate on the merits and disadvantages of

---

[4] Lionel Elvin, Friedrich Edding and Ingvar Svennilson authored the document used as a background for the OECD conference on the relation between education and economic growth held in Washington, D.C., October 16-20, 1961 (OECD, 1962).

[5] This quotation is taken from the address by Lionel Elvin, which can be found in the proceedings of the Kungälv conference: *Ability and Educational Opportunity,* edited by A.H. Halsey (1961), p. 15.

[6] The Organization for Economic Cooperation (OECD) superseded the Organization for European Economic Cooperation (OEEC) in December, 1960.

the comprehensive school, which began in earnest in the late 1950s and early 1960s, has remained highly topical in education. [7]

4. "World Crisis in Education"

It would also seem appropriate in the context of this report to reflect briefly on the significance of the Williamsburg conference in 1967. Leading educators from all over the world were invited to Williamsburg, Virginia, in order to identify pressing educational problems shared by most countries, to take a hard look at some of these, and to come up with recommendations about how they could be resolved. [8] The Williamsburg conference was a landmark, in particular in terms of international exchange in education. The conference also meant a breakthrough for a comparative and cross-national study of education, as the extent of global interdependence in education had become evident to all concerned with the objective of improving school education. This was in part due to the main working document prepared for the conference by Philip Coombs. A revised version was published a year later as a full-sized report under the same title as that of the Williamsburg conference, namely "The World Crisis in Education". [9] This book has had a strong

---

[7] A review of empirical investigations with a bearing on the debate in the 1950s on comprehensive versus selective school systems is presented by Ekström (1959) and N.-E. Svensson (1962). An update of the relevant research studies conducted in Western industrialized countries until the mid-1970s is given in Husén (1974). Fend (1982 and 1984), Mitter (1984) and Creemers and Scheerens (1988) describe trends in the early 1980s with respect to middle-school education, i.e. a common school of lower secondary education (grades 7 to 9) for all children in a given catchment area.

[8] If we glance through the list of participants we find that it includes all categories of experts in education. We find present or prospective Ministers of Education, such as Hans Leussinck and Gabriel Betancour, Presidents of leading universities, such as Clark Kerr, Eric Ashby and James Perkins, reform planners such as Friedrich Edding and Philip Coombs, and intellectuals who had orchestrated cross-national debates, such as Raymond Aron along with Jean-Paul Sartre and Maheu, who were top graduates from the *École Normale Supérieure*. There were also pioneers in educational research, such as Ralph Tyler, and in adult informal education, Bertrand Schwartz.

[9] P.H. Coombs' book *The World Crisis in Education* (1968) carried as a sub-title "A systems analysis". The intellectual origins of systems theory, which had rapidly come into vogue during the 1960s, can be traced to Durkheim who had developed a theory of the mechanistic and organic forms of society in *The Division of Labour*. The central point in the theoretical perspective known as systems theory is that a society can be regarded as representing a complex system composed of many interrelated parts. Concern is with the inputs into the social system, the way inputs are processed, and the outputs that are produced. After a series of blistering attacks during the 1970s, systems theory faded from view, at least to a degree. However, there are signs that sociology seems to be heading towards a revival of systems theory in the years to come.

international impact by making educators and policymakers aware of certain incipient problems in education – less than a year before the 1968 crisis at universities in many countries. [10]

The overriding problem dealt with in Williamsburg was the imbalance between the supply and demand of education in relation to educational services, teacher training and manpower. The central question was, on the one hand, how to reconcile the hopes of individuals and the needs of society and, on the other, the capacity of the educational system to meet these. Another concern was the widening gap in educational provision between the rich and the poor countries. The summary report from the conference also pointed out the need for improving the effectiveness and quality of educational programmes by adapting the structures and management of education to new social conditions. Other problem areas touched upon were the rapidly rising costs of education and the relevance of the school curriculum. The importance of non-formal education and community education, which was a new idea at that time, was also emphasized.

In retrospect, it can be concluded that the meetings referred to above set much of the stage for discussion on educational policy during the 1970s and 1980s. The major themes still have relevance. One example is the problem mentioned previously, that of ensuring equality of opportunity. The objective was not only to create a "fair" society but also to maximize the utilization of talent in the system and, hence, to boost the economic advantage of investment in education. [11] The threatening "enrolment explosion" in formal education was another matter of concern. The overriding question was whether, and if so how, students in the secondary school were to be differentiated into "academic" and "vocational" programmes of study. The content of vocationally oriented programmes was also controversial, especially in view of the demand that such studies ought to be linked with both post-primary education and the world of work. The expansion of secondary schooling created a pressure for increasing the intake to higher education. Universal tertiary

---

[10] The Berkeley campus of the University of California was at the centre of a riot by students protesting against the administration during the autumn of 1964 (Boyd and King, 1977). The State of California has in certain respects been a trendsetter in educational policy and practice. For example, the California Master Plan for education (1960) and others like it may well have had a rationalizing effect in the long run on educational policy in Europe. That student protests began at the University of California a few years before they did in Europe is therefore not surprising.

[11] Michael Young warned against over-emphasizing the economic function of school education in the seminal book *The Rise of the Meritocracy*, 1958.

education had begun to be contemplated within the framework of a "lifelong", "recurrent" or "permanent" system of education. [12]

The economic basis for the rapid expansion of school systems and the concomitant upsurge in public expenditure on education in the 1950s and 1960s was provided by the efficient and profitable production system. The blueprints for reform drawn up in European countries in the 1960s show that faith in continuously strong economic growth was prevalent. The concern was for establishing minimum standards in education, ensuring equality of opportunity, promoting social mobility and with redistributing education, work and leisure over the life-span, as would be the case in an integrated system of lifelong or recurrent education. The ideal of establishing a "learning society" began to be discussed in the late 1960s. The educative society, offering to all citizens multiple access to education and the cultural heritage, was regarded as the main challenge in the development of educational systems.

Those familiar with the present policy debate in education will recognize some of the problems listed above. Indeed, it would seem as if the school debate has in certain important respects turned full circle and has, by 1990, arrived at a point where the major concerns in Western school education, as perceived in the 1960s, are being revived. [13]

5. School Education in Question

But problems not foreseen, or to which not enough attention was paid, were emerging. The enrolment explosion at many universities in Europe and the lack of teaching and guidance provided by the teaching staff had begun to stir up unrest. The writing on the wall had begun to be visible,

---

[12] In the late 1960s and early 1970s, lifelong education was considered an alternative to the perceived shortcomings of the formal "front-end" system of education. Central was the idea that the rôle of the school would change completely in a system of lifelong education, because there would be no reason to cram the minds of children full with facts that would soon become outdated anyway. Another contributing factor was the belief that schools, instead of motivating the students for further education and self-directed learning, were often instilling a hostility to education in children. Thus the school was perceived as achieving the opposite of what was intended. In this critical perspective, because schooling would be a prelude to lifelong learning, the objectives of initial education would have to be directed to the mastery of study skills, perseverance in self-directed learning, the mastery of mother tongue and foreign languages, and the development of the faculties of concentration, observation and exploration, and critical reflection. The Edgar Faure Report (1972) and the study by the Council of Europe on Permanent Education (1970) were influential in spreading the concept of lifelong education in Europe.

[13] This argument does not, of course, negate the fact that important differences in school policy and practice exist.

but not enough to be regarded as a major issue at Williamsburg in 1967. The trade-off between quality and quantity was not yet identified as an overriding issue, since the predominant concern was how to expand enrolment so as to get as many children as possible into the secondary school. A change in mood occurred in the early 1970s. This is documented in the report emanating from the OECD conference on educational policy held in Paris in 1970. [14]

Whereas formal education had been described in the proceedings of the OECD meeting in Washington as a powerful instrument for promoting economic growth, social welfare and change, the report published a decade later from the Paris conference is less optimistic in several respects. The rapporteurs point out that the goals set for educational policy a decade earlier had not been achieved. For example, increased enrolment had neither increased effectiveness nor contributed to the equalization of educational opportunity to the extent expected. The then Minister of Education in Sweden, Ingvar Carlsson, acknowledged in his keynote address that experts and planners had been too optimistic about the role of formal education as an instrument for achieving equality, social progress and economic prosperity. [15]

Moreover, doubts as to the value of education as investment soon began to emerge. It was realized that, as a consumption good, it was expensive. By the early 1970s the economic arguments advanced at the OECD conference in Washington a decade before were put in question, and tools such as cost benefit analysis and manpower planning were viewed with some scepticism, and are only now being revived. Indeed, what nobody could have predicted a decade earlier had actually happened. Misgivings were aired about the worthwhileness of formal schooling. Some went further by calling into question the school, particularly secondary school, as a social and educational institution. The critical question was raised whether schooling "made any difference", and whether it would be advantageous if society was "deschooled". [16]

When attempting to explain and understand the criticism of the school as an institution in the 1970s, it is necessary to keep in mind that the pressure of numbers resulting from the post-war "baby boom" along with the increasing social demand for schooling had made a rapid expansion

---

[14] The reference is to the OECD conference on policies for educational growth, Paris, June 3-5, 1970 (OECD, 1971). The Standing Conference of European Ministers of Education is also gaining significance in the European policy context.

[15] Ingvar Carlsson, in OECD (1971).

[16] The questions whether schooling "makes a difference" and whether society ought to be "deschooled" were debated at some length during the early 1970s. P. Goodman (1962), I. Illich (1970), R.C. Rist (1973) and C.S. Jencks et al. (1972) were at the centre of the discussion.

necessary in European educational systems. A change took place whereby the élitist structure of secondary education was transformed into an institution for all, or almost all, children. By the time the reforms were well underway, however, the birth rate began to slow down and, as a result, enrolments in primary and, later, to a lesser extent, in secondary schools were dwindling in many European countries. This transformation and the mis-match between demographic developments and provision of places put the school under pressure.

An interesting case is Germany, where not the school but the school administration came under strain. Given the *Beamtentum* (the civil servant system) the governments of the *Länder* could not simply dismiss the surplus teachers, much as they would have liked to cut down on staff size. All they could do was to shift the teachers to other positions which, of course, met with resistance from both individual teachers and teachers' unions. Schools which became too small, however, could be closed and their clientèle could be fused with that of other schools. In many places it was impossible to maintain the differentiated system with one *Hauptschule*, one *Realschule*, one *Gymnasium* and even one *Gesamtschule* (comprehensive school.) Moreover, when the number of children decreased the *Gymnasien* and *Realschulen* began to persuade every single child leaving the elementary school to come to them, lest they fall below the minimal size. Thus their selective function was no longer effective; the German *Gymnasium* in many parts was called "Volksgymnasium", people's Gymnasium, and difficult to distinguish from a comprehensive school in terms of clientèle, achievement, and climate.

6. A Summary of Criticisms of the School

The educational systems in a pluralist democracies encounter great difficulties in handling the contradictions arising from divergent perspectives on religious, ethical and moral issues, and meritocracy. A common argument has been to claim that schools do not have to deal with these issues, which are considered to fall outside their mandate, nor to provide solutions for them. Yet when it comes to answering the challenges of building the "European house" proposed by Mikhail Gorbachev and of living in the larger unit, as stated in the previous chapter, mere quantitative and organizational changes may not suffice to provide the new attitudes, the widened sense of responsibility and the interiorisation of the criterion by which knowledge and skills are to be used.

This falls in line with much of what the reform movements in various European countries have been claiming for a long time. Their main complaints are:

(1) Schools are generally engaged in the preparation of their students for solving yesterday's problems, not for coping with those problems which today's children will encounter as adults in advanced European societies.

(2) Schools no longer explicitly teach ethics. Ethics and moral education have become part of what is often called the "hidden" curriculum.

(3) Schools do not even achieve the traditional goals they have set for themselves: the mastery of elementary skills, the acquisition of basic knowledge, the attainment and maintenance of attitudes generally approved of and called for in a democratic European culture, namely, respect for the rights, the needs, and the differences of other people, a sense of responsibility for their community, non-violent forms of action, personal self-direction, and the search for rational understanding.

(4) Schools are educationally inefficient and wasteful, in part because the school is faced with powerful competitors in modern society, some of which are undoing what schools have achieved or obstructing what they strive to achieve. Problems are encountered because there are tendencies in our modern, technological and meritocratic society which are at cross-purposes with the idea of the school.

(5) Schools are essentially boring, alienating, disliked by those who go there – students and teachers.

(6) Schools are burdened with many tasks some of which are in conflict with one another and with the primary purpose of the school. This primary purpose, the critics say, is not to serve *in loco parentis*, to select and sort the children into occupational careers and social strata, to prepare for jobs and how to get them, to keep the children from the streets or from deficient homes, to test, evaluate, forecast their development, and to "cool out" their aspirations so as to make successive populations fit the limited opportunities and rewards available in a society.

(7) Social pathologies have entered the school, which has therefore become a troubled institution. Certain indicators, such as absenteeism, the turnover of teaching staff, and the number of youth disliking school, point to the problems.

Though not all of these complaints are justified, they reflect a feeling of crisis and frustration which may be epitomized as follows: Firstly, the very principles of pedagogy necessitate a "transformation of the school" [17] and, secondly, the social, economic, political and cultural demands on the school are best answered when the school is doing well pedagogically.

All attempts to reform, i.e. to improve schools and other formal educational institutions, must start with asking certain basic questions: How do children become self-responsible persons? What values do we want our children to be committed to? What conditions are conducive to this process of becoming? This is important because other questions usually undergird attempts at implementing change in the school system: How can we cope with a foreseeable lack of scientists? How can schools help in diminishing the mismatch on the labour market between the supply and demand for appropriately skilled manpower? How can vandalism, juvenile delinquency or the breakdown of school discipline be effectively responded to? It follows that what may be most needed is not be a drastic change in the organization or content of basic education, but rather a revision of the aims and functions of schooling.

In this perspective, the transformation of the school into a better place for children to learn is likely to succeed only if priority is given to the pedagogical functions and problems of schools rather than to alleged organizational deficiencies. This argument is heard especially among teachers, who tend to be suspicious of the usefulness of applying managerial practices and criteria of assessment borrowed from manufacturing industry to diagnose and remedy school problems. (Cf. Chapter 3)

7. Constraints on Secondary Education

Education is increasingly perceived as an instrument for enhanced life chances and career prospects, which is at the core of "social demand". Enrolment in secondary education increased at a rapid pace in most European countries during the 1970s. The result was that competition for entrance into higher education rapidly became stiffer. Accordingly, as the number of study places did not suffice and not all young people could continue in the system, the aspirations of some were thwarted. Other students, who entered the job market after having passed through certain programmes in higher education successfully, discovered that the market value of their qualifications had substantially diminished compared to the previous period, which was contrary to their expectation. In a sense, they

---

[17] Cf. L.A. Cremin: *The Transformation of the School. Progressivism in American Education 1876-1957,* New York, 1961.

had become "over-educated". [18] This shows that the school, and indeed the educational system in its entirety, had become transformed and was given objectives it was not designed to meet. Other sources of discontent with the school system can also be pointed out. Prominent among these was the conflict between the hierarchical model of governance applied in the system and demands for increased participation in decision-making by students and parents.

The price shock and 'stagflation' in the wake of the oil crisis and the inertia of industry and public institutions in adjusting to structural change were among the factors shaping economic recession in the last part of the 1970s. Economic austerity put several new issues on the agenda. First, the provision of extended formal education was questioned. [19] Whereas huge investments in school education had been backed by a confidence in the validity of the concept of investment in education and a solid consensus about the social and private benefits of education, the theories were no longer held to be as generally valid in the mid-1970s. As the economic crisis intensified and unemployment, particularly among youth, rose to higher levels, the social and private benefits of formal education were increasingly called into doubt. [20]

In conclusion it can be noted that, by the mid-1970s, a sombre mood had replaced the strong commitment to education of the mid-1960s. The former consensus about the benefits of traditional schooling and the conviction that education always represented an intrinsic good were gone. In the 1970s problems in education were increasingly seen in terms of institutional shortcomings, both internal ones and those that pertain to the relationships between formal education and other institutions, not least work. These misgivings were, and still are, particularly strong about secondary education.

Factors such as the heavy increases in the "unit costs" (i.e. per pupil and per year) of school education, the scarcity of financial resources, doubt as to the worthwhileness of investing in education and the sliding of school education down the political priority scale led to budgetary cuts or at least austerity measures in the industrialized countries of Western Europe. The need for examining the allocation of scarce resources in the context of political priorities for national development was emphasized. As elaborated in a subsequent section, policymakers sought to increase

---

[18] The related problems of credentialism and over-education are discussed in R. Dore (1976) and R.B. Freeman (1976).

[19] This criticism can possibly also be explained because the period of compulsory schooling was being lengthened in France and the United Kingdom in the early 1970s.

[20] For a general review of the situation until the mid-1980s, see J.-C. Eicher (1984 and 1989) and G. Psacharopoulos and M. Woodhall (1985).

the efficiency and cost-effectiveness of formal education. With it came an emphasis on "value for money" and performance evaluation.

In place of financial optimism came the new and harsh doctrine of educational *accountability*. It arose in part from the new "consumerism" of the 1970s; the "consumer" of education – the parent, and in particular his proxy, the government – demanded value for money. Maurice Kogan, writing in the British context, states that accountability took a number of forms. [21] One is for tighter "public control" – and this today forms the basic of policy in most European countries –, for "the resolution of problems", "profit and loss accounting", or for "private contract" purposes (i.e., an agreement between the teacher and the individual parent.) Public control can be exerted of course in a number of ways, from bearing down oppressively upon the teacher or from merely ensuring that certain criteria of performance are met. At school level it may also be used as an instrument for explaining to parents not only what teachers are achieving, but what they are aiming at.

The theme of accountability is therefore closely associated with that of the demand for *participation*: all the actors in the educational process – governments, parents, teachers, interest groups such as employers, trade unions, the churches and, last but not least, the pupils themselves – demand a say in the educational enterprise. From about 1970 words such as "participation" in France, and "Mitbestimmung" in Germany began to be bandied about. Thus the argument as to whether the government or the parents have prescriptive rights in education was revived. In modern Europe it has expressed itself politically as one where the right-of-centre political parties champion the cause of the parents, as with the Conservatives and Christian Democrats in the Federal Republic of Germany and the Netherlands, and the Left that of the welfare state, as in the countries of the former Eastern bloc, France, Italy, Norway and Sweden. Between this Scylla and Charybdis the British have steered an uneasy course.

8. The Idea of the Comprehensive School

A "comprehensive" school system, such as the Swedish one developed in the 1960s, provides publicly supported education for all children of mandatory school age in a given catchment area. All programmes are offered, at least in principle, under the same roof. Another essential feature of comprehensiveness is that no organizational differentiation or grouping practices are employed that decisively give direction to the

---

[21] The concept is developed from a paper by M. Kogan (1990) written for OECD.

educational careers of the students. [22] By contrast, a selective system employs organizational differentiation at an early stage in order to allocate children to different types of schools or to sharply divided programmes. Furthermore, grouping practices are employed at an early stage with the main aim of spotting students who are considered to be academically oriented. Apart from selective access and internal grouping, the system is characterized by a high attrition rate in terms of grade-repeating and drop-out.

At the core of the differentiation problem, still stirring up much controversy, is the question as to how early the separation of the academically gifted children from their peers should take place. Furthermore, in what type of institutions − selective or comprehensive in enrolment and programme coverage − should students of secondary school age be accommodated? The debate among educators and psychologists in Europe about the relative advantages and drawbacks of comprehensive schooling, continuing from the 1940s until now, was, in the beginning, conducted almost entirely in pedagogical and not in social terms. The problem had been conceived as one of how far up in the grades it would be pedagogically feasible, and defensible, to keep all children of an age group "with their spread in academic ability" together in a common school. Later arguments about social justice and equality began to be advanced in favour of postponing the organizational differentiation of children into separate educational career tracks.

Élitist and selective structures of secondary education have mainly been supported by conservatives, whereas those adhering to liberal and socialist views have favoured the comprehensive model. Besides arguments of a pedagogical kind, the main arguments in favour of the comprehensive system have been those of social justice and equity, since it is expected to enhance the educational opportunities for students from the lower social strata, and efficiency, as the comprehensive system is expected to make better use of the 'reserve of talent'. [23] In the debate on the relative merits and drawbacks of the two systems it has been maintained, on the one hand, that the most able students in a comprehensive system will suffer by having to be taught together with their slow-learning peers. This will impair their standard in comparison with students of equal intellectual potential in selective systems. The adherents of comprehensive education, on the other hand, maintain that

---

[22] In a six-volume work, S. Marklund (1980-1989) provides the most thorough account yet of the school reforms carried out in Sweden from 1950 to 1975.

[23] The 'reserve of talent' was the main subject of discussion at the OECD meeting in Kungälv in 1961 (cf. footnote II-5).

the better students will not suffer as much in that system as do the great mass of less academically-oriented students in a selective system.

The main argument in favour of selecting students early and putting them into separate institutions or programmes has been that such a system caters better for able students and on the whole is conducive to the preservation of standards at all levels of ability. Those in favour of the élitist model maintain that a system of selection based on fair and uniformly employed criteria of excellence will automatically open the avenues to high-status occupations to everybody who "deserves it", that is to say possesses the necessary talent and energy. The comprehensivists counter by claiming that the selective system is beset with a greater social bias than the comprehensive, since at the upper secondary and higher education levels the proportion of lower-class students consistently tends to be lower in a selective than in a comprehensive system. They were emphasizing the evidence that in organizationally selective systems the high standard of the élite is bought at the price of limiting the opportunities of the mass of students.

The main pedagogical argument in favour of the comprehensive school is that developing and learning are highly individual processes which take place at a different pace with each child and require as many approaches as there are children in the group. A system which claims to have sorted the pupils into homogenous groups tends to forget this and regiments the learning processes. The comprehensive school, which admits all children living in one neighbourhood, cannot ever hope to have a common answer to all of their needs and will, therefore, be the least uniform among the various schools – in direct contrast to the assertions of its adversaries.

9. Main Trends in Secondary School Reforms

It is important to realize how different European secondary education systems were at the beginning of the 1960s. At least three different models could be distinguished in Western Europe. France had retained a select *lycée* that was still redolent of its Napoleonic origins, with the bulk of pupils remaining in primary schools until the school leaving age. West Germany, after the Nazi era, had reverted to a pre-1933 set-up that had its roots in the early nineteenth century, with an élitist school, the *Gymnasium* (classical, modern language, and mathematical-scientific), a *Realschule* with some vocational bias, and a *Hauptschule* which was really the old upper primary school for the majority of adolescents. Britain, in wartime 'idealism', had proclaimed "secondary education for all", instituting in 1944 a tripartite system consisting of a grammar school (for some 20 per cent of the secondary population), a secondary technical, and a secondary modern school the latter attended by the majority. Most secondary systems in Western Europe were cast in a similar mould to these.

Sweden, however, was to prove the Western European exception – even constituting what has been labelled the "Swedish Model". Its government led by the social democratic party, continuously in power with only a few exceptions since the early 1930s in a country untouched by the effects of war, was eager to achieve social equality, and considered education as one means of realizing a fairer society. Backed by psychological views on the nature of ability, it first opted to introduce, on a try-out basis over a decade, a comprehensive system to cover all compulsory education, thereby replacing a secondary system that was "parallel" as elsewhere. Through organizations such as OECD and the Council of Europe its reforms became widely known, and had a greater influence than the American model of the "common school".

By 1960 Western Europe had largely recovered from the effects of war, and was able to focus more attention on education. The view was by now widely held in the European democracies that all children deserved an equal start in life, and this could best be given through a more equitable secondary education. This promotion of the individual went hand in hand with another view that after 1961 gained credence: economic growth demanded the development of the abilities of everyone. It was emphasized that countries that prevented a large proportion of children from going on to secondary education, among which were the nation's "reserve of talent", would eventually pay a high economic and social cost in terms of wasted human resources and missed opportunities. Therefore the interests of the individual and the collectivity uniquely coincided.

In some countries developments followed on this realization. Thus in France the return to power of de Gaulle in 1958 prompted educational idealists, active ever since 1918 and supportive of the Langevin-Wallon Plan formulated after the War in an effort to secure greater educational equality, to persuade the new Fifth Republic to act. In the short interval in which the General ruled by decree – a law on educational reform would never have been passed by the National Assembly without endless debate – the proposals known as the Berthoin reforms were adopted. These stipulated that the school leaving age was to be postponed from age 14 to 16, and that the post-primary schools would become "*Collèges d'enseignement général*" (general secondary schools.) This sparked off successive reforms which culminated in 1975 with the "*réforme Haby*" by which the "*Collèges d'enseignement général*" became the unified middle school for all pupils, comprising grades 6 to 9.

Britain, however, had to wait until 1965 before much change took place in secondary education. Taking counsel from Sweden, Crosland, a Labour Minister, realizing that "parity of esteem" between the existing types of secondary school was a myth, introduced – by circular 10/65 to the Local Educational Authorities, because as in France a law would have given rise to argument and delay – six different schemes whereby local authorities could "go comprehensive". [24] Although today most secondary pupils are in some form of comprehensive school, however, a few of the old élitist grammar schools still survive.

By contrast, in West Germany the situation remained stagnant until about 1970. Many, but not all, Germans considered that the existing tripartite secondary education system had served the "economic miracle" well. [25] That this "Wirtschaftswunder" was achieved in spite of the massive destruction caused by the Second World War was to no small extent due to the fact that educated human resources were available. Hence "keine Experimente" was the watchword. [26] In 1959 an Outline Plan (the *Rahmenplan*) had put forward a justification of tripartism. According to this, there were three types of psychological ability (theoretical, technical and practical); three types of occupation ("academic", technical, and manual); and three types of social class (upper, middle, lower.) All these fitted together in a kind of matrix that justified three types of school.

Although the German system has since changed out of all recognition, even today the Gymnasium, where it has not been supplanted by a modified form of comprehensive education, retains its prestige, although access to it is much more open and no longer confined to a very small section of the relevant student population. It will be interesting to see whether the educational thinking of former East Germany will significantly influence the West in the long term, although the chance

---

[24] An excellent background to the debate on the pros and cons of comprehensive schooling in England at that time is provided by E. Boyle, A. Crosland and M. Kogan in *The Politics of Education* (Penguin Education Special, 1971).

[25] The economy has become a major if not predominant concern in Germany during the 1970s and 1980s. Some people, for example Hartmut von Hentig, have put forward the criticism that the German school reforms, if they could be called reforms at all, were not concerned with the development of pedagogy and new integrated teaching and learning devices, as one would hope, but with the one-sided and superficial objectives of economizing on time, breaking up subject-matter into constituent components, streamlining school management and organization, and meeting various economic and labour market demands on the school.

[26] For a discussion, see H. Fend: *Pädagogik des Neokonservativismus* (1984); L. von Friedeburg: *Bildungsreform in Deutschland. Geschichte und geschichtlicher Widerspruch* (1989); and K.-J. Tillmann: *Comprehensive Schools and Traditional Education in the Federal Republic of Germany* (1988).

that this will happen does not seem very high at the moment. The general approach to the comprehensive school taken in Germany resembles that of the Netherlands, where a few small-scale experiments were half-heartedly conducted but never evaluated in a satisfactory way.

All in all, everywhere in Europe the move towards a common secondary school has been slow in coming. However, in general, selection and differentiation in secondary education has been delayed from the age of 10 or 11 to that of 14 to 16, and even then the intent is that no choice should be definitive. However, this goal has proved very difficult to achieve in practice, particularly in "parallel" school systems such as those of Germany, the United Kingdom and the Netherlands, where barriers impeding student mobility across school types at secondary level have not yet been fully removed.

In Central and Eastern Europe the situation as regards secondary education was not so clear. As the Iron Curtain descended between East and West it looked as if the theories of Soviet pedagogues would prevail. Indeed the polytechnical principle, in which no distinction is made between general and vocational/technical education, had been adopted by about 1965 in many of the then socialist countries. However, in some, such as Hungary, there was no definitive break with tradition, although lip-service was paid to the ideas of Marxism-Leninism and the principles laid down by early Soviet revolutionaries such as Krupskaya, Lenin's wife. Nevertheless, earlier than in most Western countries, socialist states such as the German Democratic Republic had attempted to establish the *Einheitsschule* (which was a historical name from the period just before the First World War, but a wrong one), and to make secondary schooling more relevant to the demands of the workplace. This, and the great advances that were made in the teaching of mathematics and the sciences, may prove to be the most lasting contributions given to secondary schooling in a more united Europe.

10. Equality of Opportunity and the Meritocratic Tendencies

The strongest theme, the most persistent principle, stretching indeed back to 1945 and still surviving, although today in a more attenuated form, was that of *equality of opportunity*. Educational literature over the past quarter of a century is full of references to it in many languages, from "*démocratisation de l'enseignement*" to "*Chancengleichheit*". Democratic governments have been forced to react to this demand, if only because to assert its opposing tenet, elitism and the maintenance of privilege, was a certain vote-loser. A welter of facts supports this contention. One may cite as an example the introduction, slowly but surely, of comprehensivization in secondary education. The massive expansion of higher education, which involved the creation of para-university institutions, also owed as much to the demand for access to some form of post-secondary schooling as to the need for a more highly qualified workforce. For a while some even advocated a further form of egalitarianism that represented not only equality of opportunity but equality of outcome: children should be so treated at school that by the time they left they should have reached as near as possible the same level of attainment and the same access to post-secondary education.

Many different aspects of educational equality have been noted by researchers and brought to the attention of policymakers. At first the concern was mostly with factors of inequality between individual learners and social groups. The focus shifted to factors causing inequality between classrooms and schools during the 1960s. Inequality in the effectiveness of teachers constituted a third factor in which researchers and policymakers were interested.

One particular facet of equality of opportunity has been that of inequality between the sexes. *Sexism*, interpreted as the main cause for the under-achievement of girls in the upper secondary and post-initial levels of the educational system, was a problem that came to the fore in the 1960s, although in Britain as late as 1963 the Newsom Report cast girls in a largely domestic role, and advocated a different kind of education for them. However, the rise of feminist movements and the massive entry of women into employment from the 1960s, quickly brought about changes. Although women are still under-represented in various branches of the labour market, facilities for girls enjoying the same education as boys have vastly improved. However, a certain polarization still occurs: for example, in upper secondary education the majority of those studying science are boys. Eastern Europe, where greater educational equality between the sexes has been achieved, may well have something to offer the West in this respect.

At the other end of the scale was the attempt to use education to break into the "cycle of deprivation" by initiating a policy of positive discrimination. The creation of educational priority areas in Britain and Sweden may be seen as one facet of what came to be designated as *social engineering* through education. However, such "engineering" is now generally frowned upon, as it has been demonstrated to have been rather unsuccessful. The improvement of educational facilities has largely benefited the middle classes. Education as a tool for the redistribution of power and wealth has been seen to be the light that failed. A.H. Halsey, the British sociologist, has summed up the position: [27]

> "The essential fact of twentieth century history is that egalitarian policies have failed ... . Liberal policies failed basically on an inadequate theory of learning. They failed to notice that the major determinants of educational attainments were not schoolmasters but social situations, not curriculum but motivation, not formal access to the school but support in the family and the community".

The rising tide of meritocracy is worthy of attention. Most young people nowadays in schools are keenly aware of the fact that the educational system is used as a sorting and screening device by employers, whose demands and expectations have increased. They know that the longer they stay on in schools the better their employment possibilities are likely to be. This situation has created a paradox. Even though a significant proportion of young people do not enjoy being in school and have a generally low motivation, they understand that it is better simply to "stick it out". The other side of this is that the school is in danger of becoming a "cooling-out" institution. [28] Youth in many European societies derive little intrinsic motivation and reward from going to school. Instead, young people stay on because they realize that education is the only means of entering the "good life". This development signifies that the fight for social survival has entered the school itself. What are the implications of this for the functioning of schools in Europe? Another paradox could also be mentioned here. Namely, the more affluent a country is, the less children seem to like school. In the rich countries, the school competes with other activities for the attention of children. In less affluent and developing countries, the only realistic alternative to schooling is work, usually of an unskilled type.

A decade ago, an article by Braddock [29] entitled "The issue is still equality of educational opportunity" appeared in a leading educational periodical. It received attention internationally because it addressed a concern with the consequences that a policy of encouraging deregulation and privatization in school education could have for the distribution of learning opportunities among children. This concern is gaining currency in Europe at present, as is demonstrated by the recent flow of studies

---

[27] Cf. A.H. Halsey: "Political Ends and Educational Means", in A.H. Halsey (Ed.), 1972, pp. 3-12.

[28] The educational and occupational expectations of students and parents are not realistic under all circumstances, and public demand for education cannot always be met. Accordingly, the school is not only a place for teaching and learning, but also an institution where expectations get thwarted and ambitions 'cooled-out'.

[29] J.H. Braddock II, "The Issue is Still Equality of Educational Opportunity", 1981.

addressing the issue of the advantages and disadvantages of public vs. private schooling in European countries. Braddock concludes after reviewing a seminal study on private schooling conducted in the United States by Coleman and his colleagues: [30]

> "Not since the publication of the Equality of Educational Opportunity (Coleman *et al.*, 1966) [study] has debate on the issue of school productivity reached such a high level of intensity as it has with *Public and Private Schools*. Equality of educational opportunity should remain a major focus in these debates, and we need to understand clearly the degree to which this goal depends upon a strong public school system and how various proposals for changing the structure of public education would affect equal opportunities". [31]

This quotation refers to a significant development previously mentioned. The emphasis of public commitment to education has been shifting in the early 1980s from the provision of educational equality and the role of the interventionist welfare state to school effectiveness and the contribution of private initiatives to improve school quality. Thus the main focus of debate on educational policy has moved away from concerns aired in the 1960s, when belief in the positive and beneficial effects of formal schooling was paramount and equality of opportunity was a panacea for solving social problems. During the 1980s the idea was accepted that high-quality schooling and educational effectiveness, assessed in terms of the knowledge, skills and abilities of school leavers, were of paramount importance. Concern shifted from equal opportunity to policy intended to optimize learning. The present position seems to be that policy promoting equality of opportunity may conflict with the objectives of increasing the qualitative output of the educational system and making it responsive to the demands of consumers, primarily the workplace but also students and parents. Hence some "no-nonsense" managers of educational systems have expressed a radical idea, one that would hardly have found currency two decades ago. They essentially argue, although the case is generally not presented in such blunt terms, that the notion of equality of opportunity and, indeed, the very idea of comprehensive schooling itself, are potentially harmful – for example, in that they serve to direct attention to the educational needs of the disadvantaged students rather than to the talented, who may be

---

[30] The reference is to J.S. Coleman, T. Hoffer and S. Kilgore: *Public and Private Schools*, 1981a. See also J.S. Coleman *et al.*: "Questions and Answers: Our Response", *Harvard Educational Review* (1981b); and *High School Achievement: Public, Catholic, and Private Schools Compared*, Basic Books, 1982.

[31] J.H. Braddock II, *op. cit.*, p. 495.

mistreated as a consequence. M. Apple [32] recapitulates the argument in the quotation below:

> "Equality, no matter how limited or broadly conceived, has become redefined. No longer is it linked to past *group* oppression and disadvantagement. It is simply now a case of guaranteeing *individual choice* under the conditions of a "free market". Thus, the current emphasis on "excellence" (a word with multiple meanings and social uses) has shifted educational discourse so that underachievement is once again increasingly seen as largely the fault of the student. Student failure, which was at least partly interpreted as the fault of severely deficient educational policies and practices, is now being seen as the result of what might be called the biological and economic marketplace".

In facing the dilemma between equality of opportunity and the meritocratic demands of European societies pressed hard by international trade competition, the scales would seem to have tipped in favour of the latter. The writing on the wall is visible in this respect. [33] Comprehensiveness tends to be "outmoded" in debate in some European countries. Concern for the education of gifted students, privatization of the school, and the setting up of national monitoring programmes for assessing school and student performance – these are all reflections of the same policy orientation, namely that the efficiency and cost-effectiveness of school systems must be increased. What is in store for the European school systems in the 1990s and at the beginning of the next century in this respect?

11. Social Demand and Problems of Access

Closely linked to equality of opportunity is the concept of *social demand*. As the OECD Report on education in the Netherlands puts it generally: [34]

> "If a sufficiently qualified citizen stands at the door of any type of school he must be admitted and it is the responsibility of the appropriate government authorities to anticipate his request".

Educational reforms in Europe have over the last few decades been propelled by three main forces: the quest for equality of educational

---

[32] M.W. Apple: "Redefining Equality: Authoritarian Populism and the Conservative Restoration", *Teachers College Record,* 1988, p. 170.
[33] Cf. J. White: "Educational Reform in Britain: Beyond the National Curriculum", International Review of Education, 1990.
[34] OECD: *Reviews of National Policy for Education: The Netherlands,* 1990. See also Netherlands Ministry of Education and Sciences: *Richness of the Uncompleted. Challenges facing Dutch Education,* 1989.

opportunity, the increased trade competition on the international market, and the increased "social demand" for education. Since the first two were considered above, the present section is in the main concerned with the factors of social demand and access.

In Europe the level of general education in the population has in recent decades been increasing considerably. A four- to six-year compulsory schooling was common a century ago. This period has gradually been lengthened to eight or nine years. Yet the problems of access to education, particularly problems arising in the "pressure of numbers" in the system, have not been entirely solved. It may be true that primary and secondary education throughout the period of compulsory schooling have become universal in all European countries, but this statement conceals a host of difficulties.

The demands for more education have been met in the first place by making more years of formal education available for an increasing proportion of the relevant age groups. But the structure of the school system had in many cases to be changed in order to accommodate the growing enrolment. One solution has been the establishment of a comprehensive basic school where children are kept together longer than was the case in the parallel system, where some of them were shunted off to the academic secondary schools already after four or five years of primary schooling.

Unnoticed by many is the fact that the compulsory school period has not only been lengthened, but that it has also been extended towards the other end, by gradually developing and extending links with kindergarten. Experiments are going on with respect to a flexible school starting age. The ties between primary and pre-primary schools have generally grown stronger; in some cases primary schools have been opened up to very young children. Pre-primary schooling now precedes primary level education in practically every European country. Not only is day care and socialization offered, but increasingly also mother tongue, simple arithmetic, and even science are taught.

The statistics shown in Table 2.1 indicate that enrolment at the pre-primary level has soared in recent years. In many countries, for example Belgium, Denmark, France, Ireland, the Netherlands, and Spain, more than 95 per cent of the children in an age group are enrolled in pre-primary institutions at least one year before they start compulsory schooling. In France, even as much as one-third of the cohort of two year-olds is registered with a pre-primary institution, whereas in Germany there is a more gradual escalation, as Table 2.1 indicates. [35]

---

[35] Source of data: *OECD Employment Outlook 1990*, Chapter 5: Child care in OECD countries, p. 130.

The general trend seems to be that kindergarten is being gradually incorporated into the school system, despite the doubts aired by those critical of this development. Naturally, numerous problems of an organizational nature arise in the process of linking pre-primary and primary institutions. However, these may be considered secondary to problems of pedagogy and curriculum development. The problems hinted at here need to be studied very carefully in the coming decade.

In the United Kingdom, children generally enter into compulsory school attendance at the age of five. However, children who in other educational systems enter at seven years of age, for instance in Sweden, do not appear to perform worse on tests taken at age 10 in elementary reading, writing and numeracy skills than the 10 year-olds in Britain, who have had an additional two years of schooling. Naturally, this raises the question of whether early schooling matters. It also brings into clear focus the necessity of questioning the appropriateness of the fixed-age admission scheme that is in use in most European countries. Such schemes neglect the fact that there exists substantial variability in cognitive, physical and emotional development among children of the same age.

Whereas preschool education is regarded as highly desirable, some countries such as Britain are lagging far behind others, such as France and Belgium where, as can be seen from Table 2.1, schooling from three to six years of age includes practically all children. Moreover, there is little agreement as to where it should take place – in an independent unit, in classes attached to a primary school, or in a *crèche* and nursery attached to the workplace. The emphases on the activities tend to vary accordingly: play and socialization, formal or informal instruction, or a mere child-minding function. It is of course true that, regardless of the institutional type, the best will combine a modicum of all three.

For the handicapped and the disabled the past decade has seen great advances. The general consensus is that, unless the physical or mental disabilities are severe, given adequate safeguards such children should be integrated with their more fortunate fellows as soon as possible. The philosophy is that there is no such thing as a "normal" child; each child is "special" and therefore requires an education according to his needs and abilities. It may well be that before the turn of the century the term "special education", with its pejorative connotation, will have disappeared from the vocabulary of education.

Problems of access to primary and secondary schooling have in the main been solved in European systems. The old rural/urban dichotomy hardly exists, thanks to strategic siting of institutions and the extension of transport facilities. However, there are some who, on social and/or educational grounds regret the passing of the village school. This, however, reflects occupational changes also. Where agriculture is a minority occupation, out-of-town communities often merely serve as dormitories for those whose employment is urban. It therefore is natural that children, like their parents, should spend the day away from their place of residence.

Table 2.1: *Enrolment Rates in Pre-primary Education in European Countries (1986-1987 data, per cent)* [36]

Country	Age 2 years	Age 3 years	Age 4 years	Age 5 years
Austria	1.2	28.4	61.4	90.4
Belgium	21.1	94.0	98.2	98.9
Finland	21.1	16.0	21.1	26.6
France	33.6	95.4	100.0	100.0
Germany (West)	12.6	38.7	72.3	85.5
Ireland	—	0.3	54.3	98.9
Norway	20.8	29.8	41.1	49.9
Spain	4.6	16.8	88.3	100.0
Switzerland	—	3.8	16.2	66.4
United Kingdom	1.3	25.0	68.1	—

Perhaps a more serious problem is posed by inequalities in income. There are still many parents who are unable to support their children beyond the age when compulsory schooling ends. This means that children who would benefit from continued schooling have to drop out and, later, that the European and national communities suffer from this wastage of talent. On grounds of alleged economic stringency, the system of maintenance grants at school level, as well as at that of higher education, would seem to have been curtailed in recent years.

---

[36] OECD *op cit.,* p. 130.

A philosophy of educational élitism once appeared to be gradually dying out, but now seems due for revival. Such élitism may take the form of segregation, either by money or by ability. In the British system the existence of expensive private schools, often providing a first-class education, has been recognized as a problem. Access, for a very small number, is granted for poor children. The problem remains of how the best of such schools should be treated: if education is a market-place commodity then they should be allowed to continue to exist as they are; if not, they should be integrated into the national system. (It must by now be apparent that the problem of access is a highly political one, which hardly impinges upon educational considerations.)

As Europe "went comprehensive" between 1960 and 1980 (as in Sweden, Norway, France and Britain), or at least widened entrance to the more prestigious forms of secondary education (as in Germany, the Netherlands and Austria) élitism of ability during compulsory schooling seems to be dying out. The total "education yield" from a single type of secondary school was reckoned in many countries to be superior to the combined "yield" from two or even three parallel types of school, access to each one of which was a function of ability. However, in recent years, and connected with the drive to quality, the creation of "magnet" schools devoted to some speciality or those aimed at high-quality technology and economic pay-off, both with specific profiles, have led to a reconsideration of the right of access. Is there scope and a future for "key-point" schools of excellence in comprehensive educational systems? This question is related to another issue, namely whether the historically unique situation in Europe at present demands deliberate action to create and cultivate talent.

The focus of the contemporary discussion on comprehensive vs. selective schools seems to be less on the democratic aspect of providing secondary education for all, and more on its alleged limitations, for example the teaching of bright children, the bias against girls, and the difficulties encountered when seeking to encourage school-based innovative experimentation in a comprehensive system.

12. Gifted Pupils and Specialist Schools

The recent swing away from emphasis on the concept of equality of opportunity has, in some countries, been accompanied by a renewed interest in the promotion of gifted and talented pupils. [37] Although in

---

[37] Contributions to thought on the education of gifted children have come from different sources and countries. For many years, study of the gifted was dominated by

some Western European countries such as Switzerland and the Netherlands the very idea of singling out such children encounters hostility as being undemocratic, it is increasingly being asserted that the talented as well as the backward – the "slow learner" – may merit special treatment. In the Netherlands, for example, where according to Monks, [38] in the 1960s and 1970 educational policy was based on the "false premise" that equity meant "all are equal and can reach the same level" (i.e. in educational jargon, there must be "equality of outcome".) Even today the term "more gifted" is preferred to "highly gifted", as having less élitist overtones. How the gifted may be promoted without disadvantaging the rest is a moot problem. One method has been the offering of "enrichment" or additional programmes. Acceleration, "jumping" a year, is frowned upon, mainly because of the emotional problems that may arise. Furthermore, there may be undemocratic side-effects. Thus in Britain, where a few experimental programmes have been mounted, they have given rise to claims of sexual bias and assertions that immigrant and working-class children have not been fairly represented.

However, in the socialist and former socialist countries, where Marxist theory had hitherto ranked equality higher than liberty, no such élitist misgivings have been entertained, and other solutions to the problem have been found. The Soviet Union was the pioneer in the development of special schools for the gifted. The idea for such institutions, initially in mathematics and the sciences, sprang from Professor Keldysch, then President of the Soviet Academy of Sciences, and Professor Lavrientiev, head of the Novosibirsk branch of the Academy. It began to take concrete shape in the mid-1960s, when so-called "mathematical olympiads" were established to select the most suitable students. In the initial stages their promoter was the University of Novosibirsk. Competitive examinations were arranged at district, and then provincial level, and finally at the regional level of Siberia itself. The selection

---

Terman's five-volume 'Genetic Studies of Genius' (L.M. Terman and M.H. Oden: *The Gifted Child Grows Up. Twenty-five Years' Follow-up of a Superior Group: Genetic Studies of Genius*, Vol. 4, Stanford, 1947). General works on the education of élites are W. Troger: *Elitenbildung. Uberlegungen zur Schulreform in der demokratischen Gesellschaft* (München and Basel, 1968); and Ph.E. Vernon, G. Adamson and D.F. Vernon: *The Psychology and Education of Gifted Children* (London, 1977).

[38] Cf. H.-G. Mehlhorn and K.K. Urban (Eds). Detailed accounts of specialized schools in the Soviet Union and the former Soviet bloc are hard to find, and information is sparse regarding their internal functioning. However, the following may be consulted: *Hochbegabtenforderung international*, Bildung und Erziehung, [Beiheft 6], Bohlau-Verlag, Köln und Wien, 1989; O. Anweiler, "Die sowjetische Schul- und Bildungsreform von 1984", *Osteuropa*, 343, pp. 839-860, passim.; and J.J. Tomiak, *The Soviet Union*, World Education Series, London, 1982, pp. 82-84, a brief but extremely useful account.

process continued with a three-week summer camp where, out of the surviving 800, the 400 best candidates were finally chosen, on the basis of aptitude tests and competitive activities. The way in which the scheme was organized is typical of what has since developed elsewhere in the Soviet Union. Today such "olympiads", particularly those of the Univeristy of Moscow for mathematics, physics and chemistry, have become internationally well-known, and attract some two to three million candidates annually. They are asked to devise answers to such questions as, "Would it be possible to calculate the weight of the moon?" The aim is to detect inventiveness, creativity and logical thinking rather than to test factual knowledge.

The exact number of these special schools is not known, although they do not comprise more than three per cent of all secondary schools. They are usually boarding institutions entered at grade 8 (age 14), and provide a three-year course. Univeristy professors and members of the Academy of Sciences are drawn in to teach the students. Classes are small – usually under 15 students. The emphasis is on university methods of working, in which students are set specific problems to resolve individually. In their final year students may even be invited to interest themselves in the research work of the local Academy of Sciences before taking the usual secondary leaving examination and being admitted to the local university. The drop-out rate from the schools is small – under 10 per cent – a low figure but perhaps to be expected in view of the intensive selection procedures. It should be stressed that, unlike the English A-Level system of specialization, all students must continue to follow the complete general secondary curriculum, with additional lessons devoted to their specialization.

Other establishments such as the Moscow Music School and the ballet schools in Moscow, Leningrad and Kiev have also contributed to the Soviet reputation in another field, that of the performing arts. Again, students follow the normal school curriculum for their age and are given extra tuition in their speciality.

"Extended high schools", as they were known, for the gifted in the former German Democratic Republic, following the Soviet example, were also impressive. Ability in mathematics, the sciences – both pure and applied, – the arts (including music and classical dancing), and physical education was encouraged. The outstanding reputation that this country acquired in world sport, for example, was one spin-off from this concentration of talent in special institutions. The insistence was on beginning early: even in the kindergarten notions of science were taught; ballet dancing schools begat at grade 3 (age 9). But the main specializations, particularly in the scientific subjects, began in grade 6 and continued through till grade 12. Again not more than three per cent of the total school population was involved.

Nevertheless, nowhere has any reliable method been found for identifying the gifted. In the Soviet Union it is held that "ability is not only expressed in activity, but is developed through activity". The judgement must in the main be on a measure of performance and not on any measure of general potential such as IQ tests. All in all, it would seem that Western Europe may have something to learn from Central and Eastern Europe regarding the education of talented pupils, if it is prepared to sacrifice equity to political and economic considerations. The discussions in several European countries on so-called "Magnet" schools, on American lines, such as 'Sports Gymnasia' in the Netherlands or the creation of City Technology Colleges in England, may be symptomatic of future developments.

13. Access to Continuing Education

Another aspect of the problem is that schooling should give access to continuing education, either in some kind of technical, vocational or academic institution. Here again there is a wide divergency of practice. France is striving to provide 80 per cent of the relevant age group with the baccalauréat qualification which, given the admission rules applied at present, would result in a marked increase in the number of students eligible for higher education. Germany is lagging not far behind France in this respect. Britain, on the other hand, is struggling towards some 25 per cent of an age cohort. Closely connected with this is the question of transferability. Pupils should be able to transfer easily between tracks either within school, or when going on to further education. This "permeability", which after 1992 will have also to be addressed in a European dimension, is clearly highly desirable. The old system of "typing" children at the age of 10 or 11 should clearly not be repeated in a similar "typing" at the age of 16 or 18.

The problem of access raises the whole question of entrance after 1992 to all European institutions. This topic is too broad to be treated here in a satisfactory way, considering that the mobility of teachers and their right to take up positions in Germany, for example, may well make it necessary to influence the traditional system of state-controlled examinations conferring Beamter-status to those who pass, something many Germans will not take lightly. Suffice to say that up to yet mainly only "paper" agreements on what is called "equivalences" between levels, courses, qualifications and countries have been achieved. The real acceptability of diplomas acquired or courses followed in one country for practical use in another i.e., in terms of being able *to do the job*, whether it be plumbing or medicine, has yet to be tackled and it should not be left to the bureaucrats. Certificates, diplomas, final tests, credits, all these primarily administrative devices, have a great impact on the educational processes. They tend to reduce them to rote learning – to a preparation for examinations rather than for coping with real life, real politics, a real job. These will always vary locally, regionally, nationwise. The more unified the standards are, the less "real" will the educational substance be. The contradictory demand for common standards to safeguard mobility and equality of opportunity, on the one hand, and variety of subject matter, method, and perspective, on the other, will require formidable efforts in educational research and development in this particular field and an enlightened view of what really matters.

Generalizability of access, therefore, will remain unfinished business on the educational agenda of a unified or federal or confederal Europe, or even a continent of discrete entities bound by economic ties alone, till understanding has been reached about the primary goals of education. (cf. Chapter 3)

14. Professionalism and Vocationalism

Developments such as accountability and participation have not always been welcomed by teachers, who have resented being treated as a labour commodity in the same way as other groups in the economy, and have aspired to the same kind of status as doctors or lawyers. Throughout the period from 1960 to 1985, the teachers' unions have put forward a principle of *professionalism*, and fought for their autonomy as well as their salaries. This has evoked a varying response from governments, which in some cases have reinforced the on the spot control exercised by headteachers and the inspectorate, but in others have allowed greater freedom to the individual teacher in such matters as content and teaching method.

To raise the question of the curriculum is also to raise that of vocationalism. The initial assumption behind the doctrine of equal opportunity was that this meant the transmission to all children of the "high culture", since this represented the royal road to university education and access to the more remunerative forms of employment. However, in the 1970s, when education increasingly became a function of the economy, the concept of vocationalism, i.e. of preparing young people for the world of work, took hold both in secondary and higher education. In the former this was expressed by the creation of new broadly vocationally oriented sections in the upper secondary school, as for example in the 22 options now available in Sweden for the 16 to 19 year-old students, and a gamut of technical qualifications that could be achieved in school, such as the dozen "baccalauréats techniques" in France that were added to the conventional range of baccalauréat options. In higher education the thrust to vocationalism was expressed in the creation within universities of technological degree courses and, above all, by the setting up of parallel technological institutions. Although at the upper level such new courses and qualification have now been widely accepted, at a lower level "technical education" still retains some of its former stigma, as being a form of education for the second-rate — a "dumping-ground" or, as the French say, a "dépotoir" for those who are rejected by the general educational system. In this field much work remains to be done, from the setting up of "bridges" between the "ladders" of different courses, to ensure the principle of "transferability", to the prevention of too narrow a specialization.

One of the most radical alternatives to the parallel system of academic and vocational education was developed in the Land Northrhine-Westfalia in Germany during the 1970s. In more than 35 experimental schools, after a comprehensive secondary level, the upper grades (16 to 19 year-olds) take courses in three general divisions: the "seminar type", the "studio type", and the "workshop type", and finish school with both qualifications: an immediately marketable vocational skill and a university-entrance admission.

15. Decentralization and Parental Choice

Whereas many governments in European countries had, during the 1960s and 1970s, adhered to a strategy of implementing change through planning at central government level, the trend beginning to emerge in the 1980s was to leave school improvement and the matching of demand and supply to the local market. This new approach to educational planning was accompanied by a trend to decentralization in decision-making. By encouraging experimentation at school level, the school itself was made the focal point for innovation. With it came a call for effective management in schools and, consequently, efforts to improve educational leadership. However, decentralization of decision-making was generally accompanied by an increase in central government monitoring of developments in school education and evaluation of aspects of student performance and school effectiveness. Consequently, some European governments became increasingly interested in the development of what in Britain has become a "National Curriculum". [39]

Decentralization was parallelled by a tendency to increase parental involvement in schooling. The latter was perceived as a means of improving not only the image but also the effectiveness of the school. The concept of *parental choice* in the school the child attends and the nature of the curriculum he follows in it is more widely observed than ever before. Moreover, parents have now much more say in the running of schools than they ever had previously. In many EC countries this has been achieved by a reconcentration of power and a partial redistribution of it at school level. Even in school systems where free choice is either not allowed or strongly discouraged, the resourceful parents can still find ways of gaining access to the school of their liking. Middle-class parents are even known to have moved their residence in order to influence this decision. Another way of achieving this is by adhering to the right to freedom of religion and the related demand for religious education. It is of interest to mention here that the problem of parental choice is not that parents prefer private over public schools. The issue at stake is that parents want access to good schools, regardless of whether these are public or private. They thus demand public access to information disclosing the quality of schools in their neighbourhood. This concern with high-quality is also behind the demand for 'educational vouchers' and teaching contracts. There are some risks involved as well, since an increase in parental influence may also have less desirable effects on, for example, equity and curriculum coverage.

During the years when most European nations were beset with economic difficulties and mass unemployment, countries on the Pacific Rim took over a significant portion of output in both labour- and technology-intensive sectors of industrial production. Global competition was mounting. The belief that the economic strength and wellbeing of Europe depended on trained manpower and the ability of the inventive mind led to a new emphasis on the output of the educational system, measured not merely in quantitative but to a large extent in qualitative terms, for example in relation to the levels of intellect, skill, literacy and training essential to success in the evolving "high tech" era of global competition.

The emphasis on the economic output of the production system was as a result transferred to higher education and even the school domain. The criticism levelled against schooling in the United States was repeated in

---

[39] The following sources may be consulted for a discussion of trends and issues as regards the national curriculum in British education: United Kingdom, Department of Education and Science (1989), White (1990) and Wiliam (1990).

several countries in Europe. It was noted that Japan did better not only with respect to average student achievement in mathematics and science, but also in the proportion of an age cohort retained in the educational system. Consequently, critics passed sweeping judgements on the state of European education, arguing that schools were lacking in quality and intellectual rigour and that standards were deteriorating. Elitist ideas were revived and a quest for specialization began.

In some European countries a case was made for instituting what in the United States are called "magnet" schools for young people with certain special interests, for example in music, fashion or audio-visual engineering. A call for "key" schools for highly gifted children was also heard in some countries. The idea of creating "magnet" institutions originated with the realization that the secondary school in modern industrial society has to offer young people a profile they are interested in. The aim is not only to motivate students but also to succeed in competing with other schools in attracting sufficient numbers. By contrast, the argument for building key schools is based on the realization that Western European countries are no longer a step ahead of other nations in putting innovations in high technology and electronics to good use in industrial engineering and manufacturing.

16. School Education in "Nations at Risk"

It can be concluded that three issues receiving much attention from the media and the general public in the early 1980s were, first, financial austerity, rising costs and budgetary restraints in education; secondly, discipline problems and negativism in the classroom; and, thirdly, the alleged decline in standards. The rising costs of schooling were perceived as a problem, not least in the rich countries in Europe. An increasing share of the former had to cover staff salaries. In a situation in which the public budget for education was subject to severe constraints, increases in the salary costs of the teaching force resulted in cut-backs in other, quality-related areas of educational expenditure, such as the maintenance and improvement of school buildings, and the provision of books and teaching materials. Thus the problem of rising costs in education was partly also a problem of making choices among the various categories of expenditure. The fundamental question asked was whether countries and citizens were getting value for money. The lower priority given to education, which characterized the better part of the 1970s, continued well into the 1980s. As the evidence on the outcomes of implemented reforms was often weak, or even not supportive, the governments of most European countries – Spain and the Netherlands being among the few exceptions – were reluctant to embark on new structural reforms. [40]

A significant development in the 1980s was that the market principle of free competition was welcomed in the educational systems of Western European countries, notably in the Netherlands and the United Kingdom. The consequent shift in school policy coincided with a feeling that the economic crisis in the Western countries was finally over and that schools needed to be made flexible and responsive to changes in demand. Competition between schools was encouraged and privatization was considered as a serious option. This was interpreted in several countries as an expression of distrust among a section of the public of the capacity of central authorities to improve the quality of school education. The emerging "new confidence" in the value of education in the latter half of the 1980s was not exclusively concerned with schooling, as was the case in the 1960s, but also with the post-initial component of the educational system. It thus differed from the oversimplistic belief of the 1960s in formal school education as the main motor behind economic growth. This development explains much of the recent interest in the policy implications of the related principles of lifelong and recurrent education which, although initially proposed in the late 1960s, seem only now to gain currency in practice in many European countries.

In 1981, President Reagan, sensing the mood of the time, appointed a National Commission on Excellence, directing it to examine the quality of education and to define problems obstructing "the pursuance of excellence" in American school education. [41] It may be worthwhile to reflect on some of the problems and conclusions raised in the report the Commission presented in 1983, as they have relevance in the European context. The report opened with the provocative statement: [42]

> "Our Nation is at risk. Our once unchallenged pre-eminence in commerce, industry, science, and technological innovation is being overtaken by competitors throughout the world. This report is concerned with only one of the many causes and dimensions of the problem, but it is the one that undergirds American prosperity, security, and civility. We report to the American people that while we can take justifiable pride in what our schools and colleges have historically

---

[40] The state of the art of school policy in Spain and the Netherlands is described in two reports published by the Organization for Economic Cooperation and Development: *Reviews of National Policies for Education: Spain* (OECD, 1986) and *Reviews of National Policies for Education: The Netherlands* (OECD, 1990).

[41] The National Commission of Excellence in Education was appointed by the U.S. Secretary of Education, T.H. Bell, in August, 1981. It is striking to note how the term excellence made a comeback. Papers on excellence were also published in the late 1950s and early 1960s. Examples are Yarmolinsky (1960) and Gardner (1961).

[42] National Commission of Excellence in Education, *A Nation At Risk: The Imperative for Educational Reform,* 1983, p. 5.

accomplished and contributed to the United States and the well-being of its people, the educational foundations of our society are presently being eroded by a rising tide of mediocrity that threatens our very future as a Nation and a people. What was unimaginable a generation ago has begun to occur: others are matching and surpassing our educational attainments".

The conclusions presented in *A Nation At Risk* signalled the issues that would come to the fore in educational reform during the remainder of the decade, not only with respect to the United States but also in many European countries. The document had an impact on the school debate on both sides of the Atlantic. This may well have been due, in least in part, to the critical role played by OECD and its Centre for Educational Research and Innovation (CERI), where educationalists advocated the necessity of taking measures to raise the quality of schooling for all. [43] Given the multitude of truly alarming findings the report contains, this is not surprising. For example, it is mentioned that about 13 per cent of all 17 year-olds in the United States could be considered functionally illiterate, and that average student achievement on many standardized performance tests conducted in the early 1980s was lower than it had been in 1957, when Sputnik was launched.

17. The Quality of European Schooling

It was realized by some policymakers in Western European countries that several of the concerns expressed in the report cited above prepared by the U.S. National Commission on Excellence could well be valid even in a European context. [44] This realization intensified the debate on the perceived weaknesses of European educational systems. [45] The question was asked whether the quality of teaching, student achievement and motivation had declined also on this side of the Atlantic. This question

---

[43] This applies to the OECD project on the development of international educational indicators in particular, in which the United States has played a major rôle.

[44] A recent comparative study of mathematics and science achievement in different school systems confirmed that American youth performed poorly compared with their peers in economically advanced countries (ETS, 1989). The findings of the ETS study are used in a report summarizing several large-scale investigations conducted under the National Assessment of Educational Progress (NAEP) (U.S. Office of Education, 1990). The impact of the "Wall Chart" on school performance in the United States is also discussed (see also Ginsburg *et al.*, 1988).

[45] See, for example, Netherlands Ministry of Education and Science (1989) and, for a review of the situation in France, J. Lesourne (1988). The national reports presented at the 42nd Session of the International Conference on Education in Geneva, September 3-8, 1990, bear witness to a deep concern with the improvement of quality standards and excellence in European school education.

was given new urgency with the publication, by the International Association for the Evaluation of Educational Achievement (IEA), of results from the comparative studies of student achievement in mathematics and, in the mid-1980s, student performance in science subjects. [46]

The aim of the IEA studies has been to find out not only how well the students at a particular age and grade level in a given country are doing compared with student populations in other countries, but also to identify the policy-relevant factors which help to explain differences between countries. Case-studies of the impact of IEA research studies of student achievement, which make use of very large representative samples of students, teachers, and schools, and employ psychometric methods of analysis, show that knowledge about these productive factors in schooling has been of great value to policymakers. The results of the studies have been used as a basis for decision-making in educational reforms in European countries, for example in Hungary, Italy, and Sweden. [47]

A comparison of average student achievement in mathematics at the grade 11 or 12 level in upper secondary school in different educational systems is presented in Figure 2.2. Estimates are presented not only of student achievement in European countries but also in three reference countries, namely Japan, Canada and the United States. The estimates are given separately for three groups of students, namely, the one per cent best, the five per cent best, and the estimated average score of the entire student population. [48] Some care must be taken in interpreting these findings because of significant differences between the educational systems in the proportion of the relevant age cohort retained in advanced mathematics courses at the upper secondary level. That significant changes have occurred in most European countries since 1980, when the mathematics survey was conducted, should also be noted here.

The data for Europe show that top students at the grade 12 level in French-speaking Belgium, Hungary and Scotland scored less than 50 per

---

[46] The findings of the IEA Second International Mathematics Study are reported in Travers, K.J. and I. Westbury (1989), Burstein (1992), and R.F. Robitaille and D.A. Garden (1989). A summary of the international results of the Second International Science Study was co-edited for IEA by T.N. Postlethwaite (1988).

[47] Several case-studies have been conducted of the impact of IEA research on international and national policymaking in education. Examples are found in a report edited by A.C. Purves: *International Comparisons and Educational Reform*, 1989. Another example is T. Husén: "Policy Impact of IEA Research", *Comparative Education Review*, 1987.

[48] Source: *The Second IEA International Mathematics Study*. Informative is a report by the Swedish National Board of Education: *Swedish Schools in an International Perspective* (Newsletter 1989:6).

cent correct answers on the international mathematics test administered to all students. The same applied to students in Canadian British Columbia and the United States. By comparison, the Finnish, English, Swedish and Flemish-speaking Belgian students scored at the 50 per cent level or above, as did the Japanese. These results have provoked demands for reforms in mathematics teaching in the low-scoring European countries such as Hungary and Scotland. Reforms were carried out also in the United States.

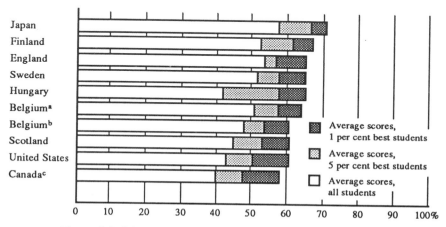

Figure 2.2: Mathematics Achievement at the Terminal Grade of Upper Secondary Schooling in Different Systems (1980 IEA data).[49]

Figure 2.3 shows the percentage of primary schools in different countries with lower average scores in science than the least achieving school in Japan, a country where the highest overall level of student performance was measured. [49] The tests were given to representative samples of 10 year-old children in 1983-1984. The findings show that there were only three systems where less than 20 per cent of the schools scored lower in science then the least scoring school in Japan. More than 60 per cent of all primary schools in England and Poland scored below this benchmark level. In these cases the results can be explained, at least in part, as a consequence of the phasing of the curricula, as science was introduced at a later age compared with other school systems.

---

[49] The data refer to the core-test only. Source: T.N. Postlethwaite and D.E. Wiley, *Science Achievement in Twenty-Three Countries*, Oxford, 1992.

The results at the 10 year-old level obtained in comparative surveys of mathematics and science achievement conducted by IEA indicate that elementary school students in European systems such as Finland, the Netherlands and Sweden perform at a lower, but largely an acceptable level, compared with Japanese and Korean children of the same age. One implication of the results is that the early exposure to science and mathematics seems to have a lasting effect, as children at 10 years of age in European systems teaching science in primary school can use basic skills in different situations. Children at this age have acquired a rudimentary knowledge that is a prerequisite for becoming scientifically literate.

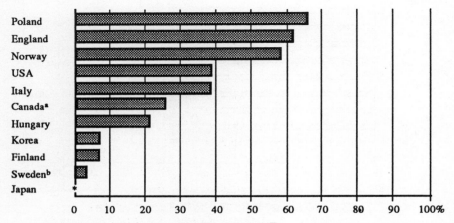

Figure 2.3: Percentage of Primary Schools in Various Countries with Lower Average Scores in Science than the Lowest Scoring School in Japan (1983-1984 IEA data).
[a]Only Canada–English. [b]Grade 4 level.

Table 2.2 shows the results of 14 year-old students from different countries on the science core-test administered in 1983-1984. The Hungarian students were at the top, followed at a short distance by students from Japan and the Netherlands. The data also indicate that 36 per cent in Italy scored lower than the lowest performing school in Hungary. This result may be compared with that of the United States, where 30 per cent of the schools scored below the benchmark level. The evidence emerging from the IEA science survey was in some countries, notably in Italy and the United Kingdom, interpreted as indicating that all was not well with the quality of school teaching. On the whole, however, there was an impression that the performance of students in the various European systems did not quite reach a level as low as that of students in the United States and that, by implication, the conclusions drawn by the U.S. Commission on Excellence in Education could be discarded as being not fully relevant in a European context.

Table 2.2: *Science Achievement at the Lower Secondary Level (14 year-olds) in Different Systems (percentage of schools scoring lower than the lowest scoring school [49.4] in Hungary, 1983-1984)* [50]

School system	Grade level	Schools (N)	Overall score (Mean)	Deviation (s.d.)	% schools below 49.4
England	9	3,118	55.8	16.4	19
Finland	9	2,546	61.7	13.8	2
Hungary	8	2,515	72.2	15.7	0
Italy	8	4,622	52.4	16.4	36
Japan	9	7,610	67.3	16.8	0.5
Netherlands	9	5,025	65.8	16.9	16
Norway	9	1,420	59.8	15.8	1
Poland	8	4,520	60.4	17.3	14
Sweden	8	1,461	61.4	16.3	2
United States	9	2,519	54.8	16.7	30

That the average performance of European students in science was, on the whole, not as alarmingly low as that of comparable students in some other world regions is also shown by the estimates presented in Figure 2.4. The results, which are reported by Keeves, [51] can be used to answer the fundamental question whether the average standard of performance in science of students in European educational systems has changed over time. The first international science study was conducted by IEA in 1970. By the time the second study was planned in the early 1980s, many countries had carried out reforms and changes in science curricula had been introduced. It was therefore of interest to examine the magnitude and direction of change in the educational achievement of similar target populations between these two points of measurement in time. An achievement scale was developed that enabled the performance of

---

[50] Source: IEA, and T. N. Postlethwaite and D. E. Wiley, Oxford, 1992, Table 3.3.
[51] J.P. Keeves: *Changes in Science Education and Achievement: 1970-1984*, Oxford, 1992.

students on the science achievement tests of 1970 and 1983-1984 to be compared. [52] The advantage of the scale was that the achievement of students could be defined to correspond to five standards: basic, elementary, intermediate, advanced and specialist. The intermediate level of achievement was set at the zero point of the test taken by the 14 year-old students in 1983-1984. Hence the mean scores of students in the different countries are located between the intermediate and advanced level.

The results recorded in Figure 2.4 generally indicate that small to moderate gains had occurred in the levels of science achievement of 14 year-old students in the educational systems of most countries investigated. [54] Changes over time in the achievement of students in six European systems could be examined with the available data: England, Finland, Hungary, Italy, Sweden and the Netherlands. Sizeable gains are recorded in Hungary. The Dutch case must be viewed with some scepticism because the populations are not fully matched. An interesting observation is that the achievement Some growth in average science achievement is also found in the remaining four systems. Hence, the data in Figure 2.4 provide some justification for the claim that the standards of achievement in European school education are not lagging behind those of competitor countries. Quite the contrary. In the chapter of the volume reporting the comparative analyses of the IEA science data over time, the authors conclude: [55]

> "The evidence presented in ... this report ... led to expectations that gains in science achievement at the 10 year-old and 14 year-old levels might be observed. The findings are generally in agreement with these expectations. Moreover, the gains at the 10 year-old level are generally larger than those recorded at the 14 year-old level. Indeed, ... it might be argued that in some countries the gains at the 14 year-old level are due to a significant extent to gains made during the primary school years".

---

[52] Adjustments were made in the comparative analysis of data for the differences in ages of students in the samples.

[54] Consistent with the findings reported in *A Nation At Risk, op. cit.*, a significant decline was found in the average standard of science achievement of 14 year-old students in the latter country.

[55] The results appear in a chapter by J.P. Keeves and A. Schleicher: "Changes in Science Achievement: 1970-1984".

School Concerns in Retrospect

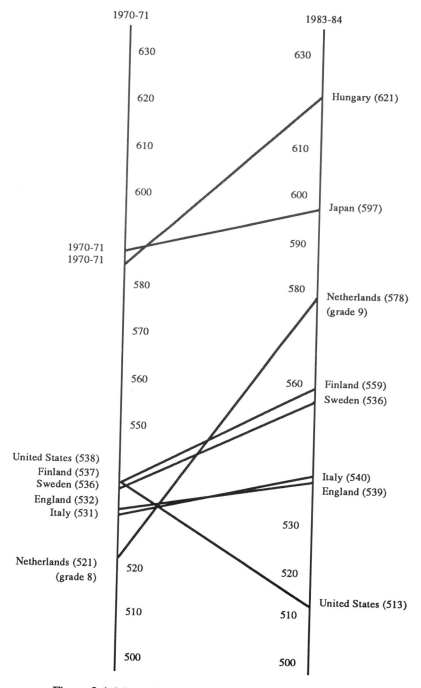

Figure 2.4: Mean Scale Scores on the Science Achievement Scale, 14 Year-Olds, Comparison 1970-1971 and 1983-1984.[53]

Findings such as those reported above contributed in the mid-1980s to a shift in concern from the improvement of teaching at the secondary level to problems of providing high-quality schooling at the primary level. Of paramount concern was the search for ways to improve the quality and effectiveness of primary education. More attention was given to the core curriculum, teaching methods, the teaching of mathematics and science and the role of leadership in school improvement. Other developments in some European countries were, firstly, that schools were encouraged to carry out experiments and, secondly, that national assessment programmes were set up and evaluation projects initiated. [56]

18. Concluding Observations

By way of summary, from a financial view point the period from the early 1960s to early 1970s may be described as one of *incrementalism* followed by a period of austerity and *decrementalism*. In the 1960s funds for school education were comparatively plentiful, and were spent on the building of new institutions and the development of new courses, etc. However, there was a "disjointment factor". Most countries lacked an overall plan for education as a whole, and there was little consideration given to the knock-on effects of expansion in one sector upon another. Thus in France the raising of the leaving age and the increase in the provision of academic secondary education had the result of swamping the facilities available in higher education, leading among other things to the "events" of May 1968. However, decrementalism soon set in, as faith in the economic effectiveness in education diminished and as the downturn in economies began with the two oil crises of the 1970s. By the early 1980s education was no longer the "priority of priorities" that it once was. The rise and fall of education may be charted in the growth and decline in educational expenditure as a proportion of the Gross Domestic Product in EC countries.

It has been noted here that major objectives have slipped down on the priority scale in Europe since the 1960s – for example, the policy of providing equality of opportunity. This may well present a risk and a dilemma, however, as the cost of failure in the school system is very high to both the individual concerned and society. [57] If equality in education is allowed to fade from view, then, as will be argued in Chapter 4, it is

---

[56] Recent trends in school effectiveness research are summarized in B.P.M. Creemers and J. Scheerens (1988).

[57] This point is made in the OECD policy document *High Quality Education and Training for All,* which was prepared as a basis for discussion at the meeting of the Education Committee at Ministerial Level, November 13-14, 1990.

likely that a "new educational underclass" will grow. However, other objectives have achieved higher priority, such as school effectiveness, quality of instruction, and the provision of public education for gifted children. This new emphasis may well be said to represent a policy shift, as it bears the hallmark of a fundamental change in the perspectives on the role of public school education in several European countries. As has been the case with all major educational reforms implemented in Europe in past decades, this "new wave" in educational policy, which would seem to mark the beginning of a period of extending the private sector in the "state education monolith" in, for example, Britain has met with strong opposition. [58] Another aspect of change well worth noting is that uniformity as a schooling objective is gradually being replaced by liberalisation and individualism, even in former socialist countries such as Czechoslovakia, Poland and Hungary. The desire on the part of parents and government officials in some countries to diversify and to extend the autonomy of the school can also be noted.

Finally, there must be mentioned the increasing *bureaucratization* and *politicization* of education. Officials, whether operating in an educationally centralist system such as in France or comparatively decentralized systems such as in Britain and the Federal Republic of Germany, increasingly intervened directly in the schools. Headteachers became their executants, managers rather than educationists. In Sweden the National Board of Education wielded its power beneficially; [59] in Britain the Schools Council was less effective; and the results achieved by the German *Bildungsrat* have likewise been characterized as abortive. Indeed, direction by officials was not accepted precisely because of increasing politicization. In countries such as France, of course, education had always been a political issue. But in Britain, where it has been said that education was seldom discussed at Cabinet meetings in the 1960s, the situation changed radically. The former consensus on education between the political parties was broken, was followed by a period of "dissensus", and then the constitution of partial agreement regarding ends but heated argument about how to achieve them. When education finally passes into the domain of the European Community,

---

[58] A summary of the criticism against the recent reform of education in Britain can be found in the publications: *Take Care, Mr. Baker* (Haviland, Ed., 1988) and *Bending the Rules. The Baker 'Reform' of Education* (Simon, 1989).

[59] The Swedish National Board of Education has been abolished from July 1, 1991. It has been replaced by a smaller institution, the School Board, with evaluation and monitoring as its main tasks. This development provides an excellent example of dual trends towards centralization and decentralization, as one part of the responsibility of the former institution has been handed over to the Ministry of Education and another part transferred to regional and local educational authorities.

the school will run the risk of becoming subject to political controversy. The fear is that the voice of the "specialists" in education, whether theorists or practitioners, will go unheeded.

This retrospective review of issues, which have dominated the educational debate from 1960 to about 1985 and have thrown up many problems still unresolved, cannot be carried further. The last quinquennium is too close to be accessible to detailed analysis. However, certain tendencies are already emerging, such as the necessity for a revision of the curriculum in the new age of information technology, the view increasingly taken of education as a lifelong process, and its corollary, the greater integration of education into the community. And there are doubtless others, of which nothing has been said in the present chapter.

Chapter 3

Emerging Issues in School Education

1. Introduction

From the past, we now turn to the future. What kind of schooling is called for in a new, enlarged Europe? What is feasible and what issues remain to be resolved? The study group has addressed itself to the task of pinpointing the problems that may arise. The order in which they are dealt with here in no way indicates any priority, nor can they be isolated and discussed sequentially, for each one has an impact on the others.

As previously noted, the main concern since the 1960s has been the restructuring of secondary education, the management of the expansion of education and the discrepancies between demand and supply in terms of study places, teachers and facilities, such as buildings. The impact of the austerity in public budgets in Western European countries in the 1970s on school education was, with a certain time lag, felt strongly during the early 1980s. At about the same time there was a change in policy orientation. Decision-makers shifted priorities and called for new strategies of improving the quality and effectiveness of school teaching, raising standards of student achievement, and promoting excellence. This emphasis on quality set much of the stage for school policies in the 1980s.

Chapter 1 concluded that the building of the new Europe may well require a new agenda for education. A future-oriented educational strategy may be required actively to prepare citizens for the "larger unit Europe". However, a reconsideration of what European education is, or what it perhaps ought to be, cannot be entirely Euro-centred. Even though national educational systems in Europe reflect unique traditions which have to be retained, the development of models of education in other parts of the world has to be followed closely. Americans have recently been pondering the lessons they could learn from the Japanese. [1] The Europeans are well advised to do the same, although critically, with respect to schooling in, for example, the United States and Japan.

2. The Role of the School

The nature of most of the problems of European cooperation in education outlined in Chapter 1 is such that not only do they challenge the content and procedure of existing basic education in the nation states but also pose the question of whether the school is the right place to deal with them. In other words: The problems of living together on this continent at this time set new goals for basic education, but the extent to which these can be attained depends on the rôle assigned to the school.

To put it bluntly: The traditional role of our traditional school makes it unlikely that it can do the job alone. But the other socializing agents cannot do it in isolation either. One has to take a critical look at the school, the home, the *polis*, and the mass media and at the actual and possible division of their functions in education.

The school, more than the other agents, has certain traditional tasks in different European nations. The pragmatism of British schools; the very notion of "Bildung" as distinguished from "Ausbildung" and "Erziehung" in the German system; the French inclination for Cartesian rationality; different patterns of organizing school: as a selective or as an egalitarian institution; the clear preference for polytechnical vocational training and schools of high specialization in some hitherto socialist states; the preference for the principles of American high school education in the Nordic countries; different affinities to the major European languages – all these variations hint at divergent definitions of the role of the school at the present time; a time when the policymakers share a common concern with the objective of raising the quality of school education.

It is therefore appropriate to begin by considering various ideal types of school. Six essentially different models of the school can be distinguished. These are:

(1) School as an institution charged with the task of contributing to the development of the *whole* individual; a place where knowledge and special skills are imparted that cannot readily be acquired solely by experience gained outside the school. This role of the school is traditional, whether it be that of the former medieval Latin school, as a preparation for medicine or theology, or another vastly different kind of vocational course today in say, motor mechanics.

---

[1] A good example is the report by M. White: *The Japanese Educational Challenge: A Commitment to Children*, London, 1987.

(2) School as a *protective* institution, sheltering the young against the outside world in which the adults are setting the wrong examples, placing them in a sort of quarantine until they are able to stand on their own feet and face society, and equipping them to do so. One feels that Rousseau, who was nevertheless against formal schooling, would have approved of this.

(3) School as the *"moulder" of human beings*. No national educational system, and, *a fortiori*, any future "internationalized" or "Europeanized" system of schooling, can be exempt from the ambition of wanting to shape the rising generation according to certain goals, which may be different depending on whether one takes a "conservative" or "revolutionary" perspective.

(4) School as an *instrument of social policy*. Society makes certain demands upon the educators, since they work with children and implement many functions of schooling that have to be fulfilled before adulthood. Such functions include the preparation of young people for further training and employment, according to the needs of the labour market. Certain habits that society deems desirable must be inculcated, and others have to be suppressed. In a democracy, the personal development and "sorting" functions must be carried out in accordance with social justice.

(5) School as an institution that *complements* the education that goes on, willy nilly, within the home and the community – the education, for which the parents usually carry the principal responsibility.

(6) School as a kind of sub-society, *a place for children to grow up in*. The school has become the focus for children for the major part of the day. In modern industrialized countries there may be less scope than previously for learning by experience and participation in the activities of the adults which increasingly are performed outside the home. However, the life problems which the children bring along to school often supersede their learning problems and have to be dealt with by the teacher. The school can therefore be seen as "a place for children to grow up in", as Paul Goodman put it. If the role of the school is widened in this sense, then it may be made responsive to the challenges of a modern, complex society: it can be organized as an "embryonic society", as suggested by John Dewey, and enter new relations with the family, the community and the media – the "conditions of life" which are still the most powerful and the least manageable agent in education.

The above models, of course, co-exist in some form or another in most schools today – in a pragmatic blend meeting the demands of various local and historical conditions. Yet, some of these mixtures are highly inconsistent; one function inhibits the other and, certainly, reforms cannot take place effectively as long as one has not decided about the proper mixture. Measures taken to improve training in special skills and those that will facilitate "self-finding" processes may be very much in each other's way, and both may not fit the idea and practice of selectivity.

As mentioned above, however, the models are mere skeletal structures that require to be fleshed out and beg the most essential questions. What kind of skills and knowledge should be taught? How far should the young person be protected from society? What goals should policymakers set for the school? If education is for the individual as well as for society, to what extent has society the right to use the school as an instrument of social policy? How far should society impinge upon the rights of the family and to what extent can it act as a surrogate for the parent? In pluralist democracies, such as those already existing in most of Europe today, where many divergent views are represented, what common frame of reference can be established?

In this context it is impossible to do more than indicate the nature of the conflicting views and showing the possibilities that may work towards a convergence.

Before moving on to a discussion of goals, it is important to note that a "European dimension" is not too difficult to identify (cf. Chapter 1), although it is at present imperfectly dealt with in schools. Thus present programmes for such subjects as civics, social studies, *éducation civique* or *Gemeinschaftskunde* – the designation, as does the content of such courses, varies from one country to another – are inadequate and may well require an overhaul, an updating to face up to the new international community of the 1990s. It can be argued that the teaching of European "commonality" should not be taught through one particular subject but should permeate all subjects. This argument holds good not only for such politically sensitive subjects as history and civics, but also, for example, in the sciences, where ecological problems and their global and European implications can be dealt with. Efforts have been made in the recent past to promote programmes such as "education for international understanding" and "peace studies". It is on such, or similar work, that new developments may be based.

The question of the proper goals of the school in the new European social order and perspective is particularly relevant at present, as it relates to that of the new European identity and the problem of managing pluralism. Contradictory developments complicate the problem of defining goals and objectives of basic education in Europe. For example, there is a trend towards increased international cooperation and economic and political interdependence, but a trend towards revived nationalism or regionalism can also be observed. How can schooling meet both ends, achieving tolerance and mutual respect as a first step towards the conception of European citizenship, while at the same time safeguarding the diverse interests of the nation-states and the different cultures they represent? There is a case for learning from the long experience of regional, cultural autonomy within the nation-state, which some European countries have gained, for example Switzerland, and the more recent experience of Germany, Spain or Italy. Education could contribute to this.

3. Goal Conflicts in Basic Education

A distinction can be made between three sets of goals for educational systems: economic, social and pedagogical. Among the economic objectives of education are the training for occupational skills, contribution to economic growth and more recently training schemes for reducing unemployment. Social objectives include satisfaction of the social demand for education, promotion of equity through education, and preparing for citizenship. Pedagogical objectives relate to curriculum reforms, educational innovations and strategies for pedagogical renewal.

However, the goals of education are usually also interpreted in prescriptive and idealistic terms. For example, many consider that the purpose of education, particularly basic education, is not confined to conferring knowledge and skills for optimizing the work potential of people. Educational systems are also charged with the aim of increasing the general ability and willingness of people, in their role as responsible citizens, to become involved in and to influence the democratic development of a society. Education is also regarded as constituting a means for realizing cultural enrichment and personal self-fulfilment, for example, through the development of enriching leisure activities.

Schooling in modern society, as has been the case in other fields of public administration, has increasingly become a "processing industry" with elements of industrial activities. This may well be said to reflect a tendency to conceive of the school as an instrument of an efficient, product-oriented manufacturing company working in a highly competitive international setting. Schools in modern societies have become an instrument of international trade competition. But it must be acknowledged that the main purpose of the school is to contribute, along with the family, to the development of the individual. The sheer size of school units leads to a fragmentation of social contacts and the formalization of elaborate controls at all levels in the educational system. The hierarchical nature of the latter particularly reflects the complexity of the highly specialized societies that have gradually affected the school. Compartmentalization, which is inherent in such control-oriented societies, leads easily to an incompatibility of norms and difficulties in socialization. Inevitably, these inconsistencies become visible in discrepancies between values, goals and functions of schooling.

Unrecognized but nevertheless built-in goal conflicts lie at the basis of the "malaise" that besets schools in a highly industrialized and technocratic society.

During the period of economic growth, educational expansion and rising levels of expectations, certain goal conflicts went either unrecognized or were simply glossed over. The objectives stated in school legislation or in curricular guidelines had in some countries, such as Sweden, quite a lot to say about education as an instrument of equalization. However, things did not, for various reasons, work out according to such expectations. First, when children enter school there are already some whose background makes it easier for them than for others to absorb what the school tries to instill. Secondly, as children move through the grades, individual differences in performance often tend to increase. By giving everybody the time and attention he or she needs, the vast differences in learning outcomes can be reduced, and those who otherwise tend to lag behind can be brought up to the level envisaged. But what cannot be controlled by individualized teaching relates to powerful factors outside the school, in first place the home and the amount of support students can enlist from their parents – if they have any and if they care.

Underlying the role of the school in the new Europe, in which economic forces are an all-important factor dictating the pace of change, is the issue of providing equality of opportunity within a meritocratic system. [2] The dilemma is that on the one hand the school provides abilities and skills that increasingly constitute an individual power base, but on the other hand school policy in Europe is based on a principle of providing formal equality of opportunity. The decision-makers answer: Because there is a meritocratic system one needs mechanisms that help to secure fair starting conditions for all children. Testing, standardization of examinations, special courses for special learners – such and other meritocratic features of school may be undesirable from a pedagogical viewpoint but they work against social bias in school education. It is for this reason that extensive testing occurs in countries, where one accepted function of schooling is to prepare the students for leaving examinations controlling access to higher education.

A related conflict inherent in schooling is the one between the functions of equality and quality. Even though the case that there may be a conflict between the equity- and quality-related goals of school systems may have been overstated, the difficulty of balancing these two goals nevertheless permeates formal education. For example, there is an obvious tension between the efficiency and economic growth objectives

---

[2] The meritocracy concept and its implications for school policy and equal opportunity in particular are discussed by M. Young in *The Rise of the Meritocracy* (1958), D. Bell in *The Coming of Post-industrial Society* (1973) and T. Husén in *Talent, Equality and Meritocracy* (1974).

of basic schooling and the redistributive aims associated with a policy of pursuing equality of opportunity and outcomes in school education. Indeed, the choice between a "comprehensive" and a "selective" approach of providing educational services constitutes a fundamental dilemma in educational decision-making. A policy promoting quality and efficiency in school education may well run counter to a policy of providing equality of opportunity.

A major goal conflict arises as a result of the tendency for differences in competencies attained by formal schooling to increase as students move up the school. This is reinforced by meritocratic tendencies in an achievement-oriented society. The rewards for good achievements in school are not seen as intrinsic but extrinsic: better life chances in terms of more desirable jobs and better pay for careers enjoying social prestige. How is learning to become appreciated for its intrinsic values if what comes out of it at the end is essentially something that determines one's life career?

Another pervasive question that should be asked is how the expansion of formal schooling relates to the economic and social needs of the European countries. Here, we are faced with a dilemma. On the one hand, future society will demand a well-educated population. The idea is that the level of educational attainment among the general population should be raised. However, this is a problem because large groups of young people are not interested in staying on at school once attendance is no longer compulsory – usually after 16 years of age. Is it really desirable to have compulsory schooling until the age of 16 years or above? Should the period of mandatory schooling be lengthened or, perhaps, should the opposite be promoted – as is argued by those who consider that problems of modern schooling could be reduced if children were given more adult responsibility, for example as would be the case if school systems were organized with a recurrent education strategy?

Tension also exists between goals such as conformity and cooperation, on the one hand, and individualism and originality, on the other. Given the social framework within which European schools operate, students are aware that they are continuously evaluated in terms of their performance, expresssed as school marks or scores on achievement tests, with the purpose of determining who will be admitted to the next level in the system and who is going to be placed at the head or at the tail of the line of job seekers.

With the "revolution of rising expectations", education has become increasingly important in determining social status. More and more, the employment system has relied on the educational system and the credentials earned there to sift job applicants. Since people realize that formal education is increasingly important as a determinant of life chances, there is an enormous upward social pressure in the system. This poses a challenge to educational systems charged with the task of educating a very heterogeneous school population. The school is supposed to educate children who like to learn and achieve good results, as well as children who dislike the institution and all that it stands for. The school cannot succeed against the will of its clients, however, so particularly "difficult" children are not tolerated. Upon failing, some children are doomed to an existence on the margin of society. It is not surprising that some of them eventually declare war on society, with disastrous consequences.

The problems indicated above exist in the separate nation-states of Europe as they are at present. This situation, as well as anticipation of change in future Europe, makes it necessary to re-assess the goals guiding educational policy and practice. A first, general observation is that the goals of schooling may well have to be defined more modestly than has so far been the case. Moreover, it is essential to place the goals of the school in a larger framework of interactions with other educative agents, such as the family and the media. A problem is encountered here, since the "consumer society" offers many distractions to teenagers: television, video, cars, hitchhiking, drugs, etc. These experiences may be taken into account under the guidance of able teachers, so as to develop critical attitudes in youth. The teacher will have to interrupt his lesson and take up what is on the students' mind: last night's TV-film – on nuclear winter or on a divorce case or on a mysterious kind of disease – thereby demonstrating that life-problems are important to him and to the educational institution, and clearing the way for his Latin grammar. But the schools with their blackboard-centred classrooms in decayed brick buildings find it hard to compete with other educative agents. A related problem is that the influence of parents on their children is weakening. Although this does not have a general validity, it can be noted that the interests of teenagers often tend to deviate sharply from those of their parents. The school now increasingly has to offer a second, or it may be first, home for young people. This requires sensitive teachers, who are able to respond adequately to the needs and feelings of their students. It also needs appropriate equipment to compete with attractions such as video games.

However, the problem is that neither parents, taxpayers nor politicians are inclined to invest resources in school maintenance and development projects commensurate with the new environment schools must offer in order to compete. The countries on the Pacific Rim, especially Japan, Korea, Hong Kong and Singapore, have made impressive advances in education. Their economy is also improving at a fast rate. And not by chance. Europe has to pay heed to those countries where children *like* to go to school.

4. The Governance and Management of School Education

In the governance and management of their school systems the countries of Europe have traditions that are even more different than those relating to content and programmes. Models range from very centralized (France) to very decentralized (United Kingdom), with every kind of intermediate ones. The term "decentralization" itself has meanings that differ from one country to another, and the same can be said about centralization.

However, what is important to stress is that these models are in the process of changing in most countries. For some years it has been noticeable that highly centralized systems are on a path of decentralization (Spain, France.) At the same time the United Kingdom has reduced the powers of the local authorities, and has legislated a common (minimum) national curriculum. Thus it seems that the processes of change tend to converge. Such a convergence by exchanges of experience appears to be an important task for European cooperation.

Beyond national divergences, it seems that the ideal model that is being pursued by the reforms under way or under discussion is that of the increased autonomy of local educational institutions and schools, together with the establishment of a new type of management.

Among the factors behind these changes, there is disenchantment with centralized planning and global reforms, which have shown the importance of school autonomy, as well as the growing inadequacy of centralized administration in the face of modern management requirements and the difficulties encountered by centralized systems in responding adequately to the needs of new categories of students attending secondary schools. Other factors are of a positive nature: the concern to facilitate the constitution of teaching teams cooperating to define the pedagogical aims suitable for their constituency of pupils, giving a certain pedagogical autonomy to the school. There is also the desire to ensure that the different forms of management work more closely together and to open the school to its environment by strengthening the participation of interest groups in school affairs.

Innovations and reforms aiming at giving more autonomy to the schools have been recommended in many recent national and international statements and documented by cooperative European programmes, such as the ISIP initiative (International School Improvement Project) run under the auspices of OECD since 1982. These changes would eventually lead each school to work out a "school plan" defining specific objectives and pedagogical strategies adapted to its environment. The result may well be a significant convergence towards a more autonomous management model for schools.

An autonomy that leads each school to work out a "school plan", that is comparable, *mutatis mutandis*, to a "business plan", implies a radical change in attitudes. It can only be realized progressively, through an appropriate training for teachers and school principals, and a modification of the mentality of all those taking part, for example *Beamter*, union representatives and other public officials. It also implies new regulatory procedures on the part of the national authorities, not only in relation to a new allocation of administrative, financial and pedagogical powers, but also to the instruments for evaluating the results achieved by schools and their pupils. Several European countries have recently undertaken to define a "minimum programme" which sets out the common core to be achieved by all pupils, and tools for assessing the real level of knowledge and know-how at differing stages in the educational career. As long as the common core is guaranteed, differences will have to be introduced, because some children will want to learn Latin or French, or specialized physics, over and above the common curriculum.

The success of school education in developing a range of student competencies and abilities, in teaching citizenship and basic human values, and in promoting literacy and problem-solving skills is singled out as a decisive factor if a prosperous Europe is to be realized. The governments of European nation-states have acknowledged that a significant investment is required in order to raise the quality of basic education for all. [3] With this recognition have come new demands on school provision. As noted above, there is a tendency in European educational systems to increase the autonomy of the individual school. In this sense a liberalization is occurring. In the past there was an attempt to "cover up" the application of meritocratic principles in education with the use of measures intended to guarantee uniform treatment. Meritocracy is now behind policy and practice in school systems in many European countries.

5. Grading and Assessment of Students

Grading and assessment are integral parts of school teaching and learning in most European countries. Grading usually refers to the activity of ranking students on a scale on the basis of their performance on mainly cognitive criteria. On the other hand, assessment is sometimes understood to have a broader connotation, encompassing not only the measurement of cognitive functions but also of "immeasurables", such as communication skills, readiness and ability to cooperate and respect for others. Yet under the influence of the "Americanization" of educational research, grading and assessment are now often regarded as synonyms, as is the case also with terms such as evaluation and assessment.

---

[3] See the report written as a background for the Meeting of the Education Committee at the Ministerial Level on "High Quality Education and Training for All", Paris, OECD, November 13-14, 1990, p. 45.

Three functions of student grading stand out as particularly important. The first is to inform students and parents about individual progress in learning. The purpose is mainly to record proven achievement. In this case, grading is of course not an end in itself. It is commonly considered as a necessary and indispensable step in raising student learning outcomes. Performance feedback is a powerful didactic tool, as several research studies show. Bloom, for example, reports a meta-analysis of studies on Mastery to Learn, showing that corrective feedback is as important a factor in raising the effectiveness of instruction measured by an achievement test as are variables such as the student's "time-on-task", graded homework and improved study skills. [4] Although other syntheses of research studies [5] report lower correlations between feedback and student learning than Bloom, there can be no doubt that grading, in so far it is being used in a "corrective" sense, is a significant, productive factor in optimalizing learning.

A second important function of grading is to inform the school system about success and failure in the educational system. At this level, the objectives of grading are to support curriculum innovation, to improve the effectiveness of instruction methods and to provide a reference standard for teachers and students.

A third purpose of grading, which in the opinion of many parents and others outside the school system may well be its major function, is to provide a basis for selecting students either for continuing education or for jobs and careers. The negative effects of grading, emphasized below, appear mainly with respect to this function.

Some teachers, but not all, have a negative attitude towards grading. This attitude can be explained by several factors. Firstly, there is a general feeling of anxiety among the teachers on how and for what purpose the evaluation results will eventually be used – for grading the student or evaluating the teacher? Secondly, grading puts extra work on the shoulders of teachers, who may have a different priority and prefer another allocation of time. Thirdly, there may be a feeling among a small minority of the teaching force that learning and achievement cannot be adequately expressed by means of school marks that are, in the main, based on test scores. They may claim, with some justification, that the writing of a detailed report on the progress a given student has made with respect to a variety of criteria may be a more valid and useful way of informing the student and his parents than by way of merely assigning a mark.

---

[4] Cf. B.S. Bloom: "The 2 Sigma Problem: The Search for Methods of Instruction as Effective as One-to-one Tutoring", 1984.

[5] Examples of meta-analyses and syntheses of research studies on the relationships between "productive" factors such as student grading and assessment, and learning outcomes are Fraser *et al.* (1987) and Kulik and Kulik (1989).

School marks, which were abolished at the primary level in the 1960s and 1970s, are now being gradually reintroduced in some educational systems. Teachers are often opposed to marking, certainly those teaching at the lower grades of the system. Marking also goes against the opinion expressed by experts in education who claim that educational systems have already become too meritocratic in orientation. The argument that extensive testing leads to a situation where schools educate not for life but for examinations is frequently heard among those critical of performance assessments.

Yet students and especially their parents, demand them. That parents require feedback may be considered a part of the meritocratic syndrome characterizing industrial and pluralist democracies, where examinations are seen by many as a way of evaluating students and anticipating their life career prospects. Employers, who often distrust diplomas, demand that achievement be assessed on standardized and readily interpretable scales, thus reinforcing public demand for marking. [6] Whether meritocracy is a defence against corruption and nepotism or not, the problem we face is that the assessment of school performance and student achievement has become a characteristic of modern society.

Teachers can be said to have a sense of responsibility towards their students that is reminiscent of the feeling of responsibility parents have towards their children. They are supposed to educate their students for the world of work, leisure and social partnership they are about to enter. But parents generally expect something in addition: they want the teacher to prepare their children to pass the school-leaving and university-entrance examinations. Full achievement of the noblest goals of school education may thus be barred, as the expectations and concerns of adult society with the sifting and sorting function of schooling are prevalent. Not only teachers but examiners too are for this reason put in a difficult situation. They are faced with the demand for assessment to take place in a strict and unambiguous way.

6. National Assessment Programmes

Fundamental policy decisions underlie the introduction of national programmes for assessing school performance in the United Kingdom, the Netherlands and Sweden. [7] In France, the assessment takes place at each transition level in the school system, as Table 3.1 indicates.

---

[6] In evaluating job applicants, employers tend to use the number of years of formal schooling as a first criterion for screening. Employers are also sensitive to the reputation of the school or the higher education institution where the applicant has been educated. School marks or written reports on the student's educational career are usually taken into consideration in the final selection.

[7] A national assessment programme was launched in the Netherlands in 1986. From the outset it was intended to describe the curriculum and measure the achievement of the

Also in other Western European countries, such as Belgium, there is a mounting public and hence political pressure for monitoring educational performance. Plans for the inception of national evaluation programmes are under way also in Eastern Europe, for example in Hungary and Poland. These evaluation mechanisms will provide the schools with reliable information concerning the knowledge acquired by their pupils, the progress achieved over the year, and the possibility of setting their effectiveness against national, regional or local averages. The international programmes of quantitative and qualitative educational indices, such as those developed under the auspices of the OECD indicator project, [8] will probably allow exchanges of experience regarding this new perspective to be deepened. Tangible results are wanted by those who finance and use the services educational systems provide. There can be no doubt that the expectancies of the users of education constitute a factor that carries political weight. This is also reflected in recent efforts to measure public attitudes towards education. Not only are parents now being surveyed in many European countries,

Table 3.1: *Stages in the National Assessment of Student Performance in France*

School level	Grade level	Test
Primary school:	1st grade	CP (cours préparatoire)
	2nd grade	CE1 (cours élémentaire 1)
	3rd grade	CE2 (cours élémentaire 2)
	5th grade	CM2 (cours moyen 2)
Lower secondary:	6th grade	6ème
	7th grade	5ème
	9th grade	3ème
Upper secondary:	10th grade	2nde
	11th grade	1ère
	12th grade	Terminale

Source: Ministry of Education, France.

---

8 and 11 year-olds and report the scores on a national scale developed for each subject. In Sweden assessments are being made of student achievement in different subjects. A first round was conducted in spring 1989 at the grade 2 and 5 levels. Sweden is one of the few countries that has made adult education a part of its national assessment.

[8] Several excellent papers on educational indicators and the various uses to which they can be put have been drafted by OECD/CERI officials. Although access to these working drafts is mostly limited, a report for general distribution will soon be published.

but also teachers. The results the polls arrive at are exerting an influence on educational policy, as can be seen in France for example. This development in Europe runs parallel to events in the United States, where demand for accountability led to the inception in the 1980s of the much debated National Assessment of Educational Progress (NAEP). [9] A recent paper prepared by the United States Office of Education [10] states that the aims of this programme are:

> "... to raise standards for student learning through such policy mechanisms as raising high school graduation and public university entrance requirements, redesign of state and local curricular guides and requirements, increased testing to monitor mastery of the curriculum, alignment of the curriculum and the testing programmes, graduation exit tests of mastery of minimal skills, [and] policy discussion and interest in higher order thinking skills and problem-solving". [11]

On both sides of the Atlantic, the ultimate aim of evaluation is to raise school quality by identifying shortcomings and implementing counter-measures. Reform is thus a corollary of educational performance evaluation. In the case of the United States, school assessment programmes have indeed been followed by attempts to implement reforms. If, however, reform followed nothing but performance evaluation the result would ultimately be disastrous. It would be like a ship sailing by a magnetized compass. What is needed is a philosophy of education resulting in a theory of the school. This may very well be formulated primarily in negative terms, i.e. in terms critical of the existing concepts, for example: not every problem rising in society should be prepared for in schools; it is not the only task of the school to develop the human mind; the human mind's main virtue is to produce, to interpret, to relate knowledge and to apply it intelligently, not to treasure it dilligently; not any knowledge is worth "having"; it certainly is not unless it has been "appropriated" (Hegel).

Elements of a modern theory of school and learning could be: learning by instruction combined with experience; learning at other places than in school buildings; learning from other people than professional teachers; learning in a time-framework different from the one in which schools are

---

[9] The various demands made on the reporting of the results obtained from educational testing carried out under the auspices of the NAEP programme in the United States, as well as their consequences for the design of educational testing and the form dissemination could take, are discussed at some length in R.D. Bock and R.J. Mislevy (1988) and Ginsburg *et al.* (1988). An excellent paper on the relationship between changing economic conditions and their consequences for school policy in terms of new demands on educational reform evaluation is contributed by J.W. Guthrie from the University of California at Berkeley: "The Industrialized World's Evolving Political Economy and the Implications for Educational Evaluation", 1990.

[10] United States Office of Education: "Educational Evaluation and Reform Strategies in the United States of America" (Report to OECD, pp. 2-3), Washington, D.C., 1989.

[11] United States Office of Education, *op. cit.*, p. 3.

organized at present; and learning by "participation" in addition to "formal" instruction.

Characteristic of the National Assessment Programme (NAEP) is that schools and local educational authorities are held accountable for their performance. The data collected in the United States are publicly reported and used as a basis for creating policy consequences, such as cash rewards for high performance and sanctions for low performance. [12] By comparison, the school assessment programmes introduced in some countries in Europe tend to serve more modest aims. Here, monitoring is often seen as an exercise with an end in itself – at least at the present stage. Even though the American school assessment programme goes a long way beyond those of European countries, [13] it may well be worthwhile to a European audience to reflect on the American example. It can be noted in passing that this example has influenced the development of some European school systems since 1945.

Large-scale evaluations of educational policy were, of course, also conducted before the 1980s. Yet the nature of educational evaluation seems to have changed during the past decade. It used to be an *ad hoc* exercise often limited to the evaluation of particular innovations and experiments. By contrast, the entire educational system has now become a target for planned, systematic evaluation. How can this development be explained? After the war, in a spirit of "social engineering", it was common to think that the information provided by educational research could be used as a basis for decision-making in education. [14] This confidence in the utility of research studies has, to some extent, diminished. Educational administrators demand information that makes the functioning and effectiveness of the system and individual schools clear to everybody. As research has generally not been able to supply this type of information, planners and policymakers have turned to other means. The large-scale evaluation studies now being carried out are often run by specialized government institutions, often in cooperation with statistical offices and experts in educational measurement.

An additional explanation is that the effects of the "high tech" revolution are now beginning to take hold. The equipment, instruments and measurement techniques needed in order to set up and run a sophisticated information system have only recently become available. A third factor may be that the public does not approve any more of evaluations of school performance conducted by the school inspector and

---

[12] E.g., by means such as the "Wall Chart" in the United States.

[13] The Rutter *et al.* (1979) assessment is a good example of a British study that takes the goals and ends of national assessment from the macro- to the micro-level.

[14] Adherence to this perspective was strong in Sweden, for example, where Tage Erlander, and Alva and Gunnar Myrdal, among others, supported the view that systematic, empirical research in the natural as well as the social sciences would provide a solid foundation for the building of the (social democratic) welfare society.

the headmaster. Yet another explanation is that different groups at the receiving end of the products turned out by the school are not satisfied. There is a concern among employers, for example, that schools do not perform as well as they should. This impression is being imprinted on the minds of the public, not least parents. Whereas in countries such as France and Sweden to publish school performance data is not encouraged, in Britain this is exactly what the public is demanding. Even in France, a similar pressure from families has led the influential monthly "Le Monde de l'Education" to publish regularly the results of the *baccalauréat* examination for each *lycée*, and a series of performance indicators for a variety of "technical" schools and higher education institutions. This reflects a significant change in attitudes.

The development of national programmes for the assessment of educational performance may well be part of a long term trend. In some European countries it mirrors a deep concern with school quality and effectiveness. Countries follow different paths to evaluation. In England, for example, national evaluation is being imposed on schools. One of the arguments was that this would establish an evaluation system at school level as a matter of course. This strategy has a long-term goal, namely, to see to it that schools gain experience in performing evaluations.

The official school system is being decentralized in Belgium, where local communities are taking over school management tasks. This has increased the demand for data on school performance. The trend in France is for the new school inspectorate to evaluate the functioning of the whole school system. Similar developments are found in the Nordic countries. In Sweden, for example, the power of national and regional boards of education is being diminished in favour of an increase in the scope of decision making at the municipal level. The German-speaking countries seem less affected by the tendency to assess school performance. It is assumed that the tripartite system in place in Germany, Austria and several cantons in Switzerland guarantees enough flexibility to respond adequately to new demands arising from technological change. However, a problem encountered is that it is not possible to gather evidence showing that adaptation to new conditions is actually taking place. Italy is a case in point. The Ministry of Education does not control or manage the functioning of the system. A lower level in the system has to be contacted if information on school performance is sought. This situation makes it difficult to carry out major school reforms.

7. Criticisms and Dilemmas of National Assessment in Education

The analytical and diagnostic functions of evaluation and monitoring in education are regarded by many administrators as integral elements in a strategy to improve the quality of teaching and learning. Yet examinations and educational testing in the Japanese and United States schools have almost become an aim, instead of a means to improve the quality of schooling, the intention at the outset. However, educational testing has become so pervasive and powerful that it has been met by growing opposition. [15] In some countries educational testing, which provides the backbone to all performance assessment, influences everything from the budget allocated to schools to the content of the curriculum and the teaching methods that are being applied. This is a matter that ought to be seriously taken into account by those European countries in which programmes for educational reform evaluation are being developed or contemplated.

Another criticism of national educational testing is that it is entirely utilitarian. It presupposes that education is something that can be used and measured. If a certain aspect is not measurable, its educational objectives and their realization tend to become neglected. If a given objective of schooling or a curriculum evades definition, then it is probably not very useful either, so the argument tends to go. Immeasurables, on this reckoning, ought to disappear. Learning to be, to enjoy, think and be critical are among the many goals of education that evade quantification. Hence they may well be threatened with obsolescence. For example, education in the humanities, which does not very well lend itself to standardized assessment, may well be given low priority in an educational system in which the school curriculum is made dependent on generalizable examination results. Classical languages will find it hard to survive the competition if the governments of European countries endorse the goal that all children should become proficient in at least two modern languages besides the national language. The utilitarian approach to defining the purposes of education can lead to the extreme situation where learning how to drive a car is given a weight in the curriculum equal to that of history or philosophy.

The history of Western Europe and, at present, Eastern Europe shows that democracy cannot be brought about overnight. Time is needed, not only for establishing an institutional and legal framework conducive to participation, but especially for the growth of democratic values. One educational task in this respect, for Europe, East and West, is to challenge racism, ethnocentrism and demagogy. Yet moral, spiritual and religious education, on which this task depends to no small extent, are not readily measurable. How can such elements retain a central place in the curriculum, if the content of schooling is increasingly being dictated by a narrowly utilitarian perspective on the purposes of the school?

---

[15] N. Bottani provides a summary of this critical discussion in a mimeographed paper: "Les nouvelles mesures de l'éducation: illusion de la recherche ou recherche d'une illusion? Le cas des indicateurs internationaux de l'enseignement" (OECD/CERI, Paris, 1990).

Thus decision-makers and managers in European school systems face dilemmas regarding the design and implementation of monitoring systems useful in assessing aspects of school performance and student achievement – unfortunately, dilemmas to which generally satisfactory answers do not seem to exist.

One dilemma is whether the results of such performance assessments of individual school units or districts ought to be made public, as is now done in parts of Britain and the United States. On the one hand, schools may be subjected to powerful incentives for improvement if their results were made known. On the other hand, this type of information, if publicly available, will undoubtedly lead to stigmatisation and considerable dissatisfaction, especially among the parents whose children attend schools that perform poorly.

A related issue is whether the choice of school ought to be left for parents and their children to decide upon, or whether teachers, administrators and guidance counsellors ought to be involved in this decision. If a country decides to publicize the outcome of a national assessment of school performance, then freedom of parental choice will obviously also have to be granted. The other side of the coin is whether schools should be allowed to test prospective students and are free to admit only those performing above a certain level of achievement.

Whether schools ought to be publicly ranked in accordance with the average achievement of their students on standardized tests and hence, whether freedom of parental choice should be granted, tie in directly with other controversial questions besetting schooling in modern societies. One of these concerns the division between public and private schooling. Another would be to ask whether Europe needs élite schools for the talented. If answered affirmatively, should such schools be allowed to receive public funding, or even to operate within the public school system, as tends to be the case with the special schools catering for children with severe learning handicaps? Much of the discussion in connection with school reforms has been concerned with the question of how to combine mass and élite education. However, serious questions of definition relate to such a discussion. For example, education for a social élite has a different connotation compared with education for the gifted children in schools set apart for them. The problem raised can be summarized by asking: How can diversity be handled in school systems working in the integrated, pluralist and meritocratic environment? Should all schools have the same grey uniform, or should they be allowed and encouraged to develop their own profiles?

Are the European countries capable of considering school quality on a European scale, not to speak of initiating a European assessment of educational progress? Given that previous attempts to arrive at a common European school leaving examination have failed, the answer seems to be no. At present many European countries consider their own evaluation strategies superior to those developed in neighbouring countries. The question that comes to mind, however, is whether it would not be better in the long run to develop a European approach to educational evaluation. It is evident that such a strategy of development cannot be initiated unless countries are prepared to share information, knowledge about methods, and resources. A pre-condition for achieving this is that all countries are given complete freedom in adapting the means and ends of such a European monitoring system to their own specifications and needs.

There are no general answers to the questions raised above even though they have been asked repeatedly. However, the complexity of the issues involved is dealt with below and in Chapter 8.

8. The Quality and Coherence of Schooling

If the main theme in the education debate during the 1960s was "equality of opportunity", slogans for the 1980s revolved round the term "quality". Building on the progress already made, policy to safeguard and improve quality provision will probably remain at the forefront during the 1990s also. Institutional and curricular restructuring has now been carried out – although equity has by no means been achieved and the curriculum constantly needs updating. Concentration has been on the achievement of higher standards generally.

However, when discussing quality in education one soon finds that the precise meaning of the concept is not clear. Quality is commonly understood to relate to standards in education. This is why, as was seen in the preceding chapter, quality is now defined in a context of educational evaluation, for example as the extent to which the objectives set for certain educational activities going on in school have been achieved. In this respect a significant change has occurred. Quality used to be interpreted in terms of the resources available to a school, for example per pupil expenditure, the size of non-teaching support staff, the provision of school meals or the attractiveness and utility of the school building. Nowadays, however, quality is interpreted by many as referring to educational outputs and outcomes, assessed by means of different indicators of school effectiveness.

It may be concluded from the above that, as yet, there is no agreed definition at policy level as to what quality implies. It is clearly related to efficiency, which can be measured in monetary terms, and even more to effectiveness, which is less quantifiable, despite efforts to do so by devices such as cost-benefit and rate of return analyses, and manpower planning, and large-scale measurement yardsticks such as programme budgeting and even assessments of student performance. The impact of education is as much long-term as short-term, and to this extent its effects are incommensurable.

Nevertheless, additional instruments can and should be developed to measure the immediate outcomes of education. The array of tools produced by the IEA over the years may well be useful in establishing common standards within Europe, the need for which will increase after 1992. Here only a few areas where quality has become a matter of public debate can be touched upon.

Concern has recently been expressed in some countries at an alleged drop in school standards in basic literacy and numeracy. If true, this presents a major problem, as these skills are fundamental to all school subjects. Information processing skills are also at the very centre of a society, affecting not only economic growth, the distribution of wealth and opportunity, and a country's capacity to attract investment, but also influencing culture and the functioning of a democracy. Table 3.2 presents data indicating that functional illiteracy is a problem in some European countries.

Table 3.2: *Estimates and Projections of Illiteracy in European Countries (population 15 years and over)*[16]

	Illiterate population (thousands)			Illiteracy rates (%)		
	1985	1990	2000	1985	1990	2000
Greece	672	548	338	8.6	6.8	4.0
Italy	1,700	1,378	966	3.7	2.9	2.0
Portugal	1,429	1,215	829	18.4	15.0	9.7
Spain	1,697	1,440	1,063	5.7	4.6	3.2
Turkey	7,689	7,046	7,459	24.0	19.3	16.4
Yugoslavia	1,614	1,342	942	9.2	7.3	4.7

Illiteracy in the adult population was, until recently, not perceived as a major problem by decision-makers in most Western and Eastern European countries. Both groups of countries had, although perhaps for different reasons, routinely reported to international agencies that illiteracy either did not exist at all or could be found only among a small fraction of the adult population. The view that illiteracy is a problem affecting only certain marginal groups in a society, particularly

---

[16] Source: UNESCO, *Compendium of Statistics on Literacy, 1990 Edition*, UNESCO Office of Statistics, Paris.

immigrants, was widely held in the Nordic countries, Germany and the Netherlands, and Switzerland. [17]

Yet, in France, the position changed in the mid-1980s when several surveys established not only that functional illiteracy affected a sizeable proportion of French-born citizens (figures of 12 to 20 per cent have been mentioned, according to the criteria used) but also immigrant workers. The "rapport Migeon" led the Ministry of Education to consider new measures for assessing and improving the reading comprehension of adults and encouraging reading practice among the young. [18] These findings corroborated those arrived at in the United Kingdom some years previously [19] and helped in drawing public and political attention to the issue. The promotional activities and studies launched by many countries in connection with the United Nations International Literacy Year in 1990 have also called attention to the idea that illiteracy is a much more pervasive and difficult problem than generally anticipated, even in the highly industrialized and affluent European countries.

The realization that adult illiteracy in fact constitutes a problem, although one that may vary in degree from country to country, has led to government action and the launching of literacy programmes in different European countries. It has also reinforced the concern with school quality, as schooling is expected to provide a basis for lifelong literacy. Of special concern is the teaching of the three R's. How reading and writing abilities, and also communication, interpretive and numeracy skills, can be improved has become a matter of urgency in a Europe

---

[17] Countries such as Switzerland and Sweden supplied this information in answer to a questionnaire distributed in connection with the 42nd Conference on Education in Geneve in September 1990: "Functional illiteracy is a serious problem mostly among women and immigrants" (Switzerland) and "... while speaking about functional illiteracy we can only refer to immigrants. No students leave school with the lack of literacy" (Sweden). The quotations are taken from Ch. Nicolitsas (1990, p. 5).

[18] A survey on illiteracy in France was undertaken in 1986-1987. Among 37 million inhabitants older than 18 years of age, 9.1 per cent were discovered to have difficulties in speaking, reading and writing, or understanding, French. Illiteracy affected about 1.4 million immigrants and 1.9 million French-born citizens (J.L. Borkowski, "L'illettrisme", in *INSEE, données sociales,* 1990, pp. 355-360). Another survey was undertaken in 1988. It established that 11.5 per cent of the adult population in France face problems in writing and 4 per cent in reading (Survey conducted by GPLI (Groupe Permanent de la Lutte contre l'illettrisme), directed by F. Bayrou, with the participation of the French Ministry of Social Affairs.)

[19] It was found in the National Child Development Study (NCDS) that about 13 per cent of the 23 year-olds in 1981 in England, Scotland and Wales reported difficulties of some kind with either reading, writing or numeracy. The results of another survey were reported in 1990 (D. Spencer, "Millions Lose Out on Numeracy", *Times Educational Supplement,* October 12th, 1990). That functional illiteracy is a problem affecting all groups in society, young and old, the employed and the unemployed, men and women, British citizens and foreigners, is confirmed by the results arrived at.

faced with the dual challenge of extending a participatory democracy and promoting economic growth.

Another area of concern in this respect is the teaching of science and technology, where the accretion of knowledge has outstripped the pace at which programmes can be updated. Moreover, quality standards are recognized as dealing also with less tangible areas such as civics. To mention a few negative traits, the rise of vandalism, hooliganism, drug-taking, alcoholism and other social scourges, and juvenile delinquency generally (in most countries the majority of criminal acts are committed by youths under 24 years of age) may well be connected in part with a failure to educate young people for social responsibility. This has gone hand in hand with a general dissatisfaction in some countries on the part of parents and policymakers as regards discipline in schools, on which successful learning must depend.

A key role in any improvement in quality is played by the teachers, whose cooperation is essential for the improvement of standards. Moreover school quality must now be defined in relation to the labour market, as it has become increasingly clear that school quality is a context-specific idea. More systematic studies of school quality in relation to the abilities employers need are called for. Furthermore, the role of parents in stimulating pupils to greater effort is pivotal, since the home, as demonstrated by many research studies, [20] is the most important single factor in school achievement.

It has been implied above that the argument about school structures has almost everywhere been settled. This does not mean, however, that the question of differentiation *within* the school by devices such as streaming and setting, or the choice of options has been resolved. A current of opinion is now tending to move once more in favour of segregating the gifted, the future key people in society, whose calibre is vital, from the less gifted. The social arguments against unstreamed classes at secondary level may well be disregarded as militating against future national or European prosperity, in which the part played by well-educated leaders is considered as paramount. But the promotion of the

---

[20] The Coleman Report on *Equality of Educational Opportunity* (1966), which was published in the same year as the U.S. Bill on International Education was presented to Congress, was among the first to establish this point. The study by Coleman and associates played a prominent role in the schooling debate in the Western industrialized countries in the late 1960s and early 1970s. The Coleman Report was matched by the Plowden Report (1967) in the United Kingdom. Similar studies were conducted also in other European countries. The results arrived at were generally not promising, in that the role of home background in determining student achievement and the outcomes of education was confirmed. These much publicized findings led to the belief, which was widespread in the 1970s, that the school did not make much difference. However, in a meta-analysis of hundreds of correlational studies, Fraser *et al.* (1987) show that, although home background factors are significant co-variates of student achievement, classroom-level and school-level variables also play an important role.

many to the highest possible standard may just as well be of equal importance. Here again one comes up against the eternal argument of quality versus equality.

In most countries the pressure is mounting in the early years of primary schooling to include, in addition to "basic" subjects, such as reading, writing, history and drawing, and elementary arithmetic, the teaching of science and foreign languages also. As with pre-primary schooling, more emphasis is put at present on the cognitive domains of learning, even though there is an awareness that this should neither cause a further dilution of learning tasks, which could have adverse effects on the quality of schooling, nor entail a reduction in the attention given to affective and physical, aesthetic and moral education. However, as the teachers of the first grades of primary school are mostly unspecialized and often lack the required knowledge to teach specific subjects, practical difficulties are encountered in the effort to upgrade the curriculum. A requirement for solving these may be to provide primary teachers with regular upgrading courses in integrated science, mathematics, and foreign language, as is done, for example, in Hungary and Sweden at present.

In the lower secondary school in most countries the curriculum includes literature, history, foreign languages and arts, mathematics and some science. Experiments are going on in the teaching of environmental education, computer education and technology.

The findings of the IEA studies cited in Chapter 2 [21] generally show that the task of harmonizing school systems in Europe is of great magnitude, since wide variations exist with respect to both the grade span in the organization of schools, general versus specific courses in different subjects, and the degree of central control over the curriculum. The examples and research results mentioned above give rise to the crucial question: what specific skills should be taught at what age, and what standard of knowledge should be achieved? Another implicit problem concerns the degree of specialization one should aim for. That there is a rich diversity in approaches and solutions in Europe can be taken as an indication that countries have much to learn from each other. This is the central issue: what can and what should we learn from each other?

Another problem concerns the organization of the school curriculum. The school timetables now in use in most European countries are based on the models developed in the 19th century. Now a new approach may be needed. Correction of the century-old curricula, as has been the normal procedure in the past, may result in overload and incoherence, particularly if diverse demands continue to be accommodated on an ad

---

[21] The reference is to the IEA Second International Science Study and the IEA Second International Mathematics Study (see Chapter 2).

hoc basis. The design of the new curriculum must take into account the fact that in many European countries more than 90 per cent of an age-group are enrolled in kindergarten at the age of four or five.

Today, the purposes of schooling are mainly seen in relation to the manifold roles as citizens and holder of occupations that people are considered to have to play in a modern industrial society. Indeed the school curricula are devised so as to impart as precisely as possible learning tasks that, it is hoped, confer the "right" skills, knowledge, and attitudes. A central problem the traditional pedagogues but also the modern curriculum developers have dealt with concerns the choice of subject matter and the amount of teaching. Yet the context within which these choices are made may change as a consequence of the emergence of new conditions in European society. There, it may be argued, many of the specific abilities and skills one needs for work can be obtained later in life, for example through effective on-the-job training. Yet an adequate foundation of basic schooling is also needed. As the demands made on the school increase in the "learning society", the role of the school itself may become even more complex than today.

9. School as a Part of Lifelong Education

Schooling tends to be extended to cover an increasing part of the life span of the individual. This development has a number of implications for school education. Firstly, the purpose of basic schooling is no longer to provide young people with the knowledge and skills for the rest of their life. As knowledge in advanced societies becomes rapidly obsolete, the goals of schooling are increasingly defined in relation to preparing not only for working life but also for educational and cultural life. Secondly, as schools are sharing their educative responsibility with other agents and institutions, the socialization function of schools must be assumed to change as well. The implication is that ways must be sought to encourage greater participation and meaningfulness in education, for example, by conferring adult responsibilities on young people in the larger societal context.

The principal objective of the so-called recurrent education strategy advocated since the late 1960s by the OECD [22] has been to increase

---

[22] The development of recurrent education in the late 1960s and early 1970s was influenced by serious doubts as to the capacity of formal education to achieve many of the goals sought by policymakers, which related to equality, efficiency, and cost-effectiveness. Mr Olof Palme, the then Minister of Education in Sweden, launched the idea of recurrent education at an OECD meeting of Ministers of Education in Versailles in the late 1960s. In his keynote address, Mr Palme envisaged recurrent education primarily as a means to guarantee democracy, equality, and the right of individual freedom of choice. The concept was subsequently worked out in detail by the

flexibility, efficiency and equity in developing and utilizing human resources. This was to be achieved by alternating education, work and leisure over the lifespan. The pursuance of alternation between education and work has a practical implication for school systems. Terminal points in the educational system, particularly at the upper secondary level, could be abolished so that all educational programmes could lead on to other programmes. Another implication is that students who have dropped out of the educational system after having completed obligatory schooling only, and who are interested in continuing their educational careers after a period of interruption, should have the right and possibility to re-enter the system.

Much has occurred since the recurrent education theory was developed in the early 1970s. Programmes of adult education and particularly job training have expanded enormously. Change in adult education and training in the 1980s has followed roughly the same course as it did in school education. The responsibility for adult education and job training has in the main come to rest with the individual and his employer. There also is an emphasis on quality control, effectiveness, and credentials.

The formal education young people receive at an early age is in many ways decisive for their life chances and careers in present-day European society, where level of formal education increasingly determines life chances, career, and social status. Shortcomings in basic education can only be compensated later in life to a limited extent, by means of recurrent adult education, retraining, and staff development programmes in working life.

The influences of home background on educational attainment and life career outcomes are well documented. Research has consistently reported that a large portion of individual differences in educational attainments are accounted for by non-scholastic factors. [23] Disparities between social groups have a tendency to continue up the educational ladder, from primary to secondary to post-school levels of education. Education gained in youth influences occupation and the opportunity of acquiring additional educational experience later in life career.

That advantages and disadvantages are accumulated over the life span can be inferred from the data presented in Table S.9 in the statistical supplement. [24] The standardized regression coefficients refer to the

---

Organization for Economic Cooperation and Development and its Centre for Educational Research and Innovation (OECD/CERI, 1973).

[23] Two interesting studies in this respect are J.P. Keeves *et al.*, 1991, and J. Peschar, 1991. Additional examples of research studies estimating the influence of home background factors on educational achievement and life career outcomes are given in Chapter 4.

[24] The estimates included in this supplementary Table are also presented in Tuijnman (1989, p. 188). The reader is referred to this source for information regarding the path model in which the effects were estimated, as well as for details concerning the

effects on occupational status and earnings of home background measured using conventional indicators such as father's education and occupation and family income, cognitive ability tested with four conventional subtests, received level of initial formal education comprising basic schooling and in some cases also higher education and received amount of adult recurrent education and job training. The estimates are based on data collected in a Swedish longitudinal study of men born around 1928, who were followed up from 10 to 56 years of age. [25] The data show the impact of home background factors on life careers, as the total effect of home background on occupational status is fairly stable from 25 to 56 years of age. It can also be seen in the supplement that the effect of home background on earned income varies substantially. It is estimated to be the strongest at the end of early career, at around 35 years of age. Cognitive ability assessed by a standardized intelligence test given at 20 years of age seems to have a constant effect on life career outcomes. It can moreover be inferred from the coefficients that the importance of home background and measured ability is, after the age of 35, mediated by both initial and post-initial education.

The results may well be interpreted as showing that a cycle of accumulation is in operation in which education received initially plays a decisive role both in mediating the effects of home background and cognitive ability, and in influencing the likelihood of receiving further education and training in working life. Hence it would seem that both schooling and further education and training are important factors in explaining occupational achievement and earnings differentials. Although the effects of home background are consistently strong, there would seem to be no place for fatalism of the kind espoused during the early 1970s, when the claim that schooling "did not make a difference" was frequently heard in debate on educational policy.

Hence the principal rationale for ensuring equality of educational opportunity in secondary education comes from the realization that inequality is reinforced by education, permeating not only the occupational and economic lives of young adults, but also exerting a clearly decisive influence on occupational and educational careers in a lifelong perspective. The implications are that the differentiation of

---

measurement properties of the variables specified and the analytical strategy taken in the fitting of the model to the data.

[25] The Malmö Study, which by its 50-year timespan may well be the oldest still active longitudinal survey in the world at present, comprises information originally gathered for all 1,542 children who attended the third grade in all public and private schools in Malmö, a city in southern Sweden, in 1938. The vast majority were 10 years of age at the time the first data were collected by Siver Hallgren. The group has been followed-up regularly since then. There have been five occasions on which questionnaires and also interviews were conducted. These are reported in Husén (1950), Husén (1969), Fägerlind (1975) and Tuijnman (1989).

students ought to be delayed and that overspecialized vocational education at the secondary level ought to be avoided. It may be noted in this respect that the mandate of the school in imparting knowledge and skills for a vocation and citizenship is diminishing relative to the influence of other educative agents operating in the society. Since the early 1970s priority has been given by governments in European countries to the creation of a so-called "learning society". The development of systems for the further education and training of the labour force, which is a high priority in Europe today, can be regarded as a first step towards the provision of education in a lifelong perspective.

The above analysis of data draws attention, once again, to the issue of equity. Europe will have to pay a high price in the long-term perspective if policy to ensure equity in school education is sacrificed for short-term gains in quality, efficiency and school effectiveness. Under the impact of changes affecting the precarious balance between equity and efficiency, it may well become necessary to reassess the basic values guiding educational policy and practice. It is possible that children and society may be better served if the content of initial schooling is broad and general rather than narrowly focused on the specific vocational skills in high demand on the labour market at a given moment. Since early experience of formal education influences both adult success and subsequent experience of education and training, the provision of equality of educational opportunity in basic education in Europe must be considered beneficial, as the overall level of education in the population as well as the flexibility of a well-trained labour force are factors determining the economic success and wellbeing of nations in the new Europe.

Longitudinal studies of how the individual's life career and earnings profile have developed over time show that those who have used the system of recurrent education later in their lives have reached occupations with higher levels of responsibility, social prestige and earnings compared to others untouched by the system. [26] But the same studies tend to show also that those who have had a relatively long and qualitatively good formal education in early life are those who make most use of the possibilities for upgrading and retraining offered by recurrent education. It may be anticipated that those who have postponed a phase of their formal education until later in their lives, instead of transferring to higher education immediately after taking the leaving examination, cannot compensate for the relative disadvantages they face early in their career by returning to formal education. The implication is that increased provision of post-initial formal education, further training

---

[26] Other examples of studies in which the long-term effects of initial education on post-initial educational and occupational careers are investigated with the use of longitudinal data are Härnqvist (1989) and Van Leeuwen and Dronkers (1991).

and retraining, does not lessen but rather reinforces the important role of basic education for career development.

There are good reasons to assume that, to a certain extent, what is beneficial for the career development of the individual is also good for the development of the nation at large. If the findings of research studies are generally valid, then they must be held to have far-reaching consequences for the structuring and organization of educational systems. This applies also to the criteria that must be employed in selecting students for university admission. One implication is that it may well be beneficial in the long run if children start school at an early age. In those countries where children begin their formal schooling at seven, one could consider whether the entry age can be lowered to six or even five years. As it does not seem to pay off, at least from the viewpoint of the individual, to postpone part of higher education until a mature age, students may be well advised to transfer to higher education immediately after finishing the upper secondary school. This applies especially to students who have taken academically-oriented upper secondary programmes with an accent on mathematics and science subjects.

10. Conclusion

In the competition between regions or continents, excellence counts. Europe cannot be competitive in the long run if the talented are receiving a mediocre education. The future cannot be predicted but invented, as the saying goes. In this endeavour the contribution of different people is needed. Therefore, more than just one route to academic studies has to be offered, as well as alternatives both in the school and the outside world. But giving special consideration to the educational needs of children with different talents ought not to result in a situation where children considered disadvantaged in comparison with their more gifted and fortunate peers are barred from prospects in European society because of an educational system that does not meet their needs. It follows that Europe cannot focus attention mainly on the education of its academic élite. Every child, regardless of nationality, creed, cultural background or intelligence should be offered a route to, and a chance to reach, the optimum of its potential.

One has begun to realize that the goals of schooling have to be more limited than was believed in the optimistic period of the 1960s. At the same time, though, it is essential to place these goals in a larger framework of interaction with other educative agents, since schools by themselves furnish an incomplete context for maturation. The question of what the goals of basic schooling are or should be in the new Europe was considered in this chapter. It can be concluded that children are best equipped for life in the European society if they are eager to keep on learning after leaving school. One purpose of the school is therefore to foster the idea and desire for lifelong learning. This may be achieved only if the school can be made into a place where children are eager to go.

Recognition of the nature of the problems spelled out in the preceding paragraphs may well be crucial for overcoming failure and developing strategies for realizing the aim of providing a high quality basic education for all children in Europe.

Chapter 4

Home and School Relationships

1. Introduction

The school is often regarded as the main, if not the only, educative agent in modern society. Education, therefore, tends to be conceived as equivalent with schooling. Nothing could be more wrong, particularly in the highly industrialised and urbanised society today where the family's educative role, as we shall see, has been considerably reduced. Survey research conducted during the last few decades has repeatedly shown that home background accounts for more between-pupil variance of school achievements than school resources and methods of instruction. This implies by no means that "schooling does not make any difference", but tells us that the "input" to schooling, represented by the pupils' home background is a major factor in accounting for the outcomes in terms of the cognitive competence the school is expected to instill.

The school is but one educative agent in our society in addition to the family, the peer groups, the neighbourhood, the media and various organizations, religious and other. However, social science research has so far focused mainly on the school in studying attainment and achievement and in trying to unravel the conditions for learning. The family as an educator has been given much less scholarly attention, which is deplorable because of the changes in the structure and role of the family, which we shall deal with in the paragraphs that follow.

Problems of school education in Europe vary from country to country. This was manifested by the rich diversity of issues and problems raised in the national reports the ministers of education submitted to the 42nd Session of the International Conference on Education in Geneva in 1990. [1] Therefore, even though an attempt is made in the present chapter

---

[1] This impression arises from the various reports prepared by the ministries of education of the European countries for the 42nd Session of the International Conference on

to generalize across countries, the applicability of the general trends and tendencies must be assessed separately and flexibly for each national geographical area in Europe.

2. Statistics on Changing Household Structures

Fundamental changes in the conditions for child rearing have taken place in industrialised countries in Europe during the latter half of the 20th century. Functions of the socialization of children have increasingly been removed from the family and delegated to specialised institutions and agents. This process started when legislation on universal elementary schooling was passed and children were sent to schools which were expected to provide literacy and some basic orientations in geography, science and history. More recently, with increased urbanization, day care and youth centres have emerged. One consequence of this has been a fragmentation of contacts between adults and young people both in terms of number and continuity. Child rearing has increasingly become a specialised and professionally licensed task carried out in institutions with specialised categories of teachers, social workers, administrators and other 'care takers', each category with its own guarded territory of competences and prerogatives. As the institutions and their organizational framework have grown, their services have tended to become increasingly impersonal and bureaucratised. The background of this development is, of course, fundamental socio-economic changes that have taken place in modern society.

The information provided in Table 4.1 has been extracted mainly from the population studies that have been conducted over the period 1986-1990 under the auspices of the Council of Europe. [2] (Additional data on demographical developments in the European countries are provided in the statistical annex to this report.) Table 4.1 compares a number of selected European countries with respect to indicators, such as the marriage rate, divorce rate, fertility rate, age of mother at birth of first child, and the average life expectancy for men at birth.

A secular downtrend in the fertility rate in most European countries has taken place since the turn of the century. A dramatic acceleration of this trend occurred in the 1960s. In 1965 the birth rate per woman in the majority of countries was between 2,3 and 3,0 children. Twenty years

---

Education, which was held in Geneva, September 3-8, 1990. Examples are not only found in the reports of EC countries but also in those of Cyprus, Finland, Norway, Poland and Turkey.
[2] The demographic data and the statistics on the changing structure of households and labour markets in European countries presented in this chapter are derived mainly from reports prepared by working parties of the Council of Europe and the Organisation for Economic Cooperation and Development. The sources are indicated in the bibliography.

later, in 1985, the rate had dropped to below replacement level, namely between 1,3 and 2,0.

Table 4.1: *Marriage Rate, Divorce Rate, Fertility Rate, Modal Age of Mother at Birth of First Child, and Male Life Expectancy (years) in European Countries*[3]

Country	Marriage rate[a] 1970 1987	Divorce rate[b] 1970 1987	Fertility rate[c] 1987	Modal age of mother at birth of first child[d] 1970 1987	Life expectancy for men[e] 1987
Austria	0,91 0,58	0,18 0,29	1.43	24 26	72,0
Belgium	1,05 0,74	0,10 0,28	1.54	24 24	70,5
Denmark	0,65 0,59	0,25 0,46	1.50	24 26	72,0
France	0,92 0,53	0,12 0,32	1.82	24 26	72,0
Germany	0,97 0,60	0,15 0,31	1.36	24 27	71,8
Ireland	0,98 0,66	— —	2.35	25 26	70,0
Netherlands	1,06 0,61	0,11 0,27	1.56	24 27	73,4
Sweden	0,62 0,56	0,23 0,43	1.84	24 27	74,2
Switzerland	0,87 0,72	0,15 0,30	1.50	25 27	73,8

Notes: (a) Total first marriage rate for women below 50 years of age. It should be noted that the rise of age at marriage, which is a general phenomenon in Europe, is associated with the increase of consensual unions, particularly in situations where these precede formal marriage. The postponement of first birth will also influence the age of first marriage. (b) The total divorce rate summarizes the divorce experience of a fictitious marriage cohort and thus represents the number of marriages which are expected to end in divorce without any distortions due to past fluctuations of the marriage rate. When examining changes in divorce rates, one should bear in mind the likely connection between declining divorce rates and increasing non-marital cohabitation. (c) Total fertility rates are calculated on the basis of data for periods. A total fertility rate less than 2.10 implies that a country is experiencing below replacement fertility. The total fertility rate is the transverse sum of age-specific fertility rates, giving the number of children who would be born, on average, to a group of women experiencing throughout their child bearing life the fertility conditions prevailing in a given year. (d) The figures are rounded off. (e) Life expectancy at birth for men.

During the same period the propensity for getting married was going down. The first marriage rate among men below 50 years of age declined by 40 to 50 per cent. The divorce rate doubled in Europe from 1970 to 1985. The typical marriage ended in a divorce in 30 per cent of the cases. This figure was even over 45 per cent in Denmark and Sweden. The increase in so-called cohabitation reflects a changing attitude towards marriage. Cohabitation among 20 to 24 year old women in Sweden and

Denmark rose from 29 to 44 per cent over the short period from 1975 to 1981.

The aging of the population has also affected the size and structure of the households. Around the turn of the century older persons made up only three to four per cent of the total population. In 1986 persons above 65 years of age were 18 per cent in, for instance, Sweden and 16 per cent in Norway.

The demographic statistics cited above mean a drastic reduction of the size of households. The three-generation family has for a long time been a rare species. The average size of the household was much smaller in the late 1980s than it was late as in the 1960s. A striking feature of the household situation is that single-person households have increased tremendously over a short period. They have, across Europe, risen by 50 per cent from 1960 to 1980. These tendencies have been much stronger in the big cities than in the smaller towns or in rural areas. In some big cities one-person households are even more frequent than "normal" ones with two adults and at least one child.

Single-parent families should be mentioned in this connection. In highly industrialised countries, their number has been going up by between 30 and 50 per cent during a few decades. By the mid-1980s, single-parent families comprised 10 to 15 per cent of all European households. [4] In most cases they consisted of one mother with one or several children. As can be seen from the data included in Table 4.2, which have been culled from OECD sources, the number of single mother families increased by 29 per cent in the United Kingdom over the period from 1971 to 1984. In France, their number increased by as much as 45 per cent from 1968 to 1982. The number of families headed by a divorced or separated mother also increased dramatically. Table 4.2 shows that this number went up in Belgium by 126 per cent from 1970 to 1981, compared with 86 per cent in Germany over the years from 1972 to 1982. The number of single father families went down in some countries and increased in others. For example, in France and Belgium a decrease is noted, whereas their number increased in Germany and the United Kingdom.

As will be seen below, more single mothers were working outside the home in the mid-1980s than was the case among married women. However, even among the latter, the majority were at work in many European countries in the mid-1980s. However, the percentage varied considerably between countries. In 1985, 79 per cent of women in the age group 16 to 64 were in employment in Sweden, compared with 75 per cent in Denmark but as few as 15 per cent in Spain. [6] Women

---

[4] Council of Europe *op. cit.*
[6] Council of Europe *op. cit.*

increasingly tend to keep their jobs when they have small children. In the Nordic countries, by 1965, roughly two thirds of women with young children had no employment outside the home. This figure went down to nine per cent in the 1980s.

Table 4.2: *Percentage of Net Absolute Increase in the Number of Single-parent Families (by each sex and marital status category)* [5]

Country	Divorced or separated mother	Single mother	Father	Time period
Germany	86	13	26	1972-1982
France	118	45	-9	1968-1982
Belgium	126	12	-12	1970-1981
United Kingdom	71	29	4	1971-1984

3. The Changing Role and Structure of the Family

Parallel to the changing household structures and related demographic developments described above, institutions which the sociologist James S. Coleman refers to as "corporate actors" [7] – business enterprises, public agencies and organised interest groups – tend to become increasingly influential. The corporate actors possess by their very nature a limited purposive structure, which means that they are limited to one or to very few tasks. The family, however, traditionally serves and performs a wide range of inter-related activities: production, consumption, procreation and leisure. The "primordial" family is the dominant child-rearing agent in a society with an agrarian economy of a subsistence character. The family can be viewed as a unit of production and consumption, and child rearing is part and parcel of a wide range of activities going on in that unit. Corporate bodies with their specialization and narrow purposive activities, however, have no place for child rearing, which in the family are an *integrated part* of all other ongoing activities.

The first big wave of industrialization and urbanization brought corporate actors, enterprises and voluntary associations on to the scene. Prior to that or simultaneously, public agencies at the central, regional and local levels became more powerful. The "natural" person, the single

---

[5] Source: *OECD Employment Outlook,* 1989 and 1990.
[7] Cf. J.S. Coleman: *Foundations of Social Theory,* Harvard University Press, 1990.

acting individual, lost relatively in influence *vis-à-vis* the public and private corporate actors. Industrialization made the husband move out of the family to a separate workplace, but the majority of wives stayed at home. As has been shown by Coleman, the decline of husbands working in agriculture went parallel with the increase in children going to elementary schools. [8] The latter could take over certain tasks when the children were no longer needed to perform certain chores at home anymore. The schools were expected to specialize in providing systematic teaching first and foremost in the three R's. In addition, the school served the State in exercising social control, not least in countries with a state church. It also helped in establishing a common national language. Not least in urban areas, it also served as a baby-sitting institution. One should, however, not exaggerate the first of these tasks. [9]

The second fundamental change in the role of the family has occurred much more recently, when the women have moved out to workplaces outside the family. Changes in the "ecology" of upbringing can be inferred from data recording the enormous increase in the number of women going out to work outside the home from 1960 to 1985. The countries included in Table 4.3 are cases in point.

Table 4.3: *Female Labour Force Participation Rates, 1967-1988, European Countries (per cent)* [10]

	1967	1977	1983	1988
Denmark	55	65	74	78
Spain	30	33	33	39
France	47	53	54	56
Italy	33	38	40	44
Portugal	25	55	59	59
Sweden	55	70	77	80

Note: The female labour force participation rate is defined in this table as the female labour force of all ages divided by the female population aged 15 to 64 years.

The result of the general increase in the female labour force participation rate has been that institutions were needed to take over

---

[8] J.S. Coleman: "Families and Schools", *Educational Researcher*, Vol. 16, pp. 32-38.
[9] The impact of legislated universal elementary schooling in terms of enhanced literacy during the first time of its existence was in some Nordic countries rather marginal.
[10] Source: OECD, *Employment Outlook*, July 1990, and OECD, *The Teacher Today*, Paris, 1990.

important child rearing tasks from the family, particularly at the preschool level. The extent of enrolment in pre-school institutions in selected European countries was shown previously in Chapter 2. Time-series data were lacking in Table 2.1, however. This is remedied in Table 4.4 below, which shows how the number of places in regulated child care for pre-school children, as defined in each country, have increased from 1975 to 1989. Even though the data do not cover the entire post-War period, it can be concluded that the character of child rearing and its setting in Europe has changed radically since 1950.

Table 4.4: *Number of Places in Regulated Child Care for Pre-school Children (rate of coverage of all children, per cent, 1975 to 1989)* [11]

Country	1975	1979	1983	1987	1989
Canada	3.1	4.1	5.6	8.9	--
Japan	13.7	18.2	20.3	20.6	--
Norway	—	18.3	24.9	30.6	32.0
Netherlands	--	12.2	13.8	--	--
Sweden	12.3	28.9	39.9	47.1	--

Note: Excluding kindergartens and nursery schools but including day-care centres.

It is clear from what has already been said that the *structure* of the family – or rather the household in which the children grow up – has changed considerably since 1950. The three-generation family has almost entirely been replaced with the two-generation family, a change that may well entail a loss to the children in terms of their sense of historical continuity. Pair relationships between adults of the opposite sex have become less formal and, in many cases, less stable. Formal marriages have declined in number. Cohabitation without marriage has increased considerably and is easier than marriage to dissolve. The divorce rate has increased and the number who marry more than once has also increased. Single-parent families have doubled in number over a short period. The "normal" household with two adults and at least one child is no longer the most frequent one in urban areas, as was the case a few decades ago. In big cities one finds more households with one or two adults without children or one adult with child than those with two adults

---

[11] H. Goulet: "Child Care in OECD Countries", in OECD, *Employment Outlook*, July 1990 (Table 5.4, p. 131.

and at least one child. The birth rate has been declining since the post-war so called "baby boom". As can be seen from Table 4.5, Sweden had below two children per woman in 1985, well below the number required for full reproduction. These demographic changes, as well as the fact that more women with children spend the day at workplaces outside the home, have affected the interaction patterns between adults and children. The latter tend to spend less time with their parents and more in collective settings in institutions, particularly pre-school ones.

Table 4.5: *Indicators of Household Conditions, Sweden, 1900-1985* [12]

	1900	1930	1960	1985
Mean persons per household (N)	—	3,5	2,8	2,2
Single person households (%)	23,6	18,9	20,8	36,2
Households with children (%)	60,7	62,9	39,9	25,8
Mean children per family (N)	—	3,28	1,74	1,68

In the early 19th century some 80 to 90 per cent of the labour force in Europe derived livelihood from agriculture and related work. In the 1980s some countries relied for their food production on an agrarian sector which employs three to four per cent of the work force. The majority of those who up to the mid-20th century left rural employment went into the manufacturing industry, which reached its employment peak several decades ago when the service sector expanded its share of employment. This fundamental change was of the same importance in influencing the family, as the one when the husband went out to work outside the home. But another change affected the family even more in some cases. The children had to go to school and were no longer available for the chores they had been in charge of before.

One century ago it was exceptional – it happened only in 15-20 per cent of the cases – that households were without children (young people under 18.) Now the majority of households, about 60 per cent in some of the industrialised countries in Europe, consist only of adults. One hundred years ago the number of children in households with children was on average more than four, which meant that children grew up with a much wider range of social contacts in their immediate environment than is the case today. Roughly one-fourth of all households are single-

---

[12] Source: *Statistical Abstracts of Sweden,* and *Historical Statistics of Sweden* (various years).

parent ones. The size of households has been shrinking and about two thirds of them now consist of three persons or less.

No doubt, according to crucial indicators, the family in modern society has been de-stabilised and its functions substantially reduced. Not least people employed by the Welfare system have often urged that other institutions staffed with professionals should step in and take responsibility. Nurseries, kindergartens, and pre-schools should take care of the children before they enter regular schools. The latter should extend their duties beyond class hours and take care of the children until their parents came back from work in the late afternoon or early evening.

Other agents substituting or competing with the family have also emerged. The amount of TV watching among children in some countries equals the time spent in the classroom. Given the increasing number of mothers working outside the home there is a pressure to have the school or school-related institutions take care of the children even after school hours and to shoulder minding duties that can, at best, carry only a remote relationship with formal schooling.

The welfare state, as constituted in several European countries by the mid-20th century, has in some respects developed into a "client society" where a growing class of publicly employed people are expected to take care of the rest of the citizens from the cradle to the grave. Old institutions, such as the school, have been given considerably widened scope with regard to what they are supposed to achieve. New institutions, particularly those taking care of children of pre-school age, have emerged. An enormous expansion of health and socio-medical institutions can also be noted.

As pointed out above, services provided by professional experts have become more and more specialised and no agency or individual within any given agency is taking care of the "whole client". This results in a fragmentation of contacts between care takers and clients both in terms of kind and continuity. Many institutions have a high turnover of staff which particularly has a strong effect on children in day care centres or other pre-school institutions. Studies have shown that children at that age level are afraid of "investing" emotionally in particular persons among the staff, because their experience has often been that their contacts with care takers have been of a short-time character.

The problem of fragmentation of adult contacts tends to remain at the primary school level even within the framework of self-contained classrooms. Leaving aside the turnover of class teachers during the school year, a change from almost complete containment has occurred because of the specialised teaching in, for instance, subjects such as music, art and gymnastics. Over the last few decades new categories of staff have been introduced and have increased in many urban school systems, particularly in schools with large enrolment: school nurses, school psychologists, social workers, personnel in charge of school lunches, janitors, technicians and — not least — administrative personnel assisting the headmaster, such as the assistant principal, subject area specialists and secretaries.

One major consequence of fragmented care taking with compartmentalised services conducted by professional experts, each with his or her own territory, has already been pointed out, namely, the reluctance of children to "invest" emotionally in adults. Adults in a given position change, which occurs by definition with corporate actors structured in positions instead of in "natural" persons. In such a system the climate easily becomes impersonal and the procedures bureaucratised. Hence social control is more difficult to establish. Fragmentation of adult contacts, for instance compartmentalization between home and school, leads to inconsistencies of norms and to difficulties in inculcating them. In a study of primary school discipline, teacher and parental attitudes were assessed with regard to the degree of strictness vs. permissiveness. It was found that norm discrepancies between teachers and parents led to disciplinary difficulties for the child in school. [13]

The shift of the responsibility for child rearing from the family to institutions can, in individual cases, lead to unwanted consequences, as the following episode illustrates. The Swedish author and Nobel Laureate Harry Martinson in his autobiographical novel "The Nettles Bloom", published in 1935, [14] presents experiences from his own childhood in a rural area at the beginning of the century and long before the welfare state with its *ad hoc* agencies and institutions had been established. His parents had deserted their children, who had to be taken care of by the parish agent in charge of the poor and desolate. The board did what was then common, namely, to "auction" the children to farming families in the area. The highest bidder was offered the task of taking care of the child. The book's main character escaped from one home after another. Finally, it was decided that he should be transferred to the parish home for the elderly. When he arrived there the lady in charge of the home, wanting to show her kindness and wanting to comfort, patted him on his head. Martinson's comment: "But I felt that it was the communal pat".

Institutions involved in childrearing in today's society are not extensions of the traditional family which, in Coleman's words is "the

---

[13] H. Martinson: *Nässlorna blomma*, Bonniers, Stockholm, 1935.
[14] T. Husén: "Interaction between Teacher and Pupil as a Determinant of Motivation and Satisfaction with School Work, *Proceedings of XIIIth Congress of the International Association of Applied Psychology*, Rome, 1959, pp. 471-482.

last kernel of the old social structure to remain". [15] The family tends to become a "legal anachronism", difficult to sustain in a society with growing powerful organizations in which "persons are transient and only the structure is permanent". The new structure is built around activities and not around persons. Membership in a family is what sociologists refer to as "ascribed", whereas in purposive organizations it is "achieved", an outcome of initiative on the part of the individual and the "corporate actor".

A basic characteristic of the family has traditionally been that it is tightly structured around and takes responsibility for *persons*. But the organization of corporate actors is structured around *activities*. Authority over a person, such as the one exercised by the family, means, or meant, taking responsibility for the welfare of a particular person. Both authority and responsibility are crucial to *all* the activities in which people are taking part.

New and specialised institutions involved in child-rearing take responsibility, each for a particular aspect, when the family is no longer exercising it. Children of pre-school age are usually in day care centres when the parents are not at work. Regular schools assume certain responsibilities from the age of five to seven up to the late teens. Youth centres are available for children after school hours. Each institution has narrowly defined responsibilities and is in charge of a particular segment of the person. The same applies to health and social welfare services.

The fact that various child-rearing activities are performed by specialised agencies and institutions, organizationally in separation from each other, creates not only a problem of coordination but, more importantly, pose difficulties for the "clients" – the school children. Incompatibilities between the responsibilities of the parents and those of the society represented by growing powerful institutions each in charge of one segment of the child have led to problems which as far as coordination between home and school sometimes reach crisis format.

Particular problems beset children who are growing up in disadvantaged or deprived families and environments. Upbringing has, as pointed out above, become increasingly fragmented and impersonal, in particular for children who grow up in large cities with big schools which by their sheer size could be characterised as "pedagogical factories". In big cities many children are attending schools with an enrolment well above 1,000 pupils. [16] An argument often used by administrators

---

[15] J.S. Coleman: *The Asymmetric Society,* Syracuse University Press, 1982.
[16] The Netherlands Ministry of Education and Science has recently proposed to merge small primary schools into larger ones. As is typical of the mood of the time, the debate is mainly about how big the financial benefits of this move to large school units will be, and

supportive of large rather than small school units is that education can be offered at a lower cost in the former ones. Moreover, research evidence showing that children in big schools tend to perform no worse than those in the, more costly, small schools, is also often quoted in the debate, as is the case in the Netherlands at present. Yet although economies of scale make it possible to save money without an apparent loss in student performance, there are other variables in the equation, often immeasurables, that are not included.

The interest in this chapter is focused mainly on the relationship between home and school and the problems which emerge in the wake of changes in the structures and roles of both. Before discussing this relationship and the problems one can identify in their interaction or lack of interaction, it would be justified to mention very briefly the scholarly efforts that have been made to elucidate the role of home for success or failure in school.

The main interest among social science researchers was for a long time focused on how the social status of the home affected the child's school career. Certain aspects of the home background, such as social class, parental education and occupational status, which were expected to influence the child's scholastic achievement, were subjected to intensive study. For a long time remarkably little was done in studying the family as a *support* system. How much interest do parents take in what happens to the child in school? Do they ask their children about their homework and to what extent are they eager to provide some help with it? Do they make a point in correcting their children's language use and to explain words and phrases to them? In brief: to what extent are they committed to their children? [17] Anthropologists have shown how this commitment differs considerably between families belonging to various ethnic groups, a phenomenon which has been observed in particular in the United States but which as a result of recent immigration can also be studied in Europe. The way a modern society and its world of work are organized have an effect on adults and thereby on children: they generally spend little time playing with, talking and reading to their children.

There exists an extensive scholarly literature elucidating how parental education, social class and occupational status are related to school achievement. [18] But the research on the family's role as an educator and

---

not, as one might expect, with the pedagogical, psychological and social consequences of school size.

[17] Fraser *et al.* (1987) present a meta-analysis of several hundred studies investigating the contribution of so-called productivity factors, among which several indicators of home background, to student achievement and school effectiveness in industrialized countries. See also Iverson and Walberg (1982).

[18] Among the classical research studies investigating how parental education, social class and occupational status are related to school achievement can be mentioned Coleman *et*

what specifically makes children ready to go ahead in school is scarce. The family as a support system is often related to other educative agencies close to the family, such as the neighbourhood, voluntary interest groups and religious organizations. Thus, the role of the family as an educator has to be seen in *relation* to other institutions and other formal and informal educative agencies, such as pre-schools, regular schools, peer groups and religious bodies.

4. The Changing Role of the School

Teachers, particularly those in big cities, have in recent years begun to complain about the difficulty of maintaining school discipline even at the lower grade levels. Children lack the ability to concentrate, which in turn creates a problem in attracting their attention. Emotionally disturbed children are also encountered more frequently than was the case a few decades ago. It was found in a study of school discipline conducted at the primary level in the Stockholm area in the 1950s [19] that both subjective complaints and objectively recorded disciplinary incidents were strikingly few in the first grades as compared to subsequent grades. Changes in the household structure and the educative functions of the home have taken their toll on schools.

The same development in school enrolment has occurred in many European countries over the last few decades: an increasing number of children has spent an increasing number of years at school. School attendance was generally compulsory up to 14 years of age before 1950. Since then lower secondary education has become universal in many countries and in a growing number of them the majority stay in school until the age of 18. This change alone marks an enormous widening of the role of the school. Before 1950 the great majority of youth aged 14 to 18 years entered the labour market after completing elementary school. Now they are kept in school, which serves a more competence-demanding, competitive and meritocratic society.

Irrespective of the number of years during which the school is in charge of young people its functions have been widened. By the mid-19th century the American educator Horace Mann labelled the universal tax-supported elementary schools common to young people from all walks of life the "Great Equalizer". It would at least represent the first step in removing social class barriers and could lift people out of

---

al. (1966), Blau and Duncan (1967), Husén *et al.* (1969), Jencks *et al.* (1972 and 1979) and Behrman *et al.* (1980). Examples of studies on this theme carried out in recent years are DiMaggio and Mohr (1985), Bound *et al.* (1986), Hauser and Sewell (1986), Lareau (1987), and Postlethwaite and Wiley (1992).

[19] T. Husén *op. cit.*, 1959.

poverty. It could provide the means for those who wanted to improve their conditions by acquiring improved competencies. A similar philosophy was behind the movement after World War II in several European countries to introduce a comprehensive basic school which would replace the class-divisive system of schools, private or public, running parallel to each other. Apart from discharging certain educative duties the school should also contribute to social equalization. Such an overriding objective was expected to be achieved by equalizing opportunities to enter post-compulsory education and by removing financial barriers preventing some from entering. By putting all students on a par with each other at entry one could expect that those who possessed a good deal of 'natural' ability would have an opportunity to go ahead and reach the social status they by nature deserved.

Thus, the school was seen by many as an instrument in bringing about social change in the direction of a society with greater equality not only of opportunity but of material standards as well. [20] School reforms with such goals were launched in several European countries where comprehensivization went on although at a different pace.

But the steps taken did not always yield the expected outcomes. The first OECD conference on policies in education, which was held in 1960, focused on education as an "investment in human capital" and how it could benefit both the individual and the national economy. [21] Economic growth should be promoted by taking care of the existing talent. Since participation in upper secondary and university education, according to many surveys, was strongly correlated with social background in terms of parental financial status and education, the evident policy recommendations seemed to be to enable young people from lower class homes and from rural areas without secondary schools to be given greater opportunity to proceed to further education at *gymnasia, lycées* or grammar schools in order to qualify for university entrance. However, during the 1970s, the optimism aired a decade earlier about what could be achieved in terms of equalization by means of increasing enrolment and providing financial aid had not been met by the actual change in the social composition of the enrolment in further education. To be sure, a certain levelling out of social strata had occurred, but not to the extent hoped for by those who engineered the pervasive reforms of schooling in European countries.

One began to realize that an institution, such as the school, where students are ranked and promoted according to their achievements cannot serve as the main equalizer of social opportunities. The reverse is more

---

[20] This idea was launched at the the OECD Conference in Kungälv, Sweden, in June 1961 (see Chapter 2). Reference: A.H. Halsey (Ed.), 1961.
[21] The reference is to the OECD (1962) Washington Conference (see Chapter 2).

accurate. Once having admitted students to the school on equal terms, the institution is by its very nature conducive to distinctions.

5. Discrepancies and Discontinuities Between Home and School

Modern society is highly specialised with regard to production, delivery of services and preparation for the various roles, occupational and other, which individuals are playing. It is hence very complex and requires increased competence on the part of the individual properly to orient him or herself. It is also, with its ongoing urbanization, a highly mobile society. Many families are uprooted from a stable rural life which has gone on for generations into the more ephemeral urban life pattern. Mobility is also social, and formal education tends to become increasingly important for the status achieved by the individual. International mobility means an increasing mixture of various ethnic groups within the same geographical area. Children of different backgrounds come into contact with each other in the playground and in the school.

There are certain discontinuities which are endemic to the very nature of the family and the school as institutions. Relationships within the family are what could be described as functionally diffuse. The family members are involved with each other in practically *all* respects: living at the same place, eating together, spending leisure together, *et cetera*. The school is a place with *specific* relationships which have to do with specific roles and tasks assigned to each individual, student or teacher. The child is taught in a certain grade by a teacher trained for a certain level of the system. Social and administrative services are provided by personnel where each individual has a narrowly circumscribed function to fulfil. There is crucial discontinuity between the family and the school with regard to the nature of relationships that exist between individuals. In the family the relationships are particularistic. In the school they are universalistic. The same discontinuity exists between the family and the public agencies that deliver health service or other social services. In the family "blood is thicker than water", which means that relationships between parents and children and between siblings, under normal circumstances, are determined by a loyalty which is directed toward a particular, favoured individual and not primarily to all individuals with whom the person happens to come into contact, not to speak of all the persons in the nation or around the world that the person has never met or even heard of.

The tendency in many urban areas, particularly those with many immigrants, to have residential segregation, affects the social structure of school enrolment. Catchment areas may become more homogeneous, but the variability in social background between schools may increase.

The background for this is that personal interaction in the family is determined by what the members *emotionally* mean to each other, whereas in the school or in other public institutions these relationships tend to depend on the *positions* that the interacting individuals occupy. The emotional involvement is different in the family from that in an institution or organization outside the family. If a parent benevolently pats a child on his head it invokes a feeling different from that conveyed by a "communal pat". Two things must be added here, however. Firstly, the teacher who is *liked* stands a better chance of being successful on the job. A second observation is that emotions may also be hostile, and that these play an important role in the family. Public institutions, for example kindergarten, are also different in this respect.

The distinctions made above ought to be kept in mind in discussing the problems of interaction between home and a school in the modern society of today. The discontinuities which are pervasive in all kinds of societies have become exacerbated by the changes that have occurred in the structure and roles of both institutions, family and school. The spread in wealth and welfare between groups of people and, especially, different urban communities, presents a noteworthy problem in this connection, because school heterogeneity tends to grow with an increase in social differentiation.

6. Consequences for School Education

So far we have dealt with a number of overriding changes that have occurred *within* the two main educative agencies in modern society, the family on the one hand and the school and school-related institutions on the other. The structure of the family has changed, which has led to a shift of responsibilities from the family to the school. We shall here indicate some of the more specific problems which have emerged as a result of these changes, notwithstanding the obvious and important benefits that they have brought about, such as enormously increased opportunities for further education which for the great majority were almost unthinkable half a century ago. We shall, in what follows, briefly mention some of the consequences of the changes.

Firstly, during World War II, when a great number of women with children in school went to work in industry to replace the men, one talked about "*latch-key children*", those who came home after school without finding either parent there. The relative number of women with children in school working outside the home has increased over the last few decades but at the same time institutional provisions have been made to take care of the children after school hours.

Secondly, the *common activities,* such as having meals together including the interaction between family members that goes with it, have been reduced, not least in places where the school provides a meal.

Thirdly, parents tend to *withdraw* from involving themselves in the school progress of their children, particularly those preoccupied with their own career and those who tend to take a submissive attitude *vis-à-vis* school authorities. Teachers and school administrators sometimes resist parents "meddling" in what they regard as their affairs. Some parents do not care at all about the progress of their children in school. So-called automatic promotion, a system whereby pupils are promoted from one grade to another without having reached the standards of the grade often leaves parents in ignorance about their children's progress.

Fourthly, the *truancy* problem has become a major one at the secondary level particularly in big cities where on a typical day, as many as one third of the students can be absent from class. There is sometimes a reluctance on the part of administrators and school authorities to come to grips with the problem simply because it reflects badly on the school. Furthermore, the resource allocation may be affected by the factual number of students present.

Fifth, the *size of the school* is conducive to difficulties with social control and causes an impersonal and bureaucratised atmosphere, not to speak of a fragmentation of contacts between pupils and adults. In a big organization like this the child feels that no single person is in charge of him or takes responsibility for his welfare. The bigger the school the less auspicious are the opportunities for stable emotional relationships to be established between adults and pupils.

Sixth, with strong *unionization* among teachers and school administrators, spheres of influence and relative power over what goes on in school is zealously guarded. This has sometimes led to problems in trying to increase the influence of parents upon how schools are governed and operated. Teachers and school administrators in many places resist parental em-powerment.

Seventh, the emergence of a *"new underclass"* has become a major problem in big city school systems. We are here dealing with a minority, on average some 10 to 20 per cent of the school population, whereas the "old" underclass was in the majority. It is a "new" phenomenon also in the sense that these children are not necessarily materially deprived but in many cases simply psychologically neglected due to various circumstances, such as broken homes, lack of involvement in what happens to the child in the school on the part of the parents, difficulties in settling down in a new culture or in a new residential area, ethnic discrimination, *et cetera.* Given the much higher frequency of migration within the countries and between countries many children are experiencing difficulties with their adaptation to new school settings, not to speak of new neighbourhoods. In most cases the "social capital" of the home, i.e. the psychological support it can provide, is low. School authorities often are at a loss over how to deal with these children because of the complexity deriving from fragmentation of responsibilities and the difficulties of coordinating the efforts of school and social welfare authorities to take proper measures.

Even though this is not the place to spell out the implications for the labour market of the "new underclass" it ought to be pointed out that school failures, drop-out and early leaving all contribute to create unemployability. Experience has shown during the last couple of decades in the highly industrialized parts of Europe that the majority of those who leave school at the age of 15 or 16 can hardly be employed. This applies in particular to those with bad school records. Their cognitive competence is insufficient and they are in most cases not used to the discipline of work.

7. The School and "Social Capital"

A pupil gets ahead in school and life not in the first place by his ability and willingness to work but also by having the so-called "F-connections" (Family, Friends, Firms.) These have to do with personal relationships and networks constituting "social capital", which is not a single entity but a range of entities which have in common the fact that they facilitate certain actions of individuals within a given social structure. Success is facilitated by having the right connections and an available support system. For the pupil, parents, who care and who involve themselves in what the child does in school and about his homework, represent social capital. The same applies to available grandparents, neighbours and close friends. A group of parents setting up a Parent-Teacher Association represent a social capital both to themselves and to the school. Central to such a network providing social capital to the child or to a group is reciprocity and mutual trust: "I do this for you and you do this for me".

There are many indicators which can be used in measuring social capital. In a survey of high school students in the United States Coleman used the following measures: two- versus single-parent household, the number of siblings at home, the ratio of parent to children, mother's expectation for the child's education, and a combination of these variables. [22]

---

[22] Cf. J.S. Coleman: "Families and Schools" *op. cit.* See also cf. J.S. Coleman (1961).

Table 4.6: *Effects of Various Measures of Social Capital Provided by the Parent-child Relation on Estimated Percentage of Students Dropping Out (grades 10 to 12)* [23]

	Percentage dropping out	Difference in percentages
1. *Parents' presence*:		
Two parents	13.1	6.0
Single parent	19.1	
2. *Additional children in family*:		
One sibling	10.8	6.4
Four siblings	17.2	
3. *Ratio of parents to children*:		
Two parents, one sibling	10.1	12.5
One parent, four siblings	22.6	
4. *Mother's expectation for child's education*:		
Expectation of college	11.6	8.6
No expectation of college	20.2	
5. *All measures combined*:		
Two parents, one sibling, mother expects college	8.1	22.5
One parent, four siblings, no expectation of college	30.6	

Notes: (a) All effects are significant at the five per cent level (DEF of 2.0 is taken into account.) (b) Analysis is carried out via a weighted logistic regression using nine variables. Results are calculated by subtracting from or adding to the overall percentage (14.4 %) the effect due to deviation from the mean of the independent variable or variables in question.

Other indicators which have been studied are talking at home with parents on personal matters as well as about how things are going at school. Such conversations indicate interest on the part of the parents in the child. For mothers to go out to work before the child attends school regularly represents an important factor and one subject to controversy, not least among professional women, who more frequently than women in the working class have jobs during the child's pre-school years. The

---

[23] Source: J.S. Coleman, *Foundations of Social Theory*, Harvard University Press, Table 22.2, p. 596.

decision to continue work leaves less time for the mother to spend with the child. In some instances it can also be taken as an indication that the career has a higher priority than the child. Finally, stability at home is related to how well the child can get along in school.

Some of the indicators mentioned above were tabulated against dropout in secondary school by J.S. Coleman, a Chicago sociologist. Although the results apply to the United States, they may well be significant even in a European context. The coefficient in the right-hand column indicates how much greater the dropout rate is in the one than in the other group. Thus, dropout occurs 1.45 times more often in single-parent households than in two-parent households. As can be seen from Table 4.6, households where all four measures indicated a weak network had almost four times as many dropouts as in households without these weaknesses.

8. The "Old" and The "New" Educational Underclass

As pointed out above, in the traditional agrarian society the family, as a rule a three-generation family, was a unit of production and consumption. The advent of industrialization located the father's workplace outside the family, a process closely related to urbanization. The next major change occured in most industrialised countries subsequent to World War II. In the 1960s and 1970s women in rapidly increasing numbers began to work outside the home. This change in the composition of the labour force was compounded by other key social developments, such as changes in the home and work environment as a consequence of technological advances and a concomitant increase in the degree of correspondence between the functions of the home, the school and the workplace. The growing participation of women in the labour force and the changing role of women in society generally have had an effect not only on the socialization context provided by the home but necessarily also on the functions of schooling. For example, children had at an early age to be taken care of in pre-school institutions of various kinds: day nurseries or day care centres and kindergartens.

Since the mid-20th century the changes in the working roles of fathers and mothers have been followed by other demographic changes, indicated above, which have transformed the traditional family further. All these changes taken together have brought about not only fundamental changes in the structure of the family but have also deeply affected the conditions in which children today grow up. In many European countries the majority of parents are now working outside the home. An increasing number of children are growing-up in single-parent families or with step-parents. A high percentage of children spend a considerable part of their time in pre-school institutions. The changes hinted at must be expected to have major implications for the provision of an adequate learning environment for young children in the different parts of Europe. Repercussions on the school can also be noted, as it has increasingly been called upon to shoulder the care-taking duties which previously were almost entirely performed by the family. An increasing number of children, although a minority, are growing up under psychologically deprived conditions.

Children in the traditional or "old" underclass as a rule lived under material poverty. This group has by no means been eliminated. It was estimated in 1981 that, in terms of comparative income, no less than 11.4 per cent of all households in the member states of the Council of Europe were living in poverty. [24] A recent EUROSTAT report presents the following conclusions: [25]

> "There were about 49 million poor persons [in the EC of twelve countries] in 1980, measured by national criteria. [26] On this basis their number increased slightly to 50 million in 1985. The Member States which had the highest poverty rates (i.e. percentage of the population living below the poverty line) in 1985 were Portugal (33 per cent), Ireland (20 per cent), Spain (19 per cent), and Greece and the United Kingdom (around 18 per cent.) Measured by Community criteria, poverty figures were even higher in 1980 (53 million people), but fell to 51 million in 1985 [out of a total population of about 326 million]. Although poverty seems to have stabilised in the period 1980-1985 for the Community as a whole, important changes in poverty rates were observed for individual Member States".

The report provides evidence showing that children and the elderly are high risk groups in terms of poverty. The poverty rates among them are significantly higher than for the Community population as a whole. The relative position of children does not seem to have improved from 1980 to 1985 in all EC countries. Quite the contrary. The poverty indices [27] presented in Table 4.7 show that the relative position of children in 1985 was worst in the Netherlands, followed by Ireland, Germany and the United Kingdom. One may well repeat that poverty is a relative rather than an absolute concept, and that poverty is perceived differently in different countries.

---

[24] Oreja, M.: *Social Cohesion and Dangers Facing It* (Report of the Secretary General presented to the Parliamentary Assembly on 6 May 1987).
[25] EUROSTAT: *Inequality and Poverty in Europe, 1980-1985* (Rapid Reports, 1989).
[26] A household is considered poor when it spends less than 50 per cent of the national average household expenditure per adult equivalent. The poverty line is located at this 50 per cent level.
[27] The poverty index reveals whether the poverty rate among children is above (higher than 100) or below (less than 100) the national poverty rate.

The indices in Table 4.7 are measured in relation to a national living standard. This seems to justify the conclusion that the four countries mentioned above have experienced, over the brief period considered, a significant increase in the absolute number of poor children in the population. The overall position is that, in 1985, 20 per cent of all children in the European Community were living under conditions of material deprivation. Even though hard data are not yet available, it seems reasonable to hypothesize that in excess of five million children are living in poverty in Europe today.

Table 4.7: *Poverty Incidence in EC Countries in 1980 and 1985 (per cent of group under the poverty line, poverty indices)* [28]

Country	Households[a]		Children[b]	
	1980 %	1985 %	1980 index	1985 index
Belgium	5.2	6.3	1.13	1.14
Denmark	8.0	8.0	1.10	1.14
Germany[c]	9.2	10.3	1.10	1.38
Greece	17.4	20.5	0.96	1.03
France	14.8	18.0	1.08	1.24
Ireland	17.4	18.5	1.22	1.43
Italy	14.7	12.0	1.14	0.97
Netherlands	7.9	6.9	1.36	1.55
Portugal	31.7	31.4	1.12	1.12
Spain	17.8	20.3	1.06	1.07
United Kingdom	18.9	14.1	1.38	1.32
EC[d]	14.4	14.1	—	—

Notes: (a) A household is considered poor when it can spend less than 50 per cent of the national average household expenditure per adult equivalent. This 50 per cent level constitutes the poverty line. (b) The poverty index shows whether a group is over- or under-represented among the poor. The national poverty rate forms the basis for comparison with an index value of 1.00. (c) The data refer to former West-Germany. (d) Excluding Luxemburg.

A "new" *educational* underclass, as was mentioned above, also exists, although statistical data on them have not yet been collected in European countries. [28] These are the children suffering from psychological neglect, often in combination with material deprivation. A common denominator

---

[29] EUROSTAT *op. cit.*
[28] But see Pallas *et al.*, 1989.

for most of the deprived children in modern European society is the lack of parental support and interest. Many children are victims of psychological and emotional poverty.

The children referred to here as an "educational underclass" are "new" in two major respects. The traditional underclass consisted of a majority of the population, mainly manual workers who, due to lack of material means, did not have access for their children to the educational institutions available to the upper classes, particularly academic secondary schooling preparing for university admission. The traditional underclass organised itself, educated its own spokesmen, acquired political power and demanded educational reforms that widened educational opportunities for the great majority.

By comparison, the new underclass is a minority and therefore does not have the same kind of spokesmen who have seized upward socially mobile opportunities. In many countries the new underclass consists of ethnic minorities, such as North Africans who have moved to France or people from Commonwealth countries settled in Britain. But in other cases they are children who are simply psychologically neglected. As they do not have a stable and supportive network at home, they lack an adequate learning environment and can rely only on a very small or non-existent amount of social and cultural "capital". Most children belonging to the new educational underclass have, for example, not been used to regular hot meals or firm work habits. As middle-class teachers often do not know how to handle culturally and psychologically deprived children who tend to be seen as disturbing elements causing disciplinary problems in the classroom, many children belonging to this category become truants at an early age and some eventually drop out completely from school. All too often, this sets the stage for a lifelong cycle of accumulated disadvantages (cf Chapter 3). The following quotation provides an excellent summary of the problem: [30]

> "The challenge of educating the children of poverty has long been recognized as a difficult one. Not only do many of these children perform poorly on academic tests, but also the typical school serving large numbers of such children faces a variety of problems that pose barriers to providing a high-quality education, among them: high rates of mobility among students' families, a high incidence of severe emotional or behavioral problems among students, large numbers of students with limited proficiency in English [or other European languages], low staff morale, and inadequate facilities and resources".

---

[30] The quotation is taken from an article by M.S. Knapp and P.M. Shields: "Reconceiving Academic Instruction for the Children of Poverty" (p. 753), *Phi Delta Kappan*, Vol. 72 (June), 1990.

9. The School Intervention Idea and Intervention Programmes

The realization around 1960 in the United States that there was "another America" and that a considerable portion of the population was living under poverty-stricken conditions led to the "War Against Poverty" launched by President Lyndon Johnson in the mid-1960s. The disadvantaged consisted largely of ethnic and racial minorities. Part of the "war" was a comprehensive programme of so-called compensatory education. It was expected that children who came from poverty pockets, by becoming beneficiaries of special attention at school, not least tutorials, in addition to the regular classroom instruction, would be "taught out of poverty". Abilities and skills acquired by education would make young people from underprivileged backgrounds better equipped for the labour market. Teaching staff was made available and underprivileged children were taken out of the classroom for tutorials, particularly in the three R's. Another part of the compensatory measures was the so-called "Head Start programme" which contained intensive pre-school courses for children from underprivileged areas and deprived cultural conditions. [31] The latter programme was much inspired by the research conducted by educational psychologist Benjamin S. Bloom, who studied cognitive development before school entry and the educative process going on at home through the interaction between children and parents. [32]

Apart from certain limited measures taken within the framework of special education, Europe has not had any large-scale counterpart to the American programme. Therefore, the experience gained in the latter country, not least by evaluation and follow-up studies, is of great interest in a Europe faced with increasing migration and problems of an ethnic nature. The rationale behind home intervention has been dealt with by a Task Force chaired by Thomas Kellaghan under the auspices of the International Academy of Education. [33] The Task Force points out the following propositions underlying home intervention:

---

[31] A brief description of the Head Start programme in the United States can be found in E.H. Grotberg (1985). Head Start began in 1964 under the Economic Opportunity Act on the assumption that the early provision of services and compensatory educational programmes would enable children from low-income families to deal more effectively with their environment and, particularly, to increase their opportunity to learn and be successful in school and life career.
[32] Cf. B.S. Bloom: *Stability and Change in Human Characteristics*, 1964, New York.
[33] Thomas Kellaghan was the main contributor to the report *Home Environments and School Learning* (1991), which has been published by Jossey-Bass on behalf of the International Academy of Education.

(1) Home environment has in all studies been identified as a powerful factor in determining student school achievement and interest in learning as well as the willingness to stay in school.
(2) Even though parents differ considerably with regard to level of formal education and occupational status it is what they actually *do* with their children in terms of support and encouragement that means more than status characteristics.
(3) It has been found that programmes which, for example, help parents give support and encouragement to their children, in addition to availability of early childhood education, have proven particularly effective.
(4) When home and school represent highly different approaches with regard to learning and value orientation the child tends to suffer in his or her learning.
(5) It is a clear advantage if parents are informed about the effect that home environment has on children's school learning and what they could do in order to support their children's learning.

In the modern, highly technological, information society a new category of people has emerged. It is perhaps a matter of taste whether to call them young adults or adult youngsters. Adolescence in the traditional sense (being related to puberty or to the maturity of the body) is no longer a relatively limited period of transition to adulthood after puberty. In today's society we can notice a phenomenon which we may call "evasion of adulthood". Its background is beset with a major ambivalence.

On the one hand the period of "preparation" for adulthood is prolonged − not least by staying in school an increasing number of years. This prolongs the phase in life when one is dependent on "established" adults and is prevented from taking on the responsibilities of a self-supporting, independent adult. This, being by itself experienced as something negative, can however, also be felt by the young person as an advantage, since he or she escapes being dragged into the "treadmill" of adult society with all its demands, regularity and discipline. We can today, among some young people, notice a tendency to cling to the adolescent life-style well into their late twenties or even thirties. They are "opting-out" of the "9-to-5 society" and can do so because of the relative affluence of their elders, who have worked hard to generate the resources from which the young are benefiting.

Given the growing demands in terms of competence on the labour market, as well as the substitution of labour by technical devices, it has become harder to find employment and be accepted as a fully-fledged productive worker. Young people are no longer needed in order to help the family economy. They have, in a way, become irrelevant.

On the other hand, young people in modern society experience no sense of being needed. In the traditional society at an early age they had to join their elders, parents and others, at work by taking part in the family business, such as agriculture. They had to take on functional responsibilities early in life as fully participating members of the adult society. In order to consume they had to produce.

A basic problem besetting education in modern society both at home as well as at school is the lack of functional participation. By "functional participation" it is meant that a person carries the responsibility for bringing about the conditions for one's own consumption and living. The common characteristic for lack of functional participation is that of always being "served", without the obligation of doing anything in return for the service one is enjoying. 'Spoonfeeding', goes on without the child's or adolescent's participation in any activity with either the parents or the teachers. The problem indicated here is the lack of "learning by doing" which is at the core of learning to become an independent, responsible citizen. By not having the opportunity to go through that process, a feeling of not being needed is instilled in the young. Many make a virtue out of their plight by actively trying to evade adopting adult life-style and responsibility.

10. Some Crucial Problems

Urbanization, changes in settlement patterns and the influx of foreign labour have created problems, such as clashes between values anchored in different cultures and, not least, lack of community in the new environment. There are difficulties in establishing a new setting with common values and mutual beneficence. Even within an ethnic community a kind of "new individualism" runs counter to forces of cohesion.

The task is to understand both tendencies: the longing of the individual for realms of self-fulfilment in an increasingly crowded, standardized, functionalized world, and the need of the collective for discipline, responsibility and solidarity of its individual members. In modern interdependent society individual processes and values are of necessity embedded in and conditioned by the social context.

Can the school become instrumental in forming new commitments? Can it reconcile individualism and collective solidarity? In most urbanised societies the constituents of the school, parents, teachers, students and administrators, have little or even nothing to do with each other outside the school context. But in order to create a school community centred around a common core of values one must rely on a network of bonds and contacts well beyond the school context. This is a major challenge to parents, teachers and school administrators, especially because the school is only one of several educative agents in society. Other social institutions, including the family, television, the neighbourhood and the workplace, also educate. Hence it is essential to place the goals of the school in a larger framework of interactions with other educative institutions.

This raises the question of the role of the school in establishing a common core of values and attitudes. The goals of basic education in Europe were considered in Chapter 3, which also dealt with the pedagogical challenge of providing a high quality education to all.

In talking about what the home can do we have to consider the work habits in the home, for instance the priority given to school work and the emphasis on regularity. The importance attached to correct language use, the stimulation given at home to explore ideas and events (for instance events presented in the news media) and academic expectations and aspirations are others important factors.

No doubt, the school can do quite a lot in helping to "educate" the parents so that they can better support their children. The great problem, however, is to establish the spirit of trust and confidence conducive to such endeavors.

The crucial question is: can the school serve as a substitute for the family? The answer is no. Each child needs to establish stable and continuous relationships with adults, be they parents, grandparents, or other kin or teachers. They cannot get along just with the "communal pat". Parents, neighbours, family and friends are involved people, and not institutions structured in impersonal positions.

The core of the problem is what above has been referred to as "fragmentation and discontinuities": in many cases nobody is in charge of the *whole* child. Many adults are dealing with the child in their segmented roles as teachers of special subjects, social workers and nurses. The only function left in the disintegrating family is procreation, which is also diminishing with the decreasing birth rate.

Evaluations of home interventions in cases of deprived and/or underprivileged homes have consistently shown that in order to be successful they have to be directed to the *total* situation of the family. [34] Piecemeal and fragmented interventions, such as taking the child out of class for a few hours of tutorials or visits to the parents now and then for

---

[34] The Head Start programme was surrounded by political and educational controversy right from its inception. Its various programme elements have therefore been the object of many research studies and other evaluation activities. Examples are: A.J. Mann *et al.* (1977) and E. Zigler and J. Valentine (1979).

counselling on child rearing, do not have long-lasting effects beyond the period when the interventions take place.

Lasting effects on the child's school attainments are achieved mainly by what could be called *ecological* interventions with reference to the family. Such a strategy is designed to affect the *total* situation of the family and to address the basic needs of its members. If nutrition, housing, and health care are at a level of minimum adequacy and if the parents have jobs that yield a minimum income then the auspices for adequate child rearing resulting from intervention are good.

Finally, the fact that millions of children in Europe are living under conditions of material deprivation must be noted. Poverty is not only inhuman, but it also must be held to exert a strong negative effect on student achievement. The prevalence of material and emotional poverty, as a determinant of "damaged lives" and low student achievement and school effectiveness, constitutes one of the most urgent problems European countries are faced with. To combat poverty in Europe is a challenge that cannot wait, as every day in school is important to the child. It requires us to invent effective counter-measures at a level much broader than that of school policy alone. Yet schools can certainly play a role, as they provide a platform of means and ends through which society can reach out to all children. [35]

11. A Concluding Note

As has been shown in this chapter by means of demographic and other statistics, the family in the highly industrialized countries in Europe has undergone a radical change in its structure and functions over the last decades. Other countries are likely to follow suit. Some of the consequences of this change have been dealt with here and the challenges emerging from them will be taken up in the concluding Chapter 11. We shall here limit ourselves to point out some evident consequences.

The establishment of pre-school institutions has become a priority as the majority of both mothers and fathers work outside the home. Countries with school entry at an age as late as seven years may consider lowering that age. The role of the teacher has changed because social problems have entered the school. The teacher in some respects has to serve as a substitute parent, and a social worker as well. The teacher's role is no longer confined to the one of conveying certain cognitive competencies which the school traditionally has been in charge of. All this has certain implications for teacher training.

---

[35] See, e.g., R.E. Slavin and N.A. Madden: "What Works for Students At Risk: A Research Synthesis", *Educational Leadership,* 1989.

Many of the issues raised in this chapter concern social problems and economic conditions that are external to the school environment, but that influence its functioning and set limits on what can be achieved. It is acknowledged that educators may not be in a position to do much about external factors such as violence on television, material poverty, divorce rates, and psychological and emotional disturbance in children due to parental neglect, or youth delinquency. Recognition of this limitation is important, because the school has often been blamed for social ills that, reasonably speaking, fall outside its immediate sphere of influence. This argument leads on to the next, that many of the problems schools are confronted with require to be addressed at a level over and above that of the individual school, and that intervention strategies need to be coordinated with other areas of policy.

Particular attention will have to be given to the consumption of television. It would be both pedagogically dubious and futile to try to prevent children from watching TV. But parents can be made aware of the consequences of indiscriminate and excessive consumption of films and shows on the TV-screen. They should be encouraged to watch TV together with their children and to talk with them about what they have seen. Media-education is by now a well developed discipline of pedagogy, but its results are not easily transformed into daily practice. The school as "a place for kids to grow up in" (see Chapter 3) would greatly facilitate this.

The changing role of the school and the teacher raises questions about character formation. Furthermore the conditions for expanded functions such as the size of the school 'plant' and the teacher:pupil ratio would have to be considered. But other factors, such as the conditions under which a child is raised at home, may be more important in determining student achievement and school effectiveness; factors that either fall outside the authority of government or that can hardly be changed by instruments of public policy. Two observations suggest themselves at this point. Firstly, there is a need for a closer cooperation between families in the neighbourhood and between home and school. [36]

---

[36] H.M. Levin and others at Stanford University implemented an accelerated school programme in two pilot schools in the San Francisco Bay Area early in 1987 (Levin, 1987). The purpose of the two pilot schools is to translate and implement the principles of accelerated schooling while simultaneously providing a basis for building knowledge on how to implement the changes on a collaborative basis with practitioners. According to Levin (1989, p. 27), the accelerated school is premised on a number of features which are probably common to all major restructuring efforts for disadvantaged students. The "accelerated school" idea is "based on a philosophy of empowering at-risk populations to have power over the circumstances of their lives, both educationally and in their other activities", and assumes that "leadership must be taken at all levels in order for such a movement to succeed", requires "deep community involvement", presupposes a

Secondly, since the school for an increasing number of children tends to be a substitute home it is paramount that children should like to be at school.

Other difficulties in coming of age in modern industrial societies seem to call more for political than for pedagogical action: city- and traffic-planners, architects, legislators (passing laws on employment, family, food, drugs and health). When the common "European home" is being built, there is a chance of taking these challenges into consideration.

---

curriculum and teaching methods that "build on the strengths of students rather than on exploiting and exposing their weaknesses and failings", "place great emphasis on school site responsibility", and require, beyond the pilot level, "support from school authorities", the cost of which is not high because significant savings can be made in other budgets of educational expenditure (Levin, 1989, pp. 27-28).

Chapter 5

New Challenges for Science Education in Europe

"Science is curiosity, discovering things and asking why ... . We must always begin by asking questions and not giving answers ... . Youngsters and adults cannot learn if information is pressed into their brains. You can only teach by creating interest, by creating an urge to know. The knowledge has to be sucked into the brain, not pushed into the brain. First you have to create a state of mind that craves knowledge ... . Avoid as much as possible, frontal teaching: teacher talking, students listening".

Victor F. Weisskopf [1]

1. Introduction

There have been dramatic changes in science education over the last hundred years. Science itself has always been concerned with asking why and curiosity has been the driving force behind much of its activity. But it is only within the last four decades that there has begun to be a realization that the processes of science are as important a part of science education as the acquisition of factual knowledge.

There have been other changes in recent years. There is an acceptance that Technology has an important contribution to make, that it is more than Applied Science, as it is certainly more than the acquisition of craft skills. There remains the need for such skills to be acquired, as there are also the needs to understand how to tune the television set or to meet the electrical needs in the home. Any educational system in any of our

---

[1] V.F. Weisskopf, quoted by Frank Smith in *Insult to Intelligence,* New York, 1986, p. 72.

countries must find ways to meet these needs, as it must also meet the requirement for a well-skilled labour force in each of our countries.

Perhaps the most recent development has been the appreciation that science education should show the relevance of science and technology to society. Public decision-making about matters such as energy supply, telecommunications, protection of the environment and biotechnology presupposes not only factual knowledge but also a scientific orientation. The citizen in modern European societies who is expected, not least as a voter, to participate in debate and take a stance on important issues, needs to be equipped with such an orientation. Verbal, mathematical and scientific literacy presuppose knowledge as well as certain attitudes and a way of thinking; they may assist people in researching social and personal problems and help them in arriving at motivated conclusions, which lead on to voting decisions influencing a common future.

Involved with this is the importance of attitudes towards science, a realization of its power as well as its limitations. Public understanding of science begins in school, and awareness of social implications is becoming an essential part of the curricula in use in schools at present. These are some of the issues to which we will be returning in this chapter.

2. The Beginning of Science Education

Traditionally knowledge and skills were transferred from one generation to the next through the family or the community, or the many apprenticeship 'schools' throughout Europe. The 19th century witnessed such a rapid development of scientific knowledge, often as the result of work by gifted enthusiasts, collecting evidence, studying or carrying out investigations in their own homes, that some more formal education in science became desirable. In the second half of the 19th century, physical science, together with mediaeval and modern history, began to find its place alongside mathematics as suitable for academic study in upper secondary schools and universities. The tradition of the lecture demonstration evolved in many parts of Europe and from this arose the acceptance that some science had a place alongside the traditional classical subjects in education. But the education in science was essentially factual: information was imparted.

In schools both botany and zoology had a place, together with some physics and chemistry, and a much smaller place for geology. There were links between well-established traditions in mathematics teaching and the teaching of physics, but in general the various sciences were taught in isolation from each other; it was extremely rare for there to be any attempt to provide a balance of the sciences in the education provided. Even within a particular discipline there was little attempt to coordinate different subjects. For example, physics tended to consist of a series of isolated topics: Mechanics, Heat, Light, Sound, Magnetism, Electricity, Properties of Matter, with little attempt to bring them together. Separate textbooks were provided for each.

Germany probably led other European countries in the development of demonstration apparatus. There was a fine tradition at the Royal Institution in Britain for regular demonstration-lectures on scientific topics, but, with notable exceptions, English schools lagged behind what was achieved in Germany. Throughout Europe science teaching was a matter of dogmatic assertion. The practical work which was done in Dutch and English schools − at least as far as physics was concerned − tended to reflect the state of physics in the 1880s. There was a confidence that perhaps the search for knowledge was coming towards an end; the only task remaining was to measure a whole lot of constants with a greater degree of accuracy. This emphasis on measurement continued to pervade science education in the European countries for the first half of the 20th century: children measured the thermal conductivities of good and bad conductors, and increasingly ingenious ways were devised for measuring, for example, the focal length of lenses. Measurement was what mattered and all the right precautions had to be taken to ensure accuracy; there was no scope for speculation and there were no uncertainties. Students filled notebooks with information which they did their best to memorize for reproduction in their examinations − and many of them eventually became teachers who imparted the same information to their pupils, so that they too could pass their examinations. But were all those measurements of specific heats really necessary? Was it necessary to learn all those properties of gases? Were the coloured and labelled drawings of plants and animals teaching very much about the nature of science?

3. The Beginning of the Change

The need for change was apparent soon after the end of the Second World War. The amount of scientific knowledge was increasing at an exponential rate and syllabuses could not go on being overloaded. Some rationalization was clearly going to be necessary. (It is perhaps interesting to note that the one country that did not get involved in the changes of the 1960s and 1970s, namely the People's Republic of China, now has easily the most overloaded syllabuses of any country in the world, comprising all the factual knowledge that was available in the 1930s with everything else added on top − and inevitably this affects the quality of the teaching.)

Furthermore, life in this post-war era had become more scientific, more technical than it had ever been. It has been said that the total amount of scientific information has been doubling every decade since the 1940s and the social demand for technical skills, for scientific knowledge and, most recently, computer literacy had increased concomitantly. Technology had begun to invade every village as well as every town. There were demands for new abilities and skills in the economy, and consequently a new awareness emerged that science and technology needed to be part of the education of all young people, not merely that of a limited few. Scientific knowledge and technological ability had become the principal commodity upon which the economy of European nations had to rely.

There was a further need. C.P. Snow had drawn attention to the Two Cultures, - arts and sciences - which can divide society. [2] Legislators and decision-makers needed a rational approach to anticipate future development, but one inevitable consequence of the growing body of scientific knowledge was that a widening gap was appearing between the specialists and the technocrats on the one hand, who may possess the knowledge required for effective policy, and the politicians and the administrators on the other, who were authorized by the people to devise and implement policy. If this posed a problem at that time, then it was safe to assume that it would only grow worse if there were not a change in the educational system. It is always important to remember that the children of today will become the policymakers 20 to 30 years from now. Realization of this fact underscores the conclusion, presented in Chapter 3, that children must be taught not for school but for life. An obvious difficulty here is that it cannot be known precisely how the future for which our children are educated will look. [3]

These factors led many countries, in the early 1950s, to reconsider the content of school syllabuses, but it was the work initiated in the United States, later in the decade, which began the reappraisal of what should be the aims of science education. First came the Physical Science Study Committee (PSSC), supported by the American National Science Foundation, closely followed by the Biological Science Study Committee (BSCS) and the two chemistry projects (CHEMstudy and

---

[2] C.P. Snow: *The Two Cultures and the Scientific Revolution,* New York, 1959.
[3] One is reminded here of the statement by Dennis Gabor (1962) in *Inventing the Future*: "The future cannot be predicted but it can be and should be invented". In the paper "Models in Science Education", G. Marx and E. Toth (1981): "... the ultimate goal of science education in our time does not differ from that of science itself: orientation and exploration into the unknown". G. Marx discusses some of the implications for science education in *Educating for an Unknown Future,* Budapest, 1990.

CBA).[4] It has often be said that it was the first Sputnik that led to the PSSC project, but in fact the Project began some time before the first Soviet space satellite was launched, although there is no doubt that the Russian developments in space were a considerable stimulus to what followed.

The influence of each of these American projects was enormous, as evidenced by the large number of different languages into which the various texts and supporting materials were translated and distributed throughout the world. PSSC had a careful rationale for what was and what was not included in the course. Careful attention was given to the development of appropriate experimental work, using simple apparatus, to encourage a spirit of enquiry, as well as an understanding of what were often quite sophisticated topics.

The early 1960s saw teams of American educators running seminars and conferences in Europe and elsewhere, explaining the philosophy behind the projects. It was sometimes a surprise to the Americans – and something of a disappointment to them – that more countries did not adopt their materials, excellent as much of them were considered to be. But each of the programmes referred to above were one year courses intended for use in American high schools. It was not unusual in the United States for a one-year biology course to be followed the next year by a one-year chemistry course, to be followed in the third year by the PSSC course. It was not lack of respect for what the American scientists and educators had achieved that explained why their courses were not adopted in their entirety. The reason was that they did not really fit into the established European tradition whereby a science course might extend over three, five or more years. But such was the influence of the American work and the thinking which resulted that a whole series of new science education programmes began to be developed in other countries as well.

There was another influence. In the 1960s, it was stated in the Soviet Union that "they had enough teachers, but they would never have enough good teachers, and therefore it was their policy to provide guidance to teachers to enable them to achieve a standard". It may seem strange these days when Teachers' Guides have become very much part of science education programmes, but at that time no country was giving the help to teachers that was being given, for example, in the Russian Republic. In many European countries it was considered essential to give a teacher a university degree in, say, physics and then for him or her to teach physics

---

[4] For details on these projects see J.D. Lockard, *Science and Mathematics Curriculum Developments Internationally 1956-1974*, University of Maryland, 1977. Another source is W.H. Brock: "From Liebig to Nuffield: A Bibliography of the History of Science Education 1839-1974", *Studies in Science Education*, Vol. 2, 1975, pp. 67-99.

for the next 30 or 40 years without further guidance! The guides developed in the Soviet Union ensured that the teacher had access to the advice that was needed. No school was opened without the necessary apparatus for the experiments to be done by the pupils, the demonstration apparatus to used by the teacher, the films to be shown and even the details of local factories and industrial complexes to be visited. The individual items may not have been of a standard which would have been accepted today, but the organization behind the courses and the wherewithal to support them was such that standards were achieved which might not otherwise have been possible. Awareness of this certainly had its influence on the various projects for the improvement of science education which were then developed in both Eastern and Western European countries.

The Nuffield Science Projects [5] were amongst the largest of the second-generation projects and these were developed in Britain in the 1960s. These were also separate courses in physics, chemistry and biology; they were initially five year courses for 11 to 16 year-olds, but later extended with advanced courses for the 16 to 18 year-old students. Awareness of the need to give help to teachers guided the development of the 'Nuffield' courses.

These projects were concerned to show that there was more to science education than the acquisition of factual knowledge. Even the formulae were printed on the front of the examination papers, so as to stop the students from thinking that memorizing them was what science was all about! Experimental work was devised with the objective of encouraging the students to think for themselves and prompting them to look for evidence so that, at least for a day, they could get the feel of what scientific work was like. Many of the national projects developed in Europe at that time came to accept the philosophy of the old Chinese proverb: "Hear and forget, see and remember, do and understand". In other words, it was now accepted that by doing experiments themselves the students were given the opportunity to come to real understanding of basic concepts.

Emphasis was laid on some of the uncertainties in science, to show that it was not a matter of established authority with all the answers.

---

[5] Because the courses developed in the United States were intended for use by all students, the emphasis was on the production of heavily structured curriculum materials and textbooks. By comparison, the science projects funded by the Nuffield Foundation were aimed at the academically able students. Hence the concentration on the development of "individualized" programmes and materials for laboratory work, which would encourage students to "learn by discovery" (R.F. Gunstone, "Secondary School Programmes in Science Education", *The International Encyclopedia of Education*, Vol. 8, pp. 4461-4465).

Models were given a central place. [6] The objective was to show how one model would lead on to another as further refinement became necessary and to encourage the students to understand that the latest model might in due course have to be replaced by another – that no ultimate truth had been achieved.

Particularly in the case of physics, there was a movement away from the compartmentalization of topics, each considered in isolation. The topics became a fabric of knowledge in which something studied in one part came to have a relevance and a contribution to make to other parts of that fabric. Waves, for example, were studied rather than light and sound in isolation from each other and there were links with mechanical waves, as of course with electric waves. And it was realized that important threads in the fabric concerned topics which only came into significance in the 20th century. Radioactivity, wave-particle duality, nuclear reactions and fission became parts of school syllabuses.

In the life sciences a major change was the integration of botany and zoology. Typically previous syllabuses would have an "animal" section and a "plant" section. In the new courses stress was put on biological principles which applied to both the animal and plant "kingdoms" and included a greater consideration of micro-organisms. Themes such as "cycles of matter and energy", "structure and function", "interaction between organisms and the environment" and "integration and homeostasis" provided the framework for the syllabuses. Also human biology and subjects such as genetics became more prominent. The latter was a major breakthrough because up to the 1960s genetics was considered too difficult for anyone except the brightest upper secondary students. Through innovations, such as those of the Nuffield Science Teaching project, heredity was studied even by primary school pupils. It was at this time, also, when the idea of biology as a non-mathematical science was challenged. Quantitative studies became an integral part of school biology at all levels.

4. International Cooperation in the Development of Science Education

A growing interest in the "modern" science teaching programmes described above could be noted throughout the world. This led to a

---

[6] The history of science is the history of models: geocentric vs. heliocentric universe, alchemy vs. immutable chemical elements, absolute space vs. inertial frames, valence arms vs. molecular orbitals, eternal species vs. mutation and natural selection. In science, great leaps forward were made possible by the invention of new concepts and variables, such as acceleration, entropy and wave function. However, the history of science also shows that the greatest danger for a successful model is to become a rigid dogma.

significant development in the 1960s, namely considerable international cooperation and the provision of opportunities for the exchange of ideas. This new phenomenon, which continues to the present day, owed much initially to OECD and UNESCO. Major contributions were made by the International Council of Scientific Unions (ICSU) through its Commitee on the Teaching of Science (CTS); the International Union of Pure and Applied Physics (IUPAP) through its International Commission on Physics Education; the International Union of Pure and Applied Chemistry (IUPAC) through its Committee on the Teaching of Chemistry; the International Union of Biological Sciences (IUBS) through its Commission for Biological Education. [7]

In most European countries there were associations for science teachers and an important mechanism for cooperation was established by the setting up of the International Council for Associations for Science Education (ICASE) and, specifically for physics, an international group GIREP (Groupe International de recherche sur l'enseignement de la physique) was another organization which brought educators and teachers together for the exchange of ideas. Promoted by these various organizations, conferences and seminars on a wide range of topics were held during the 1970s. Science education had become an international concern and much was gained from this collaboration.

Taking account of the experience of three decades of international cooperation in the development of science education, and considering that the principles and methods of science education have, in a sense, common validity, may lead one to conclude that there is scope for achieving a convergence in the science curricula of school systems in Europe and other parts of the world. Acceptance of a common core-curriculum in science may, in this perspective, be easier to reach in comparison with the position of subjects such as civics education and history.

5. Primary Science

Most of the developments described above were concerned with science education at the secondary level or with teacher training. During the 1970s there was an increasing awareness of the importance of science

---

[7] Some of these Commissions and conferences are discussed in *Science and Technology Education and Future Human Needs* (Oxford, 1986), edited by J.L. Lewis. Another source is J.L. Lewis: "Science in Society: Impact on Science Education", in T. Husén and J.P. Keeves, Eds, *Issues in Science Education: Science Competence in a Social and Ecological Context*, Oxford, 1991.

education at the primary level, a stage when it is easy to build on the curiosity of children, who do not tire of asking "What is it?" and "Why?"
Content is less important at this level than the *search* for understanding. Curiosity and investigative and exploratory behaviour are inclinations that come natural to children from five to 11, or 12, years. Between childish curiosity and scientific research there is the school. It may either kill or cultivate exploration. It may ask for the expected answer or encourage unexpected questions. It may prefer the comfort of "closed" problems with unique solutions, or it may raise open-ended questions as well, with unexpected outcomes. There is a consensus now that pupils should be encouraged to ask questions and then to find the answers for themselves. The teacher is there to encourage (and, of course, to initiate ideas) and should not be seen as the authority who knows all the answers.

At the elementary level the most common phrases might be:
Find out what happens if ...
Does it make any difference if ...
What happens to ...
Explain what would happen if ...
Is there any change in ... if ...

Key words will often be:
Search ...
Collect ...
Investigate ...
Describe ...
Explain ...

Much of primary science taught in Europe up to this time was concerned with the study of Nature. The new curricula introduced a significant amount of physical science and life science, the latter differing from nature study by its more experimental and quantitative approach and the inclusion of topics such as heredity, which were not normally included. However, nature study was not totally lost because many teachers justifiably considered that the interest it has for young people can establish the fascination with the living world that has led many children to become biologists and is a sound basis for environmental awareness.

It is now accepted in the United Kingdom that English, mathematics and science constitute the core subjects of the curriculum for all children from the age of five onwards. Also in other countries there is an acceptance that science education can certainly begin at the age of five.

6. Science for All and a Balance of the Sciences

Much of the activity of the 1960s was concerned with the education of "academically" inclined young people. In recent years there has been much more attention given to the requirements of those with different interests and abilities. Science-for-all has been the aim, and appropriate materials are continually under development. There is also a belief that all pupils need a balance of *all* the sciences; modern curricula usually require that there should be a balance of physics and chemistry, biology and earth sciences, not forgetting health science, meteorology and astronomy.

From the 1970s onwards UNESCO has been a strong supporter of "Integrated Science", [8] in which all the sciences are brought together in one joint discipline within the school curriculum. [9] There is no disagreement anywhere that, at the primary level, there should be no distinction between the sciences, as there is no obvious distinction for a child at that stage between one science and another. On the other hand it is hoped that the experience given to young children represent a broad spectrum. The physical sciences should be covered as well as the life sciences: science with young children should not be confined to "nature study" as it once was in many countries. At the secondary level, there are countries much in favour of "Integrated Science", others prefer to teach the sciences separately, but maintain a balance between them.

Fortunately the days when it was advocated that physics and chemistry are for boys and biology for girls have gone completely, although it has to be recognized that the "gender problem" is still with us. A considerable body of research in several countries has shown that, mainly through cultural and educational influences, girls are less likely to become scientists than boys and up to the age of 16 more boys still opt for physical sciences and more girls for biology. As shown in Table 5.1, boys also tend to excel in science over girls, particularly in the physical sciences but even in biology and chemistry. The position of women in science and technology has become a matter of much concern in Europe.

---

[8] Because of the growing interest in integrated science, UNESCO has published since 1971 a series of *New Trends in Integrated Science Teaching* (6 Vols). Several of the volumes contain papers presented to international conferences on integrated science teaching. An overview of the position until 1980 is presented by S. Haggis and P.A. Adey: A Review of Integrated Science Education Worldwide, *Studies in Science Education*, Vol. 6, pp. 69-89.

[9] The interdisciplinary and unified structure of science is emphasized in *integrated science education*. However, *integrated science teaching* refers to a specific approach to science teaching in which, for pedagogical and didactic reasons rather than epistemological arguments, the various disciplines of science are taught in an "integrated" or coherent way.

It is realized that, in seeking to revitalize its economy, the underrepresentation of girls in some fields of science education, especially in physics, signifies a major loss of talent and hence an unacceptable waste of both intellectual and economic resources. It can be seen from the data that there is, in most countries, evidence of an increased retention of girls in science courses at an advanced level. The expectation is that the position of girls has improved even further since the mid-1980s. Yet imbalances are still present. A difficult issue concerns the high performance level of boys compared with that of the girls.

Table 5.1: *Ratios (R) of Male to Female Students in Science Education and Indices (δ) Indicating Sex Differences in Science Achievement at the Terminal Grade of Secondary School (1970-1971 and 1983-1984; notes)* [10]

R = Ratio male: female δ = Effect size x 100	England		Finland		Hungary		Italy		Sweden	
	R	δ	R	δ	R	δ	R	δ	R	δ
Biology courses:										
1970-1971	0.8	17	0.7	38	0.5	15	2.7	10	0.9	47
1983-1984	0.7	14	0.9	35	0.7	26	4.5	32	1.3	32
Chemistry courses:										
1970-1971	1.6	69	1.6	60	0.7	15	3.0	29	1.5	49
1983-1984	1.8	39	0.9	45	0.6	37	2.3	28	2.3	51
Physics courses:										
1970-1971	2.3	101	1.6	119	0.5	63	0.9	74	2.7	104
1983-1984	3.3	75	1.8	87	0.8	57	4.2	72	2.4	85
Non-science specialists:										
1970-1971	0.5	62	0.3	69	3.9	83	1.8	70	0.6	60
1983-1984	0.8	40	—	—	0.3	28	0.8	52	0.5	37

Notes: (a) A value equal to or larger than 1 indicates that boys are over-represented compared with girls. (b) A positive effect size indicates superior performance of male students. (c) An index value of 20 denotes a small effect, a value of 50 refers to a moderate effect, and 80 signifies a large effect. (d) Index values less than 20 are considered to be not significant.

The generally lower interest and weaker performance of girls in science education can be explained in different ways. Some explanations

---

[10] Source: J.P. Keeves and D. Kotte: "Sex Differences in Science Education" (Tables 6.2, 6.3 and 6.5), in J.P. Keeves, Ed., *Changes in Science Education and Achievement: 1970-1984,* Oxford, 1992.

focus on biological differences between boys and girls. Others hold that sex differences are to a large extent socially and culturally determined, even though they are found across all the different social strata. Gender differences in science achievement may also arise because of differential opportunities to engage in academic learning tasks. A related perspective is that boys and girls construct their identities and gender roles differently, which give rise to differences in attitudes to science and motivation to enrol in science courses. [11] The stereotyping effect of the school on the development of boys' and girls' interest in science has also been noted, as has the influence of differential treatment of boys and girls by teachers. [12]

The closing of the "gender gap" in enrolment and science achievement is undoubtedly one of the major educational challenges facing us.

7. Science and Society

The decade of the 1980s has brought additional developments in science education. Educators in several countries – the Netherlands and the United Kingdom, Hungary and Norway – appreciated the need to show the relevance of what was taught in schools to the world beyond the classroom. In some countries there was an increasing belief amongst young people that science was responsible for many of the evils in the world – nuclear weapons and the pollution of the environment, for example – and there was a need to influence attitudes.

Awareness of environmental problems was increasing. Famine and plague disappeared from Europe in the wake of the Industrial Revolution. A belief in progress was dominant in politics, as it was in the schools, in the decades immediately following the Second World War. However, with the population explosion in the Third World and the energy crises of the 1970s and 1980s, the Western World was forced to acknowledge the finiteness of the Earth and the fact that there was a limit to natural resources. A new interpretation of "progress" became necessary, and the importance of preserving the natural environment was

---

[11] Reviews of research studies on sex differences in enrolment and science achievement are presented by A. Kelly: *Girls and Science: An International Study of Sex Differences in School Science Achievement* (Stockholm, 1978); M. Steinkamp and M.L. Maehr, Eds: *Women in Science* (Greenwich, Conn., 1984); J.B. Kahle, Ed.: *Women in Science: A Report from the Field* (Basingstoke, 1985); and K. Tobin: "Differential Engagements of Males and Females in High School Science", *International Journal of Science Education*, 1988 (Vol. 10, pp. 239-252).
[12] Cf. K. Tobin et al.: *Windows into Science Classrooms: Problems Associated with Higher-level Cognitive Learning*, Basingstoke, 1990. See also the study conducted by Wendy Duncan: *Engendering School Learning*, Stockholm, 1989.

something of which society in general was becoming increasingly aware. The publication of the Club of Rome's reports on *Limits to Growth* [13] drew attention to global problems, such as population growth and the finiteness of material resources, which put physical limits and constraints on future development, and pointed to some of the challenges to which education could make a contribution. [14] It was evident that Europe had paid a high price for its increases in wealth. The compression of areas of untouched natural beauty and the pollution of air and water were examples. Europe has, in the main, been free of war for half a century, but industrial catastrophe, acid rain and the deterioration of the ozone layer high in the stratosphere, and the greenhouse effect resulting from intensive burning of fossil fuels, have been substituted in its place.

Environmental protection is not so much a technological or economic problem, it is more an educational and moral matter. In the past, school education has paid little attention to international, global and cross-generation interests. Moral education in the European school systems focuses usually on the individual, the family and the nation-state. Now the increasing importance of global ethics and the responsibility towards future generations is such that science education should in future address moral education and also consider ethical issues in the teaching of science.

The Council of the EC adopted a statement on the objectives, means and ends of environmental education in 1988. [15] It calls for a concerted effort to develop and implement environmental education in the EC, and specifies policy measures to be taken by the member countries and the Commission. The plan gives high priority to the exchange of information on environmental education, the improvement of curriculum materials and teaching methods and the inclusion of themes on the environment in the syllabuses of all school subjects. Given the constraints of an already overloaded curriculum, this may well be a more realistic approach than creating an entirely new school subject.

A study was recently undertaken in Germany in order to find out in which subjects and grades the major themes of environmental education

---

[13] See for example, the first two studies: *The Limits to Growth: A Report for the Club of Rome's Project on the Predicament of Mankind, Second Edition* by D.H. Meadows, New York, 1972; and *Mankind at the Turning Point* by M. Mesarovic and E. Pestel, London, 1974.

[14] The educational implications are worked out in another Report to the Club of Rome: *No Limits to Learning: Bridging the Human Gap* by J.W. Botkin, M. Elmandjra and M. Malitza, Oxford, 1979.

[15] *Entschliessung des EG-Rates über Umweltbildung vom 24. 5. 1988*, EG-Rat Vereinigten Minister für das Bildungswesen, Brussels. A similar text has also been adopted by the Nordic Council of Ministers (Nordiska ministerrådet): *Recommendations on Environmental Education: Ten Years of Development*, Copenhagen, 1987.

were taught. [16] The results, which are reproduced in Table 5.2, indicate that a multidisciplinary approach was indeed taken, especially in secondary education. Environmental themes were most frequently taught in biology, chemistry and geography classes but some topics did also appear in the syllabuses of civics education, religion and home economics.

Table 5.2: *Teaching of Environmental Education Themes by Subject in German Secondary Schools*
(1987, per cent, see notes) [17]

Subject	Percentage share of all themes	Subject	Percentage share of all themes
Biology	23.5	Social studies	6.6
Chemistry	19.5	Technology	6.3
Geography	16.6	Politics and civics	6.1
Physics	9.5	Home economics	2.1
Religion	9.5	Total of above	100.0

Notes: (a) The results are from a nation-wide representative study of environmental education conducted in West Germany in 1986-1987. (b) Two populations were studied: primary schools (9-10 year-old level) and secondary schools (15-18 year-old level). (c) A list of about a dozen major themes in environmental education was prepared. An assessment was made of how many of these themes were covered in the different subjects by level.

This acknowledgement of the importance of dealing with the social, and indeed personal, implications of science in science education has, in fact, been a singular feature of biology since the 1960s. Human biology and social biology have been included either as separate courses or as part of general biology courses in several countries. Food and nutrition, health, human reproduction and the environment are the main foci for considering social, economic, ethical and other implications of the subject, and today must be added biotechnology and genetic engineering. The need to show the contribution which science could make to future well-being became apparent to many curriculum developers. The PLON

---

[16] G. Eulefeld *et al.*: "Die Praxis der Umwelterziehung in den Schulen der Bundesrepublik Deutschland. Ergebnisse einer empirischen Studie", in Hedewig and Stichmann, Eds, *Biologieunterricht und Ethik,* Köln, 1988.
[17] Source: *Umweltbildung in der EG, op. cit.* The data appear in a Chapter (Abb. 5, p. 59) by Günter Eulefeld: "Umwelterziehung als umfassende Aufgabe − eine Einführung in die Themen der Arbeitsgruppen", pp. 49-68.

project in the Netherlands [18] and the Science-in-Society project in the United Kingdom were pioneers in this field. [19] That the human side of scientific achievement — *the human element* — was increasingly becoming a worldwide concern was reflected in 1985 by a conference, "Science and Technology Education and Future Human Needs" [20] The conference examined the ways in which science education can be made more relevant to the needs of society in eight areas of concern, namely:

Health
Food and agriculture
Energy resources
Land, water and mineral resources
Industry and technology
The environment
Information transfer and technology
Ethics and social responsibility.

A conference such as this showed how much thinking about science education had progressed from the early 1960s, when UNESCO held its pioneer conferences on the teaching of physics and chemistry. Equally significant was the latest of the Danube conferences organized by George Marx in Budapest on the theme "Energy and Risk Education", [21] when the place of the Bhopal and Chernobyl disasters in education were considered as well as work on the monitoring of radon levels and acid rain. School laboratory experiments were showing signs of taking on new dimensions concerned with social problems — a long way from the routine measurements which played so large a part in school experiments

---

[18] An account of the PLON-project can be found in H.M.C. Eijkelhof *et al.*, Eds, *Op weg naar de vernieuwing van het natuurkundeonderwijs* (Towards the renewal of physics education), The Hague, 1986.
[19] Although it was at that time widely accepted by teachers in the United Kingdom that relevance to society was important in science teaching, unfortunately many teachers lacked knowledge and confidence in teaching these wider aspects. For this reason a *Science-in-Society* one-year course was developed for use in schools by students aged 16 to 18 or 19. Some years later a *Science-and-Technology-in-Society* course was produced, also by the Association for Science Education, College Lane, Hatfield. Both courses are complete with Teachers' Guide, Readers, decision-making exercises and audio-visual tapes and slides.
[20] This conference was conducted in 1985 in Bangalore, India under the auspices of the Committee on the Teaching of Science of the International Council of Scientific Unions (ICSU-CTS) and UNESCO. It was attended by over 300 teachers and science educators from at least 70 countries.
[21] A summary of the discussions can be found in *Proceedings of International UNESCO Conference on Energy and Risk Education,* edited by G. Marx, 1989.

50 years before – but still continuing the investigative work which began to be advocated 30 years ago.

The British have tended to have a more elaborate *examination system* than many other European countries and their Secondary Examinations Council ruled in the 1980s that all school science examinations must in future have a 20 per cent component concerned with social, economic and environmental applications of the science that is to be examined. Including social aspects of science in a public examination is one of the surest ways of ensuring that it is taught in school.

8. The Changing Role of Mathematics Education

Mathematics is a traditional part of the educational system. The clarity and purity of integer numbers and Euclidean geometry has been admired for centuries. Elementary mathematics has been considered the best training for abstract and logical thinking. Mathematics certainly preserves a privileged place in the school curriculum. But, as Edward Jacobsen of UNESCO has pointed out, [22] newspapers these days use other mathematical concepts, such as statistics, change, probability, significance, error, trend and risk, more and more frequently, yet school mathematics has remained too often dominated by blackboard concepts from earlier eras, for instance triangles and polynomials, which never occur in the headlines. School mathematics has to be opened up to mathematical logic and logic circuits.

European citizens have to obtain the skills of reading diagrams, pie-charts and percentage tables in order to make decisions in a democratic society. Furthermore, the greatest technological impact on modern society has been the extensive use of computers and this will inevitably have a major impact on mathematical teaching in the future and in its turn, this will influence science teaching as well.

9. Information Culture and Information Technology

In the 20th century, the radio has linked the peoples of Europe more efficiently than any political pact. Listening to trans-border broadcasting tore away the ideological veils of indoctrination. Today the exchange of musical fashions and scientific information and the conduct of world trade and politics, are transferred by electrons and electromagnetic waves at the speed of light, reaching far away places within a fraction of a second. A world football championship match is watched simultaneously

---

[22] E. Jacobsen, in G. Marx, *op. cit.*, 1989.

by hundreds of millions of people across the Earth. Even World War Three (the so-called Cold War) has been fought and won not by vast armies and nuclear weapons (as expected), but by advanced electronics. The fast economic rise in parts of the Far East is not due to richness in the soil or oil or mineral ores, but due to microchips and consumer electronics, products which incorporate more value in the form of information than raw materials. This is an important lesson for Europe, where energy and raw materials are scarce, labour is expensive, but the population is well educated. As Richard Feynman has put it: [23] "In historical perspective Maxwell turns out to be more important than Marx". The technological revolutions of the past four decades — the transistor, magnetic memory, the microprocessor, electronic mail and networking via phone line, radio and telecommunication satellites have transformed the intellectual environment and opened up new possibilities for education.

It is evident that the new information technologies discussed above, and the personal computer in particular, present various problems as well as opportunities for school education. Difficulties arise in introducing the new technologies in the laboratory or classroom. Many problems are encountered in implementing computer education. These range from problems of insufficient or inadequate hardware and software to the changing role of teachers and their lack of adequate training. However, once these 'common' implementation problems are overcome, the computers may well present an attractive means of bringing about change in the science syllabuses.

The data in Table 5.3 shows that the provision of computer education is a recent innovation in European school systems. Most secondary schools in most countries, including Japan and the United States, have little experience of using computers in science teaching.

Some of the principles upon which the new approach to science education is based are explained in a previous section. The students are now required to construct a world view of their own, based on systematic observation, experimentation and personal experience. In schools this approach presupposes an individualized learning environment. [25] New technologies, particularly computers, can help in creating such an environment, because most computer work takes place individually or in small groups. Moreover, as hinted at above, the computer presents a fascinating tool for model building, simulation, and problem solving.

---

[23] R. Feynman: *The Character of the Physical Law,* 1965, London.
[25] M.C. Linn discusses some of the essential features of traditional and 'modern' approaches to science education, with an emphasis on the role of information technology, in "Establishing a Research Base for Science Education: Challenges, Trends and Recommendations", *Journal of Research in Science Teaching,* 1987.

Teaching methods may change because the computers will shift the emphasis from the teacher to the student. An active and creative role will be required of the latter.

Table 5.3: *Introduction of Computers and Provision of Computer Education in Secondary Schools* [24]

Country/ educational system	Upper secondary science teachers, year of first use of computers (median of *using* teachers)	Lower secondary schools using computers for instruction (per cent of all schools)	Upper secondary schools using computers for instruction (per cent of all schools)	Student: computer ratio at the upper secondary level
Belgium, flemish	1982	78	98	32
Belgium, french	—	93	93	38
France	1982	99	99	26
German	1978	94	100	48
Greece	—	5	4	44
Hungary	1982	—	100	28
Japan	1982	36	94	32
Netherlands	—	87	69	34
Poland	1985	—	72	53
Portugal	—	53	72	289
Switzerland	1976	74	98	20
United States	1980	100	100	14

Notes: (a) Number of years of computer use in science teaching refers only to schools that have computers *and* use them for instruction. (b) ––: Estimates cannot be calculated because of insufficient valid cases.

10. Design and Technology

The most recent of all trends concerns the links between science and technology and the extent to which this should be reflected within the school curriculum. As already discussed, experimental work in science education has moved towards investigation, the looking for evidence in order to draw conclusions. Experimental work in technology is of a different kind: a need is felt, there is a design stage when a solution is sought and made, it is then evaluated and if necessary refined. It has

---

[24] The data are culled from several Tables in H. Pelgrum and Tj. Plomp: *The Use of Computers in Education Worldwide,* Oxford, 1991.

11. How to Teach Science

Little has been said in this chapter about the ways in which science might be taught. Perhaps the best way to discuss what should – or should not – be done is to describe the story of Erasto B. Mpemba, an African schoolboy. There is much charm in this account of his experience and no apology is given for repeating it here as there is much to learn from it. It was first published in the UNESCO Source Book: *Teaching School Physics.* [26]

> "My name is Erasto B. Mpemba and I am going to tell you about my discovery. When I was in form three in Magamba secondary school, I used to make ice-creams. The boys at the school do this by boiling milk and mixing it with sugar and putting it into the freezing chamber in the refrigerator, after it has first cooled nearly to room temperature. A lot of boys make them and there is a rush to get space in the refrigerator.
> One day after buying milk from the local women, I started boiling it. Another boy, who had bought some milk for making ice-cream, ran to the refrigerator when he saw me boiling my milk, and quickly mixed his milk with sugar and poured it into the ice-tray without boiling it, so that he may not miss his chance. Knowing that if I waited for the boiled milk to cool before placing it in the refrigerator I would lose the last available ice-tray, I decided to risk ruin to the refrigerator on that day by putting hot milk into it.
> The other boy and I went back an hour and a half later, and found that my tray of milk had frozen into ice-cream while his was still only a thick liquid, not yet frozen.
> I asked my physics teacher why this happened like that, with the milk that was hot freezing first, and the answer he gave me was 'You were confused, that could not happen'.
> After passing my O-level examination, I was chosen to go to Mkwawa High School. The first topics we dealt with were on heat. One day as our teacher taught us about Newton's law of cooling, I asked him the question: 'Please Sir, why is it that when you put both hot milk and cold milk into a refrigerator at the same time, the hot milk freezes first?' The teacher replied 'I do not think so, Mpemba'. I continued: 'It is true, Sir, I have done it myself' and he said 'The answer I can give is that you were confused'. I kept on arguing and the final answer he gave me was 'Well, all I say is that that is Mpemba's physics and not the universal physics'. From then onwards if I failed in a problem by making a mistake in looking up the logarithms this teacher used to say 'That is Mpemba's mathematics'. And the

---

[26] The quotation is taken from *UNESCO Source Book: Teaching School Physics*, Penguin Books, 1972.

whole class adopted this, and anytime I did something wrong they used to say 'That is Mpemba's ...', whatever the thing was.

Then one afternoon I found the biology laboratory open, and there was no teacher. I took two 50 cm^3 small beakers, one filled with cold water from the tap and the other with hot water from the boiler; and quickly put them in the freezing chamber of the laboratory refrigerator. After an hour I came back to look, and I found that not all the water had been changed into ice, but that there was more ice in the one that had the cold water. This was not really conclusive. So, I planned to try it again when I had the chance.

When Dr Osborne visited our school this year, we were allowed to ask him some questions, mainly in physics. I asked 'If you take two similar containers with equal volumes of water, one at 35°C and the other at 100°C and put them in a refrigerator, the one that started at 100°C freezes first. Why?' He first smiled and asked me to repeat the question. After I repeated it, he said 'Is it true, have you done it?' I said 'Yes'. Then he said 'I do not know, but I promise to try this experiment when I am back in my university'.

Next day my classmates in form six were saying to me that I had shamed them by asking that question. Some said to me 'But Mpemba, did you not understand your chapter on Newton's law of cooling?' I told them 'Theory differs from practical'. Some said 'We do not wonder, for that was Mpemba's physics'.

I asked the head of the kitchen staff at school to let me use a refrigerator for this experiment. She gave me the use of one whole refrigerator for a week. First I did the experiment by myself, because I was afraid that if I failed anyone else would tell the whole school that I was just stupid on that day when I asked the question. But my results were the same. The following day I took three boys with me from among those who has scorned my question and performed the experiment. We found that ice started to form first from the hot milk.

These three boys laughed very much and started telling the others that I was right, but that they could hardly believe. Some said it was impossible. I told the head of the physics department in my school that the experiment had worked; he then said 'it should not. I will have a go at it this afternoon.' But later, he found the same result".

There was a sequel to the above written by Professor Osborne: [27]

"The headmaster of Mkwawa High School invited me to speak to the students on physics and national development. I spoke for half an hour, but questions lasted for a further hour. There were questions of personal concern about entering the university, loaded questions about the remote possibility of relating parts of the school physics syllabus to national development, and questions that showed a considerable breadth of reading including one on gravitational collapse. One questioner raised a laugh from his colleagues with a question I remember as 'If you take two beakers with equal volumes of water, one at 35°C and the other at 100°C ... Why?' It seemed an unlikely happening, but the pupil insisted he was sure of the facts.

Now I enjoy answering questions and want to encourage questioning and critical attitudes. No question should be ridiculed. In this case there was an added reason for caution, for everyday events are seldom as simple as they seem and it is dangerous to pass a superficial judgement on what can and cannot be. I said that the facts as they were given surprised me because they appeared to contradict the physics I knew. But I added that it was possible that the rate of cooling might be

---

[27] *Ibid.*

affected by some factor I had not considered. I promised we would put the claim to the test at the University and urged my questioner to repeat the experiment himself.
Back in the univeristy I asked a junior technician to test the facts. The technician reported that the water that started hot did indeed freeze first and added in a moment of unscientific enthusiasm 'But we'll keep on repeating the experiment until we get the right answer'".

Further tests have confirmed Erasto's claim and in subsequent years various scientists have wrestled to find the precise explanation of what at first sight seems strange, although it is now accepted that convection is part of that explanation. The details of the physics, however, do not concern us here, but this delightful account enables us to draw attention to many aspects of teaching. The danger of a dogmatic approach as used by the first teacher, the evils of sarcasm in teaching, the picture of Professor Osborne asking the questioner to repeat the question in order to give him time to think and his wise handling of the situation to avoid discouraging a critical and scientific attitude, the cooperation of the "head of the kitchen staff" in allowing the use of a refrigerator for a week of investigation, are all indicators to what is good and bad teaching. And of course there is the final gem of the technician promising to go on repeating the experiment "until we get the right answer".

In recent years, there have been fundamental changes in methods of teaching. There is much less dogmatic assertion. Content is still important, but students are encouraged to think for themselves and to be critical, they are urged to seek evidence to support what they believe. There is an emphasis now on investigation, both in the life sciences as well as in the physical sciences: in fact what Erasto B. Mpemba was trying to do is exactly in line with what is considered good teaching practice these days. This might be summarized by saying that learning has become more important than teaching: the methods of science rather than just the facts of science have become increasingly important in the curriculum.

12. Enrolment in Science Education

In pursuing a strategy of revitalizing its economy and ensuring that its industries are successful in enhancing their competitiveness on international trade markets, Europe will have to overcome several major problems. Some of these are relevant to the school and science education in particular. As is argued previously in this chapter, a large looming problem concerns the methods of teaching science. A second problem arises because new information technologies and new developments in mathematical education require to be attended to. A third issue is that, compared with Japan and some other countries, a rather low proportion of students in an age cohort enrol in advanced courses in mathematics and science up to the terminal grade of secondary school. The data in Table 5.4 indicate that this may apply especially to England, Hungary and, with the exception of physics, Italy.

For England, Postlethwaite and Wiley (p. 137) comment on the results as follows:

> "The students in Grade 13 were 18 years old and studied, on average, only three main subjects. About 20 per cent of an age group was enrolled in school (colleges of further education were excluded) at this level and of these only one-quarter studied science. ... The students spent eight to 10 hours a week studying science and a high proportion of their homework was spent on science (8.4 out of 9.1 hours of all homework per week). Thus, England has a highly select group of students specializing to a great extent in science".

Table 5.4: *Participation in Science Courses at the Terminal Grade Level of Secondary Education (1984 data, per cent of an age cohort retained)* [28]

Country	Grade tested	All (per cent in school)	Age (mean)	Biology (per cent in school)	Chemistry (per cent in school)	Physics (per cent in school)	Average number of subjects at final grade
England	13	20	18.0	4	5	6	3
Finland	12	63	18.7	45	14	14	9+
Hungary	12	40	18.0	3	1	4	9+
Italy	12/13	34	19.0	4	1	13	7
Norway	12	40	18.1	4	6	10	7
Poland	12	28	18.7	9	9	9	9+
Sweden	12	30	19.0	5	13	15	9+
Japan	12	89	18.2	12	16	11	7
USA	12	83	17.7	12	2	1	5

---

[28] Sources: J.P. Keeves *et al.*: "Educational Expansion and Equality of Opportunity", *International Journal of Educational Research*, 1991 (Table 4.3, p. 70); and T.N. Postlethwaite: "Achievement in Science Education in 1984 in 23 Countries" (Table 6, p. 47) in T. Husén and J.P. Keeves, Eds, *Issues in Science Education*, Oxford, 1991.

The comparatively low rate of enrolment in upper secondary science courses in some European countries may well be a relic of the time when, in selective school systems following the predominantly "academic" traditions of, for instance, the British grammar school and the Dutch or German *Gymnasium*, the belief was prevalent that only a minority of an age group – a group with a special aptitude for mathematics and sciences – could benefit from advanced programmes in these subjects. There was also the conviction that a widening of student intake would put pressure on a standard.

A recent study seems to confirm some of these fears, because the rate of retention of students in science courses is, across countries, negatively related to science achievement. [29] In most Western European countries only five to 10 per cent of the 18 year-old students *who are enrolled in science courses* perform at a level marked "advanced" on an international science achievement scale. [30] This is a better yield compared with students in the United States, but lower than the average performance of Japanese students – even though a relatively large proportion of the 18 year-olds are enrolled in science courses in this country. [31]

Thus the question is whether the relatively high standard that tends to be achieved in school systems enrolling a select group in upper secondary science courses balances with the lower average score often obtained in countries admitting a large percentage of an age cohort. There is, of course, no easy answer to this. Yet whatever the solution, one may appropriately reflect on a related question: is the number of students specializing in science sufficiently large to satisfy the demand for such people on the labour markets of Europe? To this question the answer must be negative.

The low enrolment of students in upper secondary science education in some European school systems may cause problems, especially because the number of school leavers will continue to decrease in most countries until the turn of the century, at the very least. It may well become increasingly difficult to recruit a sufficiently large number of science

---

[29] Source: J.P. Keeves: "Specialization in Science: 1970 and 1984" in T. Husén and J.P. Keeves, *op. cit.*, pp. 82-83.
[30] Further data are presented in Chapter 2 of the present report. An interesting observation is that student performance in science was relatively low in England at the 10 and 14 year-old level but high at the 18 year-old level (see figures 1-3 in the Statistical Annex).
[31] Cf. T.N. Postlethwaite and D.E. Wiley: *Science Achievement in Twenty-three Countries*, Oxford, 1992.

teachers as a consequence. A shortage of such teachers may well have a negative effect on the quality of science teaching in our schools.

13. Conclusion

The issues for the future are very simple, but need to be answered in every country which is considering or reconsidering its national curriculum. At what stage is it appropriate to start teaching science – at five, seven, 11, 14 or even 16 years? If science is taught at the primary stage, what kind of science would be taught? And how should it be taught? How much should be concerned with the acquisition of factual knowledge, how much with the processes of science?

At the secondary stage, should it be *science for all?* Should it be the same science for all or should there be a distinction between the more quantitative (mathematical) science for the future specialist and the more descriptive for the children who do not seek to specialize? What science should be taught at the secondary stage? Should it be a balance of the sciences, the physical science and the life sciences? Should the sciences be taught together as a single, integrated course?

Above all, decisions have to be made as to the extent to which science courses should be experimentally based? This depends on the whole philosophy of the science education course. The Nobel Laureate Victor Weisskopf was quoted at the outset of this chapter: "Science is curiosity, discovering things, and asking why ... . Avoid as much as possible, frontal teaching: teacher talking, students listening". If science education is more than the mere acquisition of factual knowledge, as most thinking now suggest, then courses must be experimentally based and this has important economic implications about the size of classes, the size of laboratories and the provision of equipment.

There are other issues on which decisions will need to be made. Should there be closer links with other disciplines? No doubt cognizance will have to be taken of the changing role of mathematical education and of the increasing contributions from developments in information technology. Should there be an emphasis on the applications of science, on the contribution of science to society and on environmental and social issues? What should be the relationship with technology? What might be the links between science education and "Design and Technology", which some countries are now incorporating into their curricula?

What contributions should science education make to the public understanding of science? What contributions can be made by museums and the 'exploratoriums' which are increasing in number around Europe?

In what ways can radio and television, 'open universities' and the promotion of videos contribute to science education?

Lastly, what are the implications for teacher training – both pre-service and in-service? What resources do science teachers need throughout their careers? Motivation is perhaps the most important influence on the quality of teaching. What can be done to promote this? Perhaps somehow we have got to encourage ways by which people can realize that teaching science can be very great fun – and there is no greater satisfaction than helping a pupil come to understanding through his or her own efforts.

Chapter 6

The Teaching of Languages in Europe

1. Introduction

The new Europe will be one of accommodation: of migrants, guest workers and tourists, students and teachers. But also of science and technology, new industrial activity, and new attitudes, tastes and values. Communication, and thus language, are at the core of successful adjustment. As a result of European integration there are two overriding issues which constitute challenges for curriculum planners. The first, the teaching of modern science and mathematics was dealt with in the preceding chapter. Foreign language education is the focus of the present one.

Undoubtedly, what happens now in Europe will force us to rethink the role and the teaching of the main foreign languages spoken in Europe. This is and has for a long time been a problem in countries with small populations, but will increasingly become one in countries with large populations as well, such as Germany, France and the United Kingdom, who have national languages with the widest currency. These countries will have to pay attention to the teaching of foreign languages and give them more time.

2. Language and Literature as Expressions of Culture

At a forum organized in Paris in the autumn of 1990, Juan-Luis Cebrian, Director of *El Pais*, said: [1]

---

[1] The quotation is from the keynote (p. 30) Cebrian delivered at a forum held in Paris in the fall of 1990: Europe after the German Unification. A summary can be found in *European Affairs*, Vol. 4 (4), 1990.

> "I believe that it is necessary to know, to ask oneself: What is Europe, exactly? Is it a geographical notion, a historical idea? Is it an economic system, a culture, a civilization? Quite evidently, Europe is an ensemble of all these notions ... seen throughout its history".

William Wallace, a British scholar, [2] can be quoted as saying that Europe can be regarded as representing:

> "... a set of values as well as a collection of peoples and territories. Europe – or 'the West' – is identified both within a distinctive cultural tradition and with a set of political and cultural values which marks it out from other regions in the world".

What both perspectives have in common is that Europe is approached not primarily as a geographical concept covering a well-delineated territory but, rather, as an imagined space identified by a common history, its patterns of social, economic and political interaction, and distinctive values and cultural identities. Behind the general reluctance to define Europe in geographical terms is the realization that ideas about what Europe is, or should be, are constantly changing. The Northern and Southern, Eastern and Western boundaries of the European region cannot be traced on a map. This held true also in the past, when nation-states located at the European "core" often sought to solve their conflicts by warring both among themselves and with countries on the periphery.

International competition now goes not by weapons, but by new information technology, economic advantage, and influence on the values, tastes and attitudes of people. An interesting, partly unexpected, global change has occurred, namely, the emergence of economic powers on the Pacific Rim and the competition they represent. There can be no doubt that the European integration process has been speeded up as a consequence of the realization that firm cooperation among countries and their industrial and commercial undertakings is needed in order to meet this competition successfully. Several experts attribute the success of countries in the Far East in penetrating home markets in Europe and the United States to the superiority of their educational systems, in addition to other factors such as the family structure and support system, mental discipline, and attitude to work and duty.

Which way economic advantage and political influence will shift in the coming decades will in part depend on which peoples will meet the challenges of "high tech" first and best, and that, again, depends in part

---

[2] W. Wallace: *The Transformation of Western Europe*, London, 1990. The quotation can also be found in *EC Biblio-flash*, No. 40, 1990, p. 20.

on the educational system. But it seems that, ultimately, the price will go to those who have learnt to understand the developments – their dangers and advantages – and are able to govern them: to the wise people who have studied the history of mankind and have adopted the habit of thinking; to those who have learnt to expose themselves to others: to open-minded people; to those who have learnt to limit themselves: to disciplined people. The challenges of the 21st century will be best met by those who in their youth have been *prepared* for the world as it is without having been *subjected* to it.

One challenge has already been lost, or so it would seem. In international trade, agreements have to be made. The one who understands his adversary in business, and knows how to play this card well, has a definite competitive advantage. Evidence from the recent past would seem to suggest that Japanese, Korean and Taiwanese businessmen have had a good understanding, if not of Western culture, then surely of the strengths and weaknesses of Western business practice and of the ideas, produced in the West, awaiting further development. It is imperative that the educational systems that will operate in Europe in the coming decade pay close attention to culture – not only to the rich diversity making up "European cultures", but also to those of other regions, not least Asian and Islamic ones. As conventional, domestic sources of energy dwindle, Europe and the United States may well become increasingly dependent on the oil produced in the Gulf. Also substitutes such as nuclear fuel will have to be imported to an even greater extent than is the case today.

The discussion of foreign cultures is neglected in European schools. The danger is imminent that this situation will worsen as a consequence of European integration, as Europeans are likely to research their own common history and culture, to the detriment of studying the surrounding cultures. It follows from what has already been said that this ought to be avoided. Europe has a great opportunity here, because its people have grown accustomed to other languages and the idea that these represent other cultural norms and social values. Being less monolithic culturally than the United States with its "melting-pot" philosophy, or Japan, to take another example, it may more readily become able to communicate with the Third World as an equal partner. If successful, Europe may play a historic role in defusing the time bomb planted on the globe.

The European national educational systems have a long tradition. This is an advantage as well as a disadvantage. The advantage of expertise gained over a long time-period would seem self-evident, at least on first scrutiny. However, it also explains why our schools are conservative in orientation and procedure. Some subjects taught in school are still dominated by the distinctive but outdated aspects of the traditional nation-state, such as the national language, the glory of history, arts and literature, and not least the order of values. The great challenge of the 1990s is to reshape our school education from being overburdened with national, provincial ideas to becoming all-European and global at the same time. This may well be seen as a duty if the advocacy of democracy and human rights are taken seriously. It may also be a prerequisite if the plan proposed by Jacques Delors to create not just a "Common Market" in pursuit of economic integration but also a "Social Europe" aimed at social integration or at the emergence of a distinctive European model of culture and society, based on a blend of social-capitalist and democratic ideas, is to be realized.

This brings us back to the idea of a distinctive European culture or set of values. As Wallace [3] notes:

> "Resistance to the EC Commission's attempts to establish its legal competence over the field of education, and above all opposition to the efforts (largely under the auspices of the Council of Europe) to create a more 'European history' to set the separate national histories within a common context, reflected the awareness of both proponents and opponents of the direct links between education, values, loyalty and legitimacy. ... The argument during the 1980s over the EC's claims to competences in the educational and cultural spheres, over the appropriateness of 'measures to create a 'citizens' Europe,' revolved around the competing priorities of national, European and Atlantic identity".

The removal of border controls within the Single European Market does not mean, of course, that all boundaries between countries will cease to exist. Customs present only one aspect of a national frontier. At a deeper layer, such boundaries also represent a meeting-place of different cultural identities, often epitomized by a language division. Language is an integral component of individual and group identity. Natural boundaries presented by mountains, rivers and seas, can be overcome with tunnels and bridges, as can the territorial boundaries of nation-states. To overcome cultural boundaries is an entirely different matter.

Closely tied to culture is the *language question*. One could mention Spain, for example, where the various languages spoken give rise to a growing regional consciousness, and feelings of "otherness" and "national identity". Catalonia, the Basque country and Galicia have regained much of their own culture and language since Franco's death in

---

[3] *Ibid*, p. 21.

1975. The Spanish Constitution of December 29, 1978, establishes a united Spain consisting of juridicially autonomous regions: [4]

> *Article 2:* The Constitution is based on the indissoluble unity of the Spanish nation, the common and indivisible fatherland for all Spaniards, and it recognizes and guarantees the nationalities and regions constituting it the right to autonomy, as well as the solidarity of all.
> *Article 3:* Castilian is the official Spanish language of the State. All Spaniards have the duty to know it and the right to use it. The other Spanish languages will also officially exist in the autonomous areas in accordance with their statutes. Spain's wealth of different linguistic peculiarities is a cultural good, and as such shall be especially respected and protected".

Although the autonomous regions have no autonomy as in the case of German federalism, the problems of a redefinition of school and education in, for example, the question of regional languages, decentralization of the curricula, or a regional organization of the school administration become evident in connection with the recognition of Spain as a pluralistic State. [5] The objective of education in the autonomous areas in Spain is, in accordance with the Constitution, for all pupils to have mastered the regional language and Castilian Spanish by the end of primary school (grade 8).

Belgium presents another example of cultural and socio-economic pluralism. Three official languages exist in this country. Each of these has also official status in a neigbouring country. In driving around in Belgium one notes how language and culture go hand in hand. In passing from one language area to another, one sees not only that the names of townships are written differently, but also changes in the preferences and tastes of the people, for example, in the way houses have been built and whether the typical car is made in France or Germany. The reason why the language issue has become such an explosive problem in this country is closely connected with individual and social identity. Being deprived of your language means to be deprived of identity.

Literature is an important expression of culture. Reading the literature of another country provides a gateway to an understanding of its history, traditions and values, and the problems faced by ordinary citizens. Most countries have actively promoted the development of a national literature, and continue to do so in the new European context, in which even television boundaries are disappearing. All EC countries have

---

[4] These articles from the Spanish Constitution of December 1978 can be found on p. 26 in H.-P. Schmidtke: "Education and the Pluralistic Society," *Western European Education*, Vol. 18 (3), 25-30.
[5] *Educacion y Sociedad Plural* (Education in a Pluralistic Society) was the theme of the national congress of the *Sociedad Española de la Pedagogia* (Spanish Society for Education) held in Antiage de Compostela, Galicia between July 3 and July 7, 1984.

statements similar to the *Letterennota* in the Netherlands, [6] according to which the goals of Dutch policy are:

1. The preservation of the national literary heritage and the promotion of the development of Dutch and Frisian literature;
2. The opening of Dutch and Frisian literature to the outside; and
3. The promotion of participation in Dutch and Frisian literature.

That the interest of the European Community for cultural affairs is mounting can be inferred from the discussion at present on the development of "European media", including a new emphasis on "European literature". This shows up not only in the increased co-financing among Western countries of joint TV productions but also in the setting up of EC projects aimed at the translation and distribution to a wide audience of "literary treasures". The translation theme is at the heart of a special EC programme, as described in the text "Books and Reading: A Cultural Challenge for Europe". Although as a first beginning it may be worthwhile, this programme has the modest aim of achieving the translation of 30 literary works over a five-year period starting in 1989.

3. Developments in the Teaching of Foreign Languages

In Europe from the Middle Ages well into the 17th century Latin was the dominant foreign language of education and religion, diplomacy and commerce, a *lingua franca* of the élite, used widely by students and scholars, priests and diplomats, and even merchants. Students from wealthy families could, in principle, move from one university to another in country after country and expect to be able to attend lectures and read textbooks in Latin, and sometimes Greek. Amos Comenius in the 1630s published his *Ianua linguarum reserata*, which was used thoughout Europe. In the 16th and 17th century, however, French and Italian, and also English and German, became more important as written and spoken languages. Italian was considered a language of culture, for example in opera, and French became the language of diplomacy. English developed into the language of commerce, and German into a language used by scholars. The Reformation was another factor in the decline of Latin. The status of Latin diminished from that of a living language to that of an

---

[6] J. Honout: *Letteren-nota Nederlands Cultuurbeleid en Europese Eenwording*. Rijswijk, 1989.

"occasional" subject in the school curriculum as a consequence of these developments. [7]

Today in modern Europe, with the enormously increased mobility, face-to-face contacts, travel, and with mass media offering a wide range of radio and TV-communication, all young people and adults need to understand and to communicate in at least one but preferably two foreign languages. The first problem then, is which foreign language should be given priority in the syllabuses? This choice can change with a shift in the balance of international relations, at least in relatively small countries such as Denmark, the Netherlands and Sweden, where German was the first foreign language until the Second World War. The same may happen with the teaching of Russian as a first foreign language in Central and Eastern European countries. The reunification of Germany may give rise to a new, economically powerful Germany and a new form of German identity, albeit one of Germany in Europe. An enhanced German culture may also have implications for the use of German as a modern foreign language in Europe. In future more pupils may opt for German.

In describing some major developments in the teaching of foreign languages, particularly as regards the choice of a first and second language, it may be useful to apply the categories employed by Yun-Kyung Cha [8] in a recent study of how foreign languages became institutionalized into the school curriculum:

1. *National language:* either an indigenous official language spoken by a large proportion of the population, for example French and Flemish in Belgium, or an official language that is exogenous in origin, but that has become a mother tongue of a large proportion of the population.
2. *Modern foreign language:* either a non-indigenous language that is a mother tongue of a small proportion of the total population in a given country such as Turkish in Germany or Arabic in France; or a non-indigenous foreign language, such as French in Denmark.
3. *Regional and minority language:* an indigenous language, spoken by ethnic minorities, that may be taught or used in the school. The German and Gypsy languages in Hungary are examples. Instruction in local languages is usually offered mainly in the lower grades, before pupils make the transition to the major national language.

---

[7] A brief history of language teaching is given in Chapter 1 of J.C. Richards and T.S. Rodgers: *Approaches and Methods in Language Teaching: A Description and Analysis*, Cambridge, 1986.
[8] Y.-K. Cha: "Effect of the Global System on Language Instruction, 1850-1986", *Sociology of Education*, 1991 (Spring issue), p. 29.
[9] *Ibid*, p. 25.

Table 6.1: *Percentages of Countries Teaching Major International Languages as the First Foreign Language in the Secondary School*
(number of countries in brackets)[9]

	Period					
	1850-1874	1875-1899	1900-1919	1920-1944	1945-1969	1970-1986
Modern Foreign Languages						
English	8.3 (12)	11.1 (18)	26.5 (34)	39.6 (48)	62.2 (119)	72.0 (118)
French	45.5 (11)	50.0 (18)	54.3 (35)	47.9 (48)	33.3 (120)	17.6 (125)
German	45.5 (11)	44.4 (18)	27.8 (36)	16.3 (49)	.0 (121)	.8 (124)
Russian	.0 (11)	.0 (19)	.0 (38)	.0 (53)	6.5 (123)	6.3 (127)
Spanish	.0 (10)	.0 (18)	.0 (30)	.0 (38)	.0 (107)	6.3 (112)
Classical Languages						
Latin	100.0 (12)	81.0 (21)	74.4 (39)	70.2 (57)	33.1 (127)	12.5 (128)
Greek	.0 (12)	4.5 (21)	2.6 (39)	1.8 (57)	1.6 (127)	1.6 (128)

4. *Classical language:* mainly classical Greek and Latin.

Cha [10] has collated an impressive amount of data on the time devoted to the teaching of national and foreign languages in school curricula around the world. He discovered some very interesting facts, which seem to hold also for the European countries. Table 6.1 shows how the teaching of Latin as the first foreign language has decreased since 1850. A similar trend can be observed with respect to German and French. The teaching of German as the first foreign language came to an end in almost all countries immediately after the second World War. The decline of French began at about the same time, although it was still being taught as the first foreign language in about 40 school systems during the 1960s. By the mid-1980s this figure was down to 20, which can be compared with 16 countries then teaching Latin as the first foreign language.

The data in Table 6.1 leave no room for doubt that English has become a new *Lingua Franca*. It is now being taught as the first foreign language in at least 85 and probably more countries. It could be that 'English-English' is only a basic form in this perspective (although American-English may be harder to learn.) This begs the question as to what kind of English is or should be taught in European schools.

Not shown in Table 6.1 are Cha's data indicating, firstly, that the emphasis given to national languages in the curriculum has declined in a major way during the 20th century in countries that were sovereign nation-states by 1900, or that never experienced colonial rule, such as most Western European countries. Secondly, the study shows that there has been a global increase in the number of countries providing instruction in modern foreign languages over time. These results, i.e. a decline of both national and classical languages and a concomitant increase in the emphasis on modern foreign languages, especially English, may be interpreted as showing that a fair degree of internationalization has occurred in school education, in Europe as well as elsewhere.

In 19th-century Europe, the nation-states imposed common values, culture, a common history and a national language. The interesting observation emerging from Cha's analysis of school curricula is that the amount of time devoted to what may be considered "national language" decreased in Europe not since 1945, as one would assume, but from the beginning of the 20th century. The identities of nation-states had in the main become established by then. In this process of nation building, the teaching of national history, culture and language was paramount. The observed decline in the time devoted to national language and the concomitant increase in foreign language teaching since the early 20th century adds to an interpretation according to which Europeanization is not a recent phenomenon, to be associated mainly with the growing significance of the European Community, but should be seen as an intermediate outcome of a gradual process of integration, the beginnings of which predate this century.

4. The "Small" Languages of Europe

Two types of "small" languages are considered in this section. Firstly, the regional or minority languages of Europe, i.e. Welsh or Frisian, or Romany spoken by the Gypsies, who lack both a nation-state and a fatherland and, secondly, the (majority) languages of comparatively small countries, such as Danish and Dutch in the case of Denmark and the Netherlands.

For many years, the various bodies of the Council of Europe have been expressing a concern with "small" or "lesser used" languages. For example, the Council Assembly passed a recommendation in 1961 calling for measures to be taken to protect the rights of minorities to enjoy their own culture, to use their own language, and to establish their own schools. [11]

Resolution 192 on regional and minority languages was passed in March 1988 at the twenty-third session of the Council of Europe. [12] The Assembly of the Council thus approved of the draft charter for regional and minority languages. The text stated that the promotion of minority languages in the different countries and regions of Europe, far from constituting an obstacle to national languages, would in fact represent an important step towards establishing a Europe based on principles of democracy and cultural diversity within the framework of national sovereignty and territorial integrity. [13] Ratification of the Charter would entail that a country agrees to eliminate all forms of discrimination concerning the use of regional or minority languages, together with any practice having such discriminatory effect and, secondly, to make provisions for the teaching and learning of regional and minority languages at all appropriate stages in the educational system. It must be added, however, that the charter is only concerned with languages that are "historically European" and people who are not foreigners and who do not normally face difficulties of integration into the national

---

[10] *Ibid.*
[11] Council of Europe: *Report on the Regional or Minority Languages in Europe* (Explanatory Memorandum ACPL8PII.23), Strasburg, February 2, 1988.
[12] Council of Europe, *op. cit.*
[13] Council of Europe, *op. cit.*

society. [14] Due to the work of the Council of Europe and other agencies, there is now a growing insight among linguists, educationalists and politicians that a language is an essential part of the complex interchanges between people, and that language cannot be taught without considering the context in which language is used.

Table 6.2 shows the position of linguistic minorities in Europe in the mid-1980s. It also presents estimates of the number of speakers and the country concerned. [15] It can be inferred from the data that more than seven million people in the EC countries speak a recognized European minority language *as their mother tongue*. This may well be a conservative estimate, as the Council of Europe and its member states may not have an interest in bestowing recognition on a language or dialect whose status is being contested by linguists.

Whatever the exact figure, it is clear that a significant portion of the European electorate uses a minority language. Hence the protection and promotion of minority languages constitute topics that carry considerable political clout, especially at a supra-national level. This may be the case because history shows that, until very recently, the nation-states of Europe paid little heed to their local or regional languages. Spain and France are examples of countries where regional languages were not recognised, and hence could not be used in official communication or, for example, in schools. Countries such as the United Kingdom also demonstrate that the maintenance of a political balance between a majority and small linguistic minorities is precarious, and that the temptation of the nation-state to defer to the values of the majority population is strong. An interesting caveat is that seven of the countries in Central and Eastern Europe had national linguistic minorities in the 1930s that amounted to about 25 per cent of the aggregate population; by the late 1970s this figure had decreased to seven per cent – to a large extent due to uprooting and transnational migration. [16]

---

[14] Council of Europe, *op. cit.*

[15] The data recorded in this table must be approached very carefully. In the first place, because the figures are derived from an official EC publication, they may reflect the position of governments of member states rather than those of minority groups residing in these countries. Secondly, the data refer to language *minorities* in the EC countries. Thus the people speaking Albanian in, for example, Yugoslavia are not taken into consideration. In a similar vein, people speaking Catalan in Spain (where it is a regional rather than a minority language and where the number of speakers is the largest) are not accounted for.

[16] John Coakley: "National Minorities and the Government of Divided Societies: A Comparative Analysis of Some European Evidence", *European Journal of Political Research*, 1990.

[17] Source: Commission of the EC, *Linguistic Minorities in Countries Belonging to the European Community*, Luxemburg, 1986, p. 246.

Table 6.2: *Linguistic Minorities in EC Countries (numerical extent)* [17]

Language [a]	Country	Number of speakers	Other estimates
Albanian	Italy	80,000	
Basque	France	80,000	
Breton	France	550,000	
Catalan	France, Italy	220,000	
Corsican	France	162,000	
Croatian	Italy	2,200	
Danish	Germany, Netherlands	8,000	50,000
Flemish	France	100,000	
Franco-Provençal	Italy	48,050	59,650/75,700 [b]
French	Italy	10,300	
Frisian	Germany	409,000	
Friulan	Italy	625,000	
German	Belgium, Denmark, France, Italy	1,670,450	
Greek	Italy	12,500	
Irish	Eire, United Kingdom	789,429	
Ladin	Italy	27,500	
Occitan	France, Italy	1,550,000	
Polish	Germany	175,000	
Sardinian	Italy	158,000	1,200,000
Scottish Gaelic	United Kingdom	79,307	
Slovene	Italy	52,170	125,000
Welsh	United Kingdom	503,549	

Notes: (a) Estimates refer to minority groups outside the main country; e.g., those who speak Frisian in the Netherlands or Catalan in Spain are not accounted for.
(b) Only in the Valle d'Aosta.

There can be no doubt that a crucial factor in the preservation and development of a regional or minority language is the place it is given in the formal educational system. This explains why the question whether teaching in and of a given minority language in state schools should be allowed, or even encouraged, has tended to be the subject of much heated discussion in many countries. The tendency now is to stress the need for multilingualism. This by-passes the problem of language competition to an extent, because it is emphasized in this perspective that several languages need to be learned by everybody: perhaps a local language, certainly the national language, and definitely two modern foreign languages, one of which is English. The need to include at least two foreign languages in the curriculum arises from language interaction both within the country, as in the case of Finland, Belgium and Ireland, and between countries, as in the case of Germany, France and Britain.

The charter mentioned above stipulates that countries must offer families who so wish, pre-school and primary education dispensed mainly or entirely in the regional or minority education. [18] It is even stated that the expression "in the language" means that all or the greater part of teaching must be conducted in that language. That this provision is made for pre-school and primary education is not surprising, because it is known from research studies that 3 to 11 year-olds are particularly receptive to languages and instruction in them. [19] (The in-depth discussion on the educational implications of minority languages, for example, in terms of cultural matters and methodological problems, for example inductive versus deductive foreign language teaching, that may be encountered in using or teaching minority languages in school education, is taken up in the next chapter.)

Apart from the regional languages considered above, which are spoken by a small portion of the population within a given country, there are also national languages that can be considered minority languages, if viewed from a pan-European perspective. Examples are Danish, Portugese and Dutch. It is evident that there is a deep concern in these countries with the preservation of cultural identity – which may be more difficult in their cases compared with large countries. The teaching of the national language and, of course, literature tend therefore to be given a prominent place in the school curricula of the former. Yet it is also imperative for the citizens of small countries to be proficient in at least two but preferably *three foreign languages* – small states tend to be economically tied to larger ones, whose people tend to be unwilling to invest time in learning the less widespread language.

5. Approaches and Methods in Language Teaching

There have been several changes in language teaching methods this century. These have reflected recognition of changes in the types of skills

---

[18] *Resolution 192 on Regional or Minority Languages in Europe*, Council of Europe, *op. cit.*
[19] Reviews of research studies on language development in young children can be found in S. Krashen (1981), B. McLaughlin (1984 and 1985) and L. Arnberg (1987).

learners require, for example, an emphasis on oral proficiency and reading comprehension rather than ability to write the language as the goal of language education. Change has also been brought about as a result of theoretical developments.

When foreign languages were taught in the school in the 18th and 19th centuries, the methods applied were the same as those used in the teaching of Latin. The school was then attended mainly by an elite preparing itself for higher education and leadership. In school teaching, the emphasis was not on oral practice but mainly on the ability to read a foreign language. For a long time most of the textbooks at, for instance, Danish and Swedish universities with their small student populations, were in German, English or French. Until early in the 20th century, the main emphases in foreign language study were on grammar, the analysis and translation of selected texts, and the writing of abstract sentences constructed to illustrate the grammatical system of the language.

Opportunities for communication increased toward the second half of the 19th century. With them came a demand for a new approach to foreign language study, one in which spoken language was the central concern. Linguistic theories were developed and new methods for the teaching of foreign language created. The direct method based on what may be termed *natural principles* was influential around the turn of the century. Functional theories provided for the development of the *oral language approach* in the 1920s. Structural theories of language learning, from which *situational language teaching* and the *audio-lingual method* were derived, were *en vogue* from the 1930s to the 1950s. [20]

The assumptions underlying these approaches were increasingly called into question in the early 1960s by linguists such as Noam Chomsky, Dell Hymes and D.A. Wilkins, [21] who adhered to the view that language teaching should focus on communicative proficiency rather than on the mastery of the structure of a language. Whereas the audio-lingual approach focuses on the linguistic ability of the learner in mastering the structure and form of a language, the new "interactionist" perspectives emphasized that language learning is learning to communicate. Reading and writing are delayed in the former approach until speech is mastered, whereas reading and writing can start from the first day in

---

[20] These methods and approaches to language learning are discussed in Richards and Rogers, *op. cit.*
[21] i.e., N. Chomsky: *Syntactic Structures* (1957); Dell Hymes: "On Communicative Competence" (1972); and D.A. Wilkins: "The Linguistic and Situational Content of the Common Core in a Unit/Credit System" (Manuscript, Council of Europe, 1972); and Wilkins: *Notational Syllabuses* (1976).

communicative language teaching. [22] The latter approach has been actively promoted by the Council of Europe since the early 1970s. The Council has supported the development of first-level communicative syllabuses for foreign language teaching. According to the British linguists Richards and Rodgers [23] some of the characteristics of this communicative view of language are:

1. Language is a system for the expression of meaning;
2. The primary function of language is for interaction and communication;
3. The structure of language reflects its functional and communicative uses;
4. The primary units of language are not merely its grammatical and structural features, but categories of functional and communicative meaning as exemplified in discourse.

6. Language Teaching and the Syllabuses

Foreign language teaching for a long time took place in secondary schools. As general secondary education was restricted to a relatively small student population, foreign language study was by implication only available for an élite. This position changed with the expansion of secondary education in the 1960s and 1970s. Another factor of change was that foreign languages were introduced in the primary school curriculum in some countries. Foreign language is taught from grade 3 at the Bielefeld laboratory school. In Sweden pupils can now even take a second foreign language in grades 7 to 9 and begin with a third in grades 10 to 12. At the upper secondary school the range of options was widened in the 1960s so that pupils could take Russian, Spanish and even Chinese.

At the primary level, in the 1960s various experiments in teaching a foreign language took place. In England, the Nuffield Primary French Project, however, was eventually judged to be a failure, and abandoned. Experiments in learning English in French primary schools were also largely given up. The new National Curriculum in England explicitly rules out the teaching of a foreign language in primary school; in France, however, since September 1989 trial programmes for the teaching of English and German have been re-introduced into primary education.

---

[22] The major distinctive features of the audio-lingual method and the communicative approach to language teaching and learning are compared by Finocchiaro and Brumfit (1983, pp. 91-93) and Richards and Rodgers (1986, pp. 67-68).
[23] Richards and Rogers, *op. cit.*

Furthermore, as Table 6.3 demonstrates, at least five other countries are contemplating, if they have not already begun, the teaching of foreign languages at this level [24] (Table S14 in the statistical annex provides further information on the inclusion of foreign languages in the primary and secondary school syllabuses.)

Table 6.3: *Early Teaching of Modern Language (ETML, primary school level)* [25]

	When ML teaching to begin	No. of lessons per week
Belgium	Last two years	2
Denmark	Equivalent of last two years	2
Italy	Not yet fixed	3
Luxemburg	Throughout primary years	5-7 [2]
Netherlands	Last two years	2

The EC has consistently recommended that at secondary level at least one foreign language should be studied throughout, that the number of hours devoted to this should be increased, and that languages other than English, German and French should be encouraged. A supplementary Table in the Statistical Annex gives some idea of the gamut of languages offered, although the reality must depend upon the availability of teaching staff and the willingness of pupils to opt for a particular language, as apart from the first foreign language, a second foreign language is often optional. The authors of Table 6.3 comment as follows: [26]

> "If [in the countries included] comparative homogeneity exists at lower secondary level, great disparities are to be found at upper secondary level. The study of a modern language is practically compulsory for all at lower secondary level, with an average timetable of three hours a week. On the other hand, except in certain tracks in the German system, it is rare for a second language, even as

---

[24] Luxemburg is a special case, inasmuch as Luxembourgeois, French and German are all taught. French is the language of instruction in the secondary school. Belgium also constitutes a special case, because both French and Flemish (Dutch), as the two national languages, must be studied.

[25] This Table, as well as other Tables and much of the information given is taken from: J.M. Leclercq and Christiane Rault, *Les systèmes éducatifs en Europe: vers un espace communautaire?*, Notes et Études documentaires, No. 4899, La Documentation française, Paris, 1989, the debt to which is gratefully acknowledged.

[26] J.M. Leclercq and Christiane Rault, *op. cit.*, p. 132.

an option additional to the programme ('facultatif') to be offered. In the upper secondary level of general education, the place assigned to foreign languages varies broadly according to the options offered. This place is often precarious, as in Italy and the United Kingdom, whereas it undoubtedly enjoys priority in Germany and Denmark. As for the vocational tracks, the situation is even more diverse, with certain countries such as Italy and the United Kingdom making do with a bare minimum save in the preparation for occupations that require a good practical use of languages".

Figures for the take-up of languages seem to be the result of mere chance rather than acts of deliberate policy. [27] If and when all school education comes within the purview of the EC some common language policy would seem a high priority.

The absence of what might be termed "less common" languages from the list is striking. Not only does it mean, for example, that Dutch is ruled out for Britishers wishing to do business with the Netherlands – an important trading partner –, but the study of the rich culture of this country, even at university level, is uncommon. If one takes seriously the concept of a European heritage, then it is surely right that in every country some pupils at least should be able to enter into particular aspects of it through a mastery of the 'rarer' languages. The problem becomes even more acute with the enlargement of the concept of Europe to embrace not only the Nordic and the other non-EC countries of Western Europe, but also those of Central and Eastern Europe.

The commercial and industrial value of foreign languages is of course self-evident as a means of communication. The (unfashionable) view, however, has been advanced that the priority given to being able to *speak* them, wastes much valuable time in school. What is needed more is the ability to *understand* the foreign language and to reply in one's own language to one's interlocutor, who is equally able to understand. For most Europeans, even those who use languages in their work, it may be argued that the ability to *read* the foreign language is of great importance, even in commerce and industry. This view upsets the priorities established by reasons of teaching method given, in descending order, as speaking, understanding, reading and writing, which are now generally accepted wisdom in language teaching.

It is imperative that all of these curriculum elements be retained in future, as mastery of a foreign language depends on all of them. The emphasis could, moreover, be on the study of principles of language learning, so that one can learn a new language by means of self-directed learning in adulthood when the need arises.

---

[27] Cf. W.D. Halls, *Foreign Languages and Education in Western Europe*, London, 1970 (Table, p. 15).

That the variety of languages taught needs to be extended appears self-evident. The fact that pitiful numbers of Europeans are studying such important idioms as Russian (now further depleted as the countries of the former Eastern bloc are substituting English or German for it in their schools), Japanese and Chinese (perhaps one of the world languages of the next century), or even Spanish, with its almost universal currency in Central and South America – not to mention its commercial importance in Europe itself – is surely to be deplored. Furthermore, confining the languages of communication to one or two common languages may well lead to grave errors unless a very good knowledge of them is acquired. During the Munich crisis of 1938, the Nazis used the word *unterbrechen* to describe the talks then proceeding. Both the English and German translators of a communiqué agreed that this signified the "breaking-off" of negotiations, whereas of course it merely meant that they were 'interrupted'. The mistake was only spotted at the very last minute. Language can be a source of international misunderstanding as well as mutual comprehension.

There is certainly a remote possibility that the whole problem of languages will eventually be solved by the invention of translation machines that will be sensitive enough to discern all the nuances of a foreign language and communicate them accurately. But such a discovery seems, at the present time, decades away. Meanwhile the linguistic problem must surely be one of the highest priority on the agenda for a more integrated Europe.

7. Language Policy and Practice in the New Europe

EURYDICE, set up as an information network in February 1976, was given additional responsibilities, this time in the field of modern languages, at a meeting of the Council and Ministers of Education of the EC held in Luxemburg in June 1984. [28]

In their conclusions the ministers emphasized the key role of knowledge of the EC languages in facilitating the free movement of persons as laid down in Title III of the Treaty of Rome and for cooperation between member countries generally. In their desiderata the ministers stressed:

1. The need for pupils to be able to communicate orally and in writing in the modern language.

---

[28] Source: *The Teaching of Languages in the European Community*, EURYDICE, Brussels, 1988.

2. One modern language should be studied in depth, and the learning of others should be encouraged. One of these languages should be that of a member country.
3. The ultimate aim should be for pupils to acquire by age 16 a practical knowledge of *two* languages as well as their mother tongue, and to maintain the level of knowledge acquired in vocational, higher and adult education.
4. Measures such as the exchange of modern language teachers and students should be encouraged.
5. A central unit of Eurydice should be set up in Brussels to coordinate language teaching and exchange programmes in member states.

As reported by the Eurydice Central Unit in 1988 [29] the overall situation was as follows:

1. There was little or no modern language teaching in *primary* education. Where it exists it is on an experimental or optional basis, and the teacher may not be qualified to teach it.
2. In *secondary* education in almost all cases the teaching of a foreign language is compulsory from the first year, for a period of time varying between two and six hours a week.
3. The most usual language taught is English, although in some cases it may be French or German. Other languages may be offered. In Britain and Ireland the language offered is usually French.
4. During secondary education most pupils can opt for a second modern language, which is taught for two to four hours a week. This language is usually German, Spanish or Italian, although it can be Russian or even other languages.
5. At *upper secondary* level some pupils can take a third modern language for two to four hours a week.
6. Teachers: At secondary level they are usually specialists in one foreign language – occasionally two. Exchanges, usually optional, exist between teachers of member states.
7. Pupils: There are some pupil exchanges, usually privately arranged.

During the 1970s and until the early 1980s the United States was the first foreign destination for Western European students. [30] Since then a gradual shift towards a greater focus within Western Europe has occurred. Germany was the most important European host country in the years from 1983 to 1987. The institutionalization of the EC's ERASMUS programme has served to strengthen the position of France and Britain.

---
[29] *Ibid.*
[30] Source: UNESCO *Statistical Yearbooks*.

The aims of this action programme are to increase the number of students spending a period of study abroad, promoting cooperation among universities in Europe, developing a pool of graduates with educational experience gained in countries other than the home country, and contributing to the creation of a "People's Europe". Almost 5000 inter-university cooperation agreements were supported under the programme in four years, involving the mobility of about 80,000 students. [31] Together Germany, France and the United Kingdom received some 70 per cent of all EC students enrolled abroad in the 1989-1990 academic year. [32]

In May 1989 the EC formally adopted LINGUA, a modern languages programme designed to improve the quality and number of foreign languages taught in the various countries of the EC, and for the four-year period 1990-1994 allocated 200 million 'écus' to finance it. The programme is to be administered from Brussels by three agencies: the British Council, the Goethe Institute and the Centre International d'Études Pédagogiques, all of which have been prominent in promoting respectively the study of English, German and French, which are the main foreign languages studied in the EC, and indeed throughout Europe as a whole.

The objectives of the LINGUA-programme of the European Community are: [33]

1. To promote the continuing education of teachers and instructors of modern foreign languages, in particular the national languages of EC member states;
2. To promote the teaching and learning of modern foreign languages in higher education;
3. To promote the ability of people to use modern foreign languages in economic life;
4. To promote the exchange of students in vocational and technical education;
5. To support associations and institutions working on a European level with the aim of promoting the use of foreign languages in the media.

Unlike the massive exchanges of students and staff in higher education under the ERASMUS programme, or the EC's related programmes such as PETRA and TEMPUS, LINGUA merely allows for the study abroad of teachers of modern languages in schools, and for exchanges of young

---

[31] Roger Absalom: "Practical rather than Declamatory Co-operation: ERASMUS in 1990, An Appraisal", *European Journal of Education*, 1990.
[32] *Ibid.*
[33] LINGUA information pamphlet, EC Commission, Brussels, 1990.

people following technical or vocational programmes. Policy measures directly affecting general education at school level are still ruled out by the Treaty of Rome, so that the teaching of modern foreign languages within schools falls outside the defined scope of LINGUA.

However, one important feature that will have an undoubted knock-on effect in schools, although its effect may become visible only in the long term, is not only the improvement of the linguistic ability of teachers, but also in teaching methods, for example the development of methods using new information technologies, which will be intensively studied.

The language programme of the EC is only just getting off the ground – although similar initiatives were made in the 1960s without very good results.

What of Central or East European languages, or even those of non-EC countries in Western Europe? Are these to be left out? If so, again cultural impoverishment will occur, not to speak of the technological or commercial handicaps that will arise.

The Treaty of Rome makes it difficult for policymakers to go further. School education is not specifically mentioned. Thus any measures taken as regards schools by member states cannot be promulgated as binding. The effect of this omission on modern language initiatives in some countries – to cite only England and Ireland – may well be disastrous.

It is to be hoped that one outcome of the LINGUA programme may be to improve the methodology of language teaching as well as the competence of those that undertake it. Where, however, are the attempts to improve pupil motivation, for example in England or France?

More drastic measures should be envisaged: e.g., the compulsory teaching of foreign languages at primary level, because at this stage children are more eager to learn and less self-conscious about speaking a foreign language; why cannot lessons in some school subjects (e.g. history and geography) be given at secondary level in the foreign language? (One complaint is always that language teaching has no *content*, other than the language itself.)

8. Conclusion

A special issue in European culture is multilingualism. Indeed, it may well be said that this is the rule rather than the exception. The importance attached to multilingualism not only derives from the increasing demand for communication in the modern world, but is also related to the challenge of balancing the cultural identities and linguistic rights of both majority and minority groups in a European society. This must be done if peace is to be preserved in future, because a highly significant portion of Europeans speak a language considered a regional or minority language. That there is no systematic provision in the EC yet for the study of minority or "small" languages, without which the concept of Europe will become immeasurably poorer, ilustrates the enormous difficulty of the problem. There is some room for hope that a solution to this problem can be worked out, however, because political controversy over the teaching of a minority language is likely to become less in the new, integrated Europe than it was in the independent nation-state. The reverse of the medal is that the 'small' language countries will be confronted in the new, unified Europe with the kind of problems minorities experience in the nation-states of Europe at present.

Tailoring our secondary school curricula to comprise a minimum of three languages in every European country would ensure the recognition of the real values of European civilization and the utilization of such values to their full potential. Sacrifices may well have to be made if several foreign languages are to be given a place in the curriculum, including also the big non-European languages, mainly Arabic and Oriental languages, which do not yet seem to receive the attention they deserve. This makes it even more important to start language teaching early, as in the case of Sweden, where English is already taught in the 3rd and 4th grade of primary school. The Hungarian experiment of teaching biology and physics in a foreign language is also noteworthy.

Chapter 7

The Education of Linguistic, Cultural, and Immigrant Minorities in Europe

1. Introduction

Questions relating to national minorities and to immigrant groups are basically of the same order. A fine distinction may be drawn between a minority group that has lived for a long while within a given nation-state, and one of more recent date, initially temporary, which may be classified as an immigrant group, but which, with the passage of time may become permanent, as bids fair to be the case in Europe today. These problems may conveniently be classified as ones relating to *ethnicity*.

Ethnicity in turn may be broken down into its various components – biogenetic, territorial, linguistic, cultural, religious, economic and political. [1] It would theoretically be possible to construct an objective profile of each minority group, whether temporary or permanent, by measuring the degree to which each of these aspects were present. The profile thus arrived at could then be placed against a similar profile of the majority group, as well as the subjective profile that the latter constructs for itself of the same minority group. Where there is a lack of congruence between all three profiles problems are likely to occur, although the degree of mismatch is all-important. Thus, to take an extreme example, the existence just within the Netherlands of the tiny Belgian enclave of Baarle-Nassau poses far less 'ethnicity' difficulties than does a small North African *bidonville* on the outskirts of Lille. The purpose of this

---

[1] This concept has been developed from a paper by F.M. Ravea: "Ethnicity, Migrations and Societies"; which can be found in OECD/CERI: *Multicultural Education*, Paris, 1987, p. 107. Cf. also: E. Verne, "Multicultural Education Policies: An Appraisal", *op. cit.*, p. 32.

chapter is to examine the key components of ethnicity that relate to education, and the problems to which they give rise.

2. National Minorities in Europe

First, a brief consideration of minorities in the traditional sense, i.e. national minorities. The opening up of Central and Eastern Europe, where such minorities are particularly numerous, gives the subject renewed topicality.

In fact, no European nation presents a picture of total cultural and linguistic homogeneity. The patchwork quilt of nation-states that makes up the continent has always accommodated within national borders many different minorities and cultures. Some such groups even predate the constitution of the state itself, others result from the redrawing of frontiers, usually after wars. The existence of the European Community will hopefully avoid such conflictual settlement of boundaries in future and may indeed, according to some, eventually lead to their withering away completely, although this may well leave minorities in much the same position as before. These more permanent types of minority are characterized by their longevity, the geographical concentration of their members, and the self-assurance they possess regarding their own language and culture within a national majority.

Educational problems concerning minorities relate primarily to language and culture. How these groups fare may give some indication of the solution to similar problems arising for immigrants. We turn first to language.

There is no European country that does not include at least one permanent group speaking a language different from that of the majority. [See Table 6.2]. Within the two states that comprise the British Isles no less than six languages are current – some, however, spoken by very few: English, Welsh, Scots Gaelic, Erse, Manx and French (in the Channel Islands) – and to these some would add Lallans (Lowland Scots) – the language of Robert Burns' poetry – and even Cornish (which has been reintroduced by a few enthusiasts, after dying out in the early 18th century.) France has a similar linguistic diversity: apart from French, some also speak Flemish, Breton, Basque, Catalan and (in Alsace and Lorraine) a dialect of German. Within Spain not only Castilian, but also Catalan, Basque and Galician are spoken. In countries with smaller populations such as Sweden, Belgium and Switzerland two or more languages are in daily use in certain areas. In Eastern Europe a parallel situation exists. Thus in Hungary 3.5 per cent of the population consists of gypsies, using Romany as well as Magyar. Two million Hungarian speakers live in Romania, where up to recently their language and culture were suppressed. It is understandable that, at a time when Europe is moving towards a greater unity, minority linguistic and cultural groups cling more fiercely to their identity – whilst not necessarily repudiating the concept of a common European home.

How languages should be dealt with educationally in this context is problematic. Three solutions have so far been implemented: *Firstly,* the minority language is used as the medium of instruction. Where the minority language is sufficiently concentrated in one area, as in Catalonia, this presents no difficulty. However, if it is scattered among the majority, – as it is in parts of Wales, or as German is in parts of Poland – the provision of separate schools may present financial and resource difficulties. Moreover, since the minority will inevitably come into contact with the majority population, great prominence must also be given to the teaching of its language.

Secondly, the minority language is taught compulsorily as an alternative national language, as is the case with the teaching of Erse (Irish Gaelic) in the schools of the Republic of Ireland, or in Belgium where both Flemish and Walloons have to learn either French or Dutch as the second national language. However, since pupils have usually also to learn another European language this constitutes a heavy linguistic baggage to carry.

Thirdly, the minority language is not taught as part of the normal school timetable, but may form an 'optional extra' to the curriculum. An example of this may be in France, where the Loi Deixonne allows the teaching, for example, of Breton. [2]

Which solution should be adopted in particular cases will depend on circumstances. The teaching of minority languages may also serve broader ends, usually cultural, but sometimes political, by groups seeking enhanced autonomy or even full independence – "identity" is neither all cultural nor all political.

Here Celtic examples may be cogent. Thus Welsh enthusiasts – including one party that seeks independence for Wales – have lobbied successfully for the spread of Welsh-medium schools and a Welsh TV channel. Occasionally the demand may seem somewhat artificial. Thus in France the small *Diwan* (= 'seed', 'germination') movement in Brittany, which in recent years spearheaded a drive for Breton-medium schools, was largely inspired by a group of Parisian middle-class professionals of Breton origin who resettled in rural Brittany but found little support from an indigenous population of small farmers and fishermen. As a Scottish anthropologist who lived for a year in a Breton community puts it: "The

---

[2] For a discussion of the minority languages of Europe cf. E. Corner, "The Maritime and Border Regions of Western Europe", *Comparative Education* (Oxford), Vol. 24, No. 2, 1988.

automatisms of the socio-linguistic model and those of the local world were quite different, and it required self-conscious efforts on both sides for one to accommodate the other". [3] This statement holds good in other contexts. The artificial stimulation of Breton would seem to be borne out by the fact that, although French law authorized the teaching of regional languages (but, incidentally, not Flemish, Alsatian or Corsican on the grounds that they are variants of the languages of neighbouring countries – 'langues allogènes') in 1951, not until 1978 did some of the one million Breton speakers – most of whom also of course spoke French and used it more often than Breton – desire its implementation, and then only in parts of Brittany. It is noteworthy also that Breton, like all regional languages of France, can only be offered as an additional option, and not used as the medium of instruction. This is in conformity with historical precedent. Indeed, non-French speakers in France largely welcomed the Revolutionary edict that French should be the sole medium of instruction, realizing that this would be of inestimable benefit for their children. Another French example is also instructive. Today, in the extreme south-west of France, there are only some 1,000 pupils attending private 'seaska' schools, where Basque is the medium of instruction, as contrasted with the 70,000 pupils who attend Basque-medium schools beyond the Pyrenees in Spain. The lesson to be drawn is surely that it is important to evaluate the demands of minority groups, which may or may not represent the population as a whole.

This said, however, in a more united Europe any attempts at multilingualism are surely worthy of encouragement, as being an instrument of greater mutual understanding and the preservation of cultures that may be at risk. Singularly, at a time when nations are partially surrendering sovereignty, there is a nostalgia for the familiar, smaller entity. "Small is [sometimes] beautiful". In some cases, minority schools might be thrown open to non-minorities, with beneficial effects. One interesting example of how pupils from different linguistic backgrounds can mingle happily comes from Schleswig-Holstein. [4] In Tingleff (Denmark) on the Danish-German border, the German-medium school, that of the minority, comprises a mix of pupils from various linguistic backgrounds. On investigation it was found that only 14.8 per cent consisted of pupils whose home language was solely German; 62.2 per cent spoke Sonderjysk (Jutish); 1 per cent spoke Danish; 17.5 per cent a mixture of Jutish and German. It could be argued that this case is atypical. The Danish children (or their parents) enrolling in German-

---

[3] M. McDonald: *We are not French: Language, Culture and Identity in Brittany*, London, 1990.
[4] Cf. C. Byram: *Minority Matters and Ethnic Survival: A Case Study of a German School in Denmark*, Cultural Matters Ltd, Clevedon, England, 1986.

medium schools want to profit from the situation and learn German in addition to the language they know and speak anyway. However, what also emerges from the study is that the minority culture and the language of the home were *not* considered the affair of the school. This might well be a rational expedient for the education of immigrant children, since the cultural combinations that arise in a host country can be very numerous.

Despite the Danish example, the other educational demand from long-established linguistic minorities is for the teaching of their cultural heritage. This can occasion difficulty. The school timetable may not be elastic enough to accommodate it. In the United States, minorities may have succeeded in time being set apart for language teaching, but have recognized that initiation into the culture is not a concern of the school. A particular problem can arise with the teaching of history, where the minority's view of the past may clash with that of the majority group, since its very existence on the national territory may be the result of a war. Here an urgent need would seem to be for a broadly agreed view of history, preferably in a European dimension. One recalls that bilateral and even multilateral talks between countries to settle points of difference and interpretation of the past were undertaken in the immediate post-War years, mainly under UNESCO auspices. The moment may well have come when such dialogues between historians should be resumed both on a European scale and within countries. In the same way geography teaching can also give rise to difficulties that require resolution: one recalls, for example, the long-standing disputes between Flemings and Walloons in Belgium regarding the exact delineation of the linguistic boundary. The aim may well be a linguistic and cultural pluralism, intra-nationally and internationally.

We turn now to the much more pressing problems associated with immigration.

3. Immigration: The Overall Picture

The traditional minorities of today were either the immigrants of yesterday or people incorporated into a given nation-state by the shifting of borders. In the past the movement of people from one country to another arose largely because of persecution, political or religious. Even today there are many refugees in the European countries.

Most migrants, however, moved for economic rather than political reasons. The 1910 German Imperial census, for example, showed that there were already 1.2 million foreigners living in Germany, mainly Italians, Poles and Ruthenians. Today easier communications have increased such numbers everywhere. Although a minority are refugees seeking asylum, the main motive for immigration today appears to be economic, generating the influx of people from a poorer area of the world to a richer one. The example of the Americas is illuminating: fifteen million people have this century moved from South and Central America to North America.

Table 7.1: *Refugees and Displaced Persons in Western European Countries (1988 or 1989)* [5]

Host country	Number of registered refugees [a]
Austria	17,500
Belgium	22,000
Denmark	30,000
France	177,042
Germany	146,069
Italy	10,592
Netherlands	24,000
Norway	15,000
Sweden	130,000
Switzerland	30,195
United Kingdom	100,000

Note: (a) Displaced persons with UNHCR refugee status.

In Europe the movement has been from the developing countries – mainly former colonies of European powers – and from the Mediterranean area to the more prosperous North. Thus North Africans have installed themselves in France, Belgium and the Netherlands, where Surinamese and Indonesians have also settled. From the Indian subcontinent have come to Britain not only Indians, but also Pakistani and Bangladeshi, and from the West Indies Jamaicans and others from the Caribbean islands. Italians have settled in Belgium and Switzerland, Portuguese in France and Luxemburg, Turks and Greeks in Germany, Slovenians in Austria and Germany, Finns in neighbouring Sweden, to mention only the main migratory flows.

In recent years, however, the movement of migrants from the south to central and northern Europe has begun to tail off, in part because of the restrictive measures imposed by the "host" countries, and also because of

---

[5] Source: *United Nations Year Book 1990*.

slightly greater prosperity within what were the "despatching" countries. Moreover, former despatching countries such as Greece, Italy and Spain have now in their turn become host countries, receiving nationals from the Arab states, and in the case of the last two countries, particularly from their former colonies. [6] Turkey, however, may want to maintain emigration if only to lighten the burden on its own overcrowded labour market.

A further complication may well develop if the burgeoning democracies of Eastern Europe achieve full or associate status within the EC, when the tide may flow from east to west rather than from south to north. Soviet citizens may also become part of that movement. After 1992 it is anticipated that workers who are nationals of EC member states, whether unskilled, skilled or professional, will move freely within the Community seeking employment opportunities. Whereas those immigrants from outside the EC or from the periphery now usually wish to remain permanently, those from EC countries will probably be more flexible, the length of their stay depending upon personal circumstances and conditions in their home country. However, in the later 1980s the influx into Western Europe was still estimated at 850,000 per annum – and this takes no account of illegal immigrants, such as those coming from North Africa via Italy. [7] Whether permanent or temporary, legal or illegal, the problems of education for immigrants and their families will continue to be posed. Moreover, their length of stay and status will determine the kind of educational facilities offered. Table 7.2 shows the foreign population as per cent of the total population in selected European countries.

Such figures do not reveal, however, the large numbers of recently naturalized, or recent arrivals, in a country with a claim to its citizenship, as is the case with many migrants settled in Britain or France. Thus the British minority population numbered 2.43 million in 1977 – some 4.5 per cent of the total. A quarter of these, some 600,000, were Irish by nationality; these enjoyed from the outset exactly the same rights as British citizens and were considered to present few problems. Of those who came from the Third World another quarter of the total were West Indians, with English as their mother tongue; the rest came from the

---

[6] M. Weiss, H. von Recum and P.O. Doring: "Prospective Trends in the Socio-economic Context of Education in European Market Economy Countries", working document for UNESCO International Congress on Planning and Management of Educational Development, Mexico, March 26-30, 1990, German Institute for International Educational Research, Frankfurt am Main.

[7] I. Fägerlind and B. Sjöstedt: "Educational Planning and Management in Europe: Review and Prospects", working document for UNESCO Congress on Planning and Management of Educational Development, Mexico, March 26-30, 1990, Institute of International Education, University of Stockholm.

Indian sub-continent and were of great diversity in language and culture. But already by 1977, 43 per cent of all those of immigrant origin were born in the United Kingdom.

Table 7.2: *Percentage of Foreigners in European Countries in 1989* [8]

Country	Percentage of total population	Largest group
Belgium	8.8	Italians
France (1985 data)	6.5	N. Africans
Germany	7.3	Turks
Netherlands	4.2	Turks
Norway	3.2	Finns
Sweden	5.0	Finns
Switzerland	15.3	Italians

It is, in fact, the sheer heterogeneity of the modern migrations that presents problems. Thus Denmark, for example, has a work force consisting of only two per cent foreigners, but between them they speak over a hundred languages. The diversity of ethnic origins – not only from every quarter of Europe, but also from Africa, Asia and the Caribbean – further complicates the picture, as do the various religious allegiances – Christian, Moslem, Hindu, Buddhist. [9] In France Islam has now become the second largest religious grouping after Catholicism. It is with such a mixture of languages and culture that European educational systems have to deal.

4. Educational Problems and General Policies

Ever since 1946 international organizations and national governments have been conscious of the problems raised by the presence of minorities, including immigrants. During the 1960s UNESCO sought to guarantee minority rights in matters of education and culture. However, up to the mid-1970s European nations did no more educationally for

---

[8] Source: *SOPEMI. Continuous Reporting System on Migration*, Paris, OECD, 1990. The data are derived from Table A2, p. 113.
[9] Cf. CDCC Project No.7, *The Education and Cultural Development of Migrants*. Final Report of the Project Group, Council for Cultural Cooperation, Strasburg, 1986.

immigrants than what little previously had been done for any foreign child entering another educational system. The EC Directive of 1977 [10] dramatically changed the situation, however. It charged member states with the duty of ensuring for migrants adequate teaching in the official language, and the language and culture of origin (in cooperation with the States of origin), and with the training of teachers able to fulfill these tasks. Member states had four years in which to implement the directive. Nevertheless, the EC Commission still saw it as a preparation for migrants eventually to return home. Some of the 'States of origin', on the other hand, seized upon it as a unique opportunity to promote their own language abroad. For Western Europe as a whole these moves were reinforced by the Bordeaux Declaration of 1978, which took the form of a Council of Europe convention. This pronouncement, which was, however, no more than a statement of intent on the part of its signatories, called upon member countries to provide adequate financial support for such minority groupings. In 1981 the European Parliament approved the Arfè Report [11] setting out a charter for non-national languages and a bill of rights for ethnic minorities. Such sweeping measures caused much consternation, as countries counted the cost of their full implementation. By mid-1974 the economic situation had been such that many host countries had suspended immigration of foreign workers. However, those already in Europe had to be allowed to bring their families in, and this signified the beginning of permanent immigration: the transition was neatly summed up in France as one from an "immigration de main d'oeuvre" to an "immigration de peuplement" and finally to an "immigration sédentarisée". [12] With the arrival of dependents immigration, from being an economic asset had become a social and educational problem. Realizing the new turn towards permanent settlement that immigration was taking, governments began to offer financial inducements to migrants to return home, but found comparatively few takers. They hoped in this way to solve their own unemployment problems – there are at present 10 million unemployed in the EC – and also avoid social conflicts that were turning racist in character.

---

[10] Reference: 77/486/EEC, 25 July 1977.
[11] *A Community Charter of Regional Languages and Cultures and a Charter of Rights of Ethnic Minorities* (The "Arfè" Report), European Parliament, Brussels and Strasburg, 1981 (see also Chapter 6.)
[12] For a description of the course of immigration in France, cf. M. Silverman: "The Racialization of Immigration: Aspects of Discourse from 1968-1981", in *French Cultural Studies,* Vol. 1, No. 2, June 1990, p. 2.

It has been argued that the educational problems of immigrants may, however, have been exaggerated. Sophia Mappa, in 1987, states as a fact:

> "75-90 per cent of young people of foreign origin go to school for the first time in their host country and only rarely have lived in any other country". [13]

If this is correct, conditions are not so bad as was once feared. However, the figures for the enrolment of foreigners in selected European countries included in Table 7.3 show large numbers still in the late 1980s.

Table 7.3: *School Enrolments of Foreigners in Selected European Countries (1985 data)* [14]

Country	Per cent
Belgium (overall)	10.8
French-speaking	18.3
Flemish-speaking	5.0
France	8.7
Germany	9.0
Netherlands	4.2
Sweden	9.3
Switzerland	16.5

Complications also arise from the fact that numbers of immigrant children – the official statistics tend to speak only of 'foreigners' – as compared with indigenous children may still be growing, if the figures for 1974-1975 compared with 1982-1983 given in Table 7.4 foreshadow a trend.

The situation in Germany, for example, is that whereas the birth rate for indigenous Germans is 8.8 per thousand, for foreigners (mainly Turks) it is 15.6. [15] In 1989 one in two children born in West Berlin was of Turkish origin. By 2000 it is estimated that one in six births in [former

---

[13] S. Mappa: "Education for Immigrants' Children in the OECD Member Countries". In OECD/CERI, *The Future of Migration,* Paris, 1987, p. 247.
[14] Source: *SOPEMI, op. cit.* (OECD, 1990). The percentages are derived from the E-Series of Tables, pp. 175 ff., and represent 1988 figures except where otherwise stated.
[15] Quoted in M. Weiss *et al. op. cit.*

West] Germany will be to a foreigner. Such changes in the composition of the population will have considerable educational repercussions.

Table 7.4: *Percentage Increase or Decrease in Number of Nationals and Foreigners in School Education (1974-1975 compared with 1982-1983)* [16]

Country	Nationals	Foreigners
Belgium	- 9.6	+ 20.6
France	- 3.1	+ 13.5
Germany	- 9.4	+ 95.3
Luxemburg	- 17.7	+ 22.2
Netherlands	- 5.9	+ 67.5
Switzerland	- 3.6	- 0.4

The scope of the problem as regards levels of school education can be gauged from the percentage of foreign enrolments in various national systems, which can be seen from Table 7.5.

Table 7.5: *Foreign School Enrolment in Selected Countries (1982-1983, per cent)* [17]

	Belgium	France	Germany	Luxemburg	Netherlands	Sweden
Pre-primary	13.0	9.8	17.5	37.0	5.9	20.1
Primary	13.5	10.1	11.9	38.0	4.8	18.0
Secondary	9.9	6.4	3.8	19.3	2.3	12.8

The high numbers attending kindergarten and primary schools in some countries in 1982 will now have worked their way through to secondary education.

---

[16] OECD, *SOPEMI, op. cit.*
[17] OECD figures, 1987, quoted in M. Weiss *et al. op. cit.*

The educational measures adopted to deal with immigrants will depend upon general policy. At first, when it was realized that immigrants might remain permanently, the policy commonly applied was one of *assimilation*. It was argued that Asians would be turned into Englishmen, Turks into Germans, and North Africans into Frenchman. It was merely a process of homogenization that was required. But "making similar" as a viable policy encountered resistance on all sides: from the immigrant population, from the host community, and from the immigrants' home country. It may well be that in the distant future all the descendants of immigrants will become indistinguishable from the rest of the population, as are, for example, the descendants of the French Hugenots who settled in Britain, the Netherlands and Germany. But this is for the future and does not resolve the immediate educational problem.

An assimilationist policy brought about few education concessions, even linguistically, to the minority. Governments were forced to change tack, and embarked on a new policy of so-called *multiculturalism*. It is one that defies exact definition, because it has been variously interpreted. Generally speaking, it requires children – *all* children – to acquire a knowledge of both the host and the various immigrant cultures. The host country would acknowledge in the curriculum and in other facets of school life the need for the majority of pupils to learn the culture, even perhaps the language, of the countries from which their new schoolfellows originated. However, this somewhat crude, if Utopian, idea of multiculturalism, although it survives in a new form, as will be seen, was soon pronounced a failure. It implied the co-existence side by side of various cultures esteemed of equal worth, relevance, and importance. But in many cases the actual contacts between these different cultures that followed were rare. Thus, paradoxically, it led in extreme cases to a form of cultural apartheid.

The present policies that European governments are following are by no means clear. The terms 'integration' and 'interculturalism', the latter employed by the Council of Europe, have been used to describe them. [18] L. Taft has described it as a situation of "reciprocal role relationships". [19] It still postulates the multicultural nature of modern European society, but shuns an approach that might lead to any cultural 'separate development'. In Belgium it has been described as 'contact' education ("each-meeting-the-other-education" – "elkaar ontmoetend onderwijs"),

---

[18] Cf. *Interculturalism and Education*, Council of Europe document DECS/EGT (87) 26, Strasburg, 1987.
[19] L.R. Taft: "Coping with Unfamiliar Structures", in N. Warren (Ed), *Studies in Cross-cultural Psychology, Vol. 1,* London, 1976, pp. 121-153. See also the Chapter by L.R. Taft in T. Husén and S. Opper (Eds), *Multilingual and Multicultural Education in Immigrant Countries,* Oxford, 1983.

respecting and developing communication between the various cultural identities, but aiming at a complete integration, and recognizing that all pupils can be enriched by contact with other cultures. In 1985 the Swedish Parliament resolved formally to adopt the intercultural approach, as did the Dutch Parliament. The Swedes' policy has been described as interventionist, and contrasted with that of the British, which is one that relies upon the free interplay between communities to produce the desired results. It has been presented idealistically to the European Parliament by one of its members as "the reciprocal fertilization of cultures". [20] The Council of Europe has also had some success in convincing the European ministers of education of the merits of such a policy. However, a certain vagueness is still attached to the concept of interculturalism. Some have the image of the American 'melting-pot' in mind. Others see it as an attempt to combine *significant* elements, rather than *all* elements, of the host culture with similarly significant elements of the migrant culture, with the aim of producing a new and mutually enhancing synthesis. After all, historical parallels are not lacking. The Norman invasion of England, as Michelet reminds us, changed dramatically the indigenous Anglo-Saxon culture, although the French elements, in the minority, were eventually incorporated into the majority culture. But the process took over three centuries.

That there are negative forces at work rejecting any policy save that of repatriation is clear. In Switzerland the so-called "Schwarzenberg Initiatives" sought just this. In France the French Comité National du Patronat Français in 1978 made plain that it considered there was no place for immigrants in a period of high national unemployment. In Baden-Württemberg and the Netherlands immigrants have been offered a refund of their compulsory retirement pension contributions if they left. This stance of rejection has continued, as an EC census of November 1989 demonstrates. [21] Its findings conclude: "One European in three believes that there are too many people of another nationality or race in his own country". Their presence, it is alleged, lowers educational standards: 25 per cent of the British, 30 per cent of West Germans, and 59 per cent of the Spanish apparently hold this view. Of the countries surveyed, only Denmark emerges positively. It is plain that before any coherent policy can be pursued such attitudes must be faced.

---

[20] M. Oreja: "Social Cohesion and the Dangers Facing It" (Report to the European Parliamentary Assembly, DOC.6.5.87).
[21] *Racism, Xenophobia and Intolerance: An EC Survey*, Luxemburg, 1989 (reported in *The Times*, 14 July 1990).

5. Policies in Action: Germany and France

If these are the general policy commitments, then what is the reality on the ground? Two examples, Germany and France, highlight the background to the educational problems involved.

In Germany the chronological shifts in policy are clearly discernible. Until 1973 the immigrant problem which, then as today, mainly concerned Turks, was associated with labour market policy. From 1973 to 1978 educational measures for immigrants were still largely based on the view that their stay was temporary. In 1978 the government declared itself no longer to be a country for permanent immigration, and limits were set on numbers. By 1981 restrictions had also been applied to the immigrants' relatives, and thus on their children. However, more problems may arise because of an immigrant population increase associated with the enlargement of the "common European home" to include the countries of Eastern Europe. Meanwhile the Federal government has discouraged a 'rotation policy', whereby immigrants would settle for a few years and then, equipped with skills of benefit to their own less industrially advanced country, would return home, to be replaced by another contingent. It has encouraged immigrants to return home, but with little success. Whereas, in 1976 11.9 per cent of immigrants from Mediterranean countries, including Turkey, returned home, by 1981 this had dwindled to 4.6 per cent. [22] (For Sweden the 1981 figure was just over 2.5 per cent.) By 1985 the patterns of immigration, with immigrants then forming 7.5 per cent of the total West German population, were as given in Table 7.6.

Not all immigrants remain in the host country for ever. There are great differences from nation to nation in this respect. The Greeks for example practically all return home at some point, whereas the majority of the Turks stay. For the immigrants that remain the Federal government now seeks complete social and economic integration. It is realized that anything less is prejudicial to the welfare of all. Thus, for example, the frustration felt by youngsters from immigrant families, but often born in Germany, at not being fully accepted into the host society may have led to a higher incidence of crime. However, there is some resistance. Some Moslem Turks refuse integration. The Greek government does not wish to lose its nationals for good and has responded by maintaining a network of Greek-medium schools in Germany. [23] Nevertheless, both the central government and most of the *Länder* are now planning educational measures on the assumption of permanence.

---

[22] OECD, *The Future of Migration*, Paris, 1987.
[23] Cf. A. Schrader *et al.*, *Die zweite Generation: Sozializations und Akkulturation ausländischer Kinder in der Bundesrepublik*, Königstein, 1979.

Table 7.6: *Foreigners Working in Germany (1985, in thousands and as a percentage of the total population of foreign workers in Germany)* [24]

Country of origin	Numbers	Per cent
EC countries [a]	313	7.2
Greece [b]	281	6.4
Italy	543	12.2
Yugoslavia	592	13.6
Morocco	47	1.1
Portugal	77	1.8
Spain	154	3.5
Turkey	1400	32.1
Other countries	943	21.6

Note: (a) except Greece and Italy. (b) Data for Greece are from 1981-1982

In France a first wave of migrants consisted of workers from south- or Latin-European countries. Educationally these posed few problems and a schooling policy of assimilation was successfully pursued. However, associated with the oil crisis, a halt was called to free immigration in 1974, but male workers already in France were allowed to bring in their families. Latest figures, for 1988, show that 30,000 came in that year. For Moslems who may have several wives, and other Africans who practise polygamy, this allowed many dependents to join the head of the family. Numbers were also swollen by what has been estimated at one million illegal immigrants. In addition 60,000 foreigners ask each year for political asylum. Some 2.6 million people are largely from the Maghreb countries, where Arabic is the official language (although it is reckoned that half of those who are Moroccan and perhaps one-third of Algerian immigrants speak Berber rather than Arabic as their mother tongue.) [25] In 1987 no less than 48 per cent of the Algerians in France were under 25 years of age. Such North African immigrants, originating from what has been for over a century a sphere of French influence, are

---

[24] Bundesinstitut für Statistik, *Statistical Abstract of Germany*, 1990, Bonn.
[25] For this and other information that follows we are indebted to A.G. Hargreaves, "Language and Identity in Beur Culture", *French Cultural Studies* Vol. 1 (1), February 1990, pp. 47-58.

mainly a diffused minority, settled in or near large industrial conurbations. Interestingly, North African youngsters, although proud of their 'Arabness', are disinclined to speak their home language and are in many respects integrated with their French peer group. This would appear to validate the policy of integration being pursued by the French government, although incidents occur such as the recent 'affaire des foulards' when three Moslem girls turned up for school wearing the traditional headscarf, and were refused admission. [26]

These therefore are the educational realities that two of the largest states of the Community are dealing with in respect of immigrants.

6. Immigrants and the Majority School System

School systems vary greatly, and a problem exists when the one with which the immigrant child or his parents was familiar may be vastly different from that of the host country. The example of Turkey is sometimes quoted. Turkish parents are often reluctant to accept the German system of 12 years of compulsory schooling since in Turkey it extends in practice only over 8 years (5 years primary, and 3 years middle school, although in the villages, where there may be no middle school, it may be limited to 5 years only.) The temptation is for the more uneducated immigrants to remove their child from school at the earliest opportunity, or let them play truant. The father is so intent on his family surviving in an alien environment that he may have little interest in schooling, waiting only for his child to enter employment to help the family fortunes. Such attitudes, a result of an impoverished life style, were also not unknown among Western Europeans only a generation ago. Moreover, the underlying philosophy of the two types of school may differ: in Turkey it is allegedly more authoritarian, and the aim is to teach "obedience and order", "learning and achievement", whereas in Germany the ideal would be *Erziehung zur Verantwortung* – "education for responsibility", for independence, self-reliance and initiative. The Turkish parent may therefore feel that the discipline of the home is being undermined.

Turning now from general matters, it is evident that two key educational concerns must be uppermost. They are the teaching of language – both the majority language and the immigrant's mother tongue – and the teaching of the culture – again, that of the majority and that of the country of origin. We deal first with language.

---

[26] *Ibid.*

7. The Teaching of Languages in Intercultural Society

The school environment is all-important for the teaching of languages. The proportion of immigrant children in a school must not be too large – although the adjective begs the question – if the immigrant pupil is to be helped in assimilating the majority language. There has in fact been much discussion as to what that proportion should be. The French speak of a '*seuil de tolérance*' of between 10 and 25 per cent; for the English a 'benign quota' is one where no more than one-third of the pupils consists of immigrants. In practice this ratio has proved unworkable, because of the concentration of immigrants in certain areas. Where schools do have a majority of immigrant children it is not unknown for parents of the majority group, and also some immigrant parents, to withdraw their children and place them elsewhere. In 1987 British parents won a legal battle in Dewsbury, near Manchester, to move their offspring from a school where 90 per cent of the pupils were Asian. In another British case a headmaster had to resign because he highlighted the difficulties faced by a minority of purely English-culture children in a school made up largely of immigrants. The charge of racism is easily made, but it should be recognized that the real problem may essentially have nothing to do with ethnic origins. One solution, that of bussing, tried out in Boston in the United States in the 1960s proved unsuccessful; in Britain it has been banned since 1977. It has been found to isolate immigrant pupils from their neighbourhood and yet continue their isolation from their non-immigrant schoolfellows. Moreover, since the immigrant children often came from poorer inner city areas, parents in the receiving schools located in the more prosperous suburbs were not content. The majority community considered their own children were "held back" by the presence of less linguistically competent pupils. [27]

Other remedies have been tried to compensate for lack of educational competence in the majority language. The Dutch, whose linguistic competence is well-known, and who have over the years accepted a large number of immigrants and refugees into their society, have examined this problem in detail. [28] According to the Dutch Advisory Council, [29]

---

[27] C. Jones, *Immigration and Social Policy in Britain*, London, 1977, p. 220.
[28] The following is the picture of immigration into the Netherlands. Immigrants began to arrive in the Netherlands from 1949 onwards. Between that date and 1960 250,000 arrived from former Dutch Indonesia. In 1951 they were joined by 4,000 Malaccans (a total that rose to 12,500 with dependents.) From 1960 onwards 'gastarbeiders' arrived from the Mediterranean countries. Their ranks were swollen in the 1980s by a number of political refugees – Tamils, Vietnamese, Chileans – who joined previous refugee arrivals – Hungarians and Czechs. Another major wave of immigration came in the 1970s from the former Dutch colony, Surinam.

policy guidelines laid down in 1976 for the acquisition of Dutch as a second language had proved ineffective. The council, noting the over-representation of immigrants among low achieving students and pupils in special schools, considered that the most important element in the successful education of immigrants lay in the mastery of languages. Yet a mere half-hour a day of extra Dutch lessons gave poor results. Two years of this for the older pupils failed to achieve what was wanted. Some immigrant pupils came to school illiterate. Success depended greatly upon the role played by their own (written) language in their culture. Many children were incompetent in their own language. It was important to know whether language development in either language was sufficient to produce understanding. Teachers were not trained to teach Dutch as a second language. What has now been recommended is a 'total immersion' classes for immigrant pupils, where everything proceeds in Dutch (the so-called "indompelingsmethode"), although the warning is given that "immersion must not lead to drowning".[30] An extreme example of 'total immersion' has been tried out in Germany in North-Rhine Westphalia: immigrant children were plunged *in media res* and all pupils, German or not, were educated together in German regardless of their level of competence, with what success is unclear.

No firm typology for teaching the majority language can be established, although various other expedients have been tried. Initiation or introductory classes have been set up, where the immigrant child is gradually introduced to the common language of instruction; elsewhere reinforced language teaching has been given either in the form of additional classes held after school or by withdrawing pupils from certain lessons. Another method has been to give 'bilingual' teaching, in which some classes are given in the mother tongue, others in the majority language. As pupils progress teaching takes place increasingly in the latter. This has been tried out with Finnish immigrants in Sweden. There appears to be no hard data regarding the effectiveness of these different solutions to what is, after all, the most fundamental problem. Practices seem to vary greatly. In West Berlin no Turkish child stays longer than 14 months in a transitional language class. By contrast, in the State of Bavaria what seems to be a counsel of despair has been adopted. Considered as 'transients', Turkish children have the option of being educated in Turkish according to Turkish curricula taught by a thousand Turkish teachers who have been hired expressly for the purpose, and,

---

[29] Dutch Advisory Council for Education, *Een boog van woorden tot woorden*, ARBO, Utrecht, 1986.
[30] This *Indompelingsmethode* seems to have met with success in the Netherlands.

although living in Germany, pupils receive only eight hours a week instruction in German. [31]

The importance of early language learning has often been stressed, and with this in mind countries have opened up kindergartens to their immigrant children. In France and Belgium the move has met with some success. In Germany it has been less effective. Pre-school education in the Federal Republic is the norm, but only 38 per cent of Italian immigrants have been persuaded to allow their children to attend. Turkish children also do not attend because there is no tradition of pre-school education, and because some parents fear a different religious influence would have a detrimental effect upon their children. [32] Another useful expedient has been to set up classes in the majority language for parents, particularly the mothers, so that it can be used more in the home. In Sweden immigrant workers have even been given instruction in Swedish during working hours.

A particular problem, but one now dying out, is presented by the arrival of immigrant children when they have already started school in their native land. What the Germans call *Seiteneinsteiger* are in fact the most difficult groups to integrate linguistically into the normal pattern of schooling.

What, however, of the language of origin? Immigrants cling to every link with 'home', and language is its most tangible attribute. However, since many immigrant children have now been born abroad, their knowledge of the original language is poor, or only conversational. Nevertheless, the international consensus is that every effort should be made for them to learn their native language, and vast sums of money have been expended in order to make this possible, as various examples show. Some immigrants have pressed for the teaching of the mother tongue so that their children are capable of studying the sacred texts of their religion. In England there was a very positive response when courses in the immigrants' mother tongue were offered, partly for this purpose. In 1981-1982 a sample of pupils in key immigrant areas such as Coventry, Bradford and Haringey, revealed that the overall 'take-up' for extra classes in the mother tongue (18 languages were involved, including not only those of the Indian subcontinent, but also Vietnamese, Turkish, Portuguese, Polish and Ukrainian) was 94 per cent of the possible total. [33] On the other hand, elsewhere immigrants appear not to have been so enthusiastic. Since 1975 the French authorities have cooperated with the Algerian government in allowing Arabic classes

---

[31] R.C. Rist, p. 48 in Husén and Opper *op. cit.*
[32] Information regarding Turkish *Gastarbeiter* in Germany received from the *Zentrale für Politische Bildung* in Berlin is gratefully acknowledged.
[33] *The Other Languages of England*, Linguistic Minorities Project, London, 1985.

either as an optional curriculum subject or held outside school hours, but the take-up rate in 1975-1976 was only 37 per cent. [34] It is true that at the beginning few Arabic teachers were available, although today there are some 1,000. The German state of Baden-Württemberg in 1983 laid on over 600 courses and 232 teachers for Greek pupils, but these were attended by only 7580 pupils out of a total of 52,000 – some 19 per cent. Hesse, catering for other languages as well, had greater success, both absolutely and over the years, as Table 7.7 reveals.

Table 7.7: *Immigrants Studying their Mother Tongue in the State of Hesse (per cent)* [35]

Nationality	1976	1983
Greek	20	53
Italian	23	62
Turkish	29	66
Portuguese	42	80
Spanish	44	84

How far down this road one can go is uncertain, as the magnitude of the problem is huge. For example, there are now half a million pupils of North African origin in French secondary schools. In Denmark immigrants speak over a hundred different languages. In France the teaching of the 'mother tongue' sometimes involved the learning of a third language: Berber Moroccans were obliged to learn Arabic, and Berber and Arabic are not mutually comprehensible. In Sweden it would seem that immigrant parents themselves are now questioning the wisdom of mother tongue policy, claiming that it places an undue strain upon their children. As for England, it is stated that in London alone there are pupils speaking no less than 151 different languages, and in the country as a whole it is estimated that half a million pupils come from families where English is not the first language of the home. [36] Although the

---

[34] R. Chamouni: "Evaluation de l'enseignement de la langue nationale", in *Les Algériens et l'enseignement de l'arabe en France,* Centre culturel algérien, Paris, 1989, p. 59.
[35] J.J. Smolicz: "The Mono-ethnic Tradition and the Education of Minority Youth in West Germany from an Australian Multicultural Perspective", *Comparative Education* (Oxford), Vol 26, 1, 1990, p. 37.
[36] J. Rex and S. Tomlinson: *Colonial Immigrants in a British City: A Class Analysis,* London, 1979, p. 170.

1975 Draft Directive of the EC prescribing the teaching of the mother tongue included all immigrant languages, it will be recalled that it had eventually to be modified to include only those languages spoken in the Community.

The expensive nature of this linguistic provision can be easily perceived. At one period in Stockholm, since mother tongue instruction is laid down by Swedish law, schools were providing instruction in no less than forty languages. Swedish decentralization of education to the municipal level meant that an enormous burden of expense was placed upon local authorities. Since immigrants tend to settle in the poorer areas, the financing therefore fell upon the poorest school districts.

What are the problems that arise from what in effect is giving immigrants a double task: learning the language of the majority, and learning more about the language of the home? It should first be noted that they may also be learning a third, foreign language such as English or French. In Germany, for example, this is forced upon the bright Turkish pupil, since he must be competent in two languages other than German, and Turkish is not recognized as an *Abitur* qualification. It is certain that immigrants start with a great 'linguistic deficit' that requires to be eradicated. In terms of Basil Bernstein's theory, their overall achievement will suffer unless they can progress beyond the 'restricted code' in their use of both languages. [37] Theories regarding the effects of such 'bilingual' education abound. Taft [38] states categorically that:

> "... bilingual education leads to confusion, culture conflict, diffusion of self-identity, loss of skill in the prevailing [majority] language and even semi-lingualism [sc. semi-linguality]".

Writing of bilingualism in Canada, Lambert, who has conducted experiments with Québecois pupils, [39] has argued that a sufficient competence in the child's mother tongue is required in order to achieve a level of competence in the other language. The experience of well-known international schools such as that of Geneva, or Atlantic College in South Wales, which cater for the wealthy and for intellectual elites, in circumstances far removed from that of the struggling immigrant family, certainly bears out this contention. Logic would therefore seem to demand that at the beginning the immigrant child not born in Europe should be given a thorough grounding in his mother tongue. In any case,

---

[37] Basil Bernstein: *Class, Codes and Control,* Vol. 1, London, 1973.
[38] Taft, *op. cit.*
[39] W.E. Lambert: "The Effect of Bilingualism upon the Individual: Cognitive and Socio-Cultural Consequences". In: P.A. Hornby (Ed.), *Biligualism: Psychological, Social and Educational Implications,* New York, 1977.

much depends upon what has been termed the 'linguistic space' that separates the mother tongue from the majority language. It would seem a matter of common sense that an English-speaking immigrant set down in Germany would encounter far less difficulty in learning German than does, for example, a Croat or Slovenian. This seems to be confirmed by a study made of bilingual children in Sweden. [40] We have in fact little knowledge, save experiential, of the difficulties, say, of a speaker of Punjabi, Gujerati or Urdu, in learning English. In this particular branch of applied comparative linguistics much research remains to be done.

The same holds good for culture, with which we now deal.

8. The Teaching of Culture

The White Anglo-Saxon Protestant establishment that ruled the United States throughout the period of massive migrations from Europe used the "little red schoolhouse" to acculturate the children of immigrants, who took home the ideas they learned and passed them on to their parents. No concessions were made to these "new Americans": schooling was in English; the culture of their countries of origin found no place in the school. Only German, and then solely for a brief period, challenged this supremacy of the original settlers of the continent. As we have noted, for a while this policy also prevailed in regard to the flood of immigrants into Northern Europe. However, concessions had to be made. The American situation did not represent an exact parallel: the view at first was that migrants would eventually return home. Consequently their children would need to know not only the language of the home, but also the culture into which they would eventually be reintegrated. Since the immigrant is now more likely to be a permanent resident the need to retain the original culture has been called into question. Like the home language, some now believe that this culture should be imparted within the family. They hold that not all cultures are equally valuable, both in an absolute and a utilitarian sense; since school time is limited it is the dominant culture that is most formative for minorities.

However, as Bullivant [41] in his study of six countries has pointed out, multiculturalist policies demanded that knowledge of minority cultures should be imparted not only to the minorities but to all pupils. The more the majority knew about the minority the less would cultural conflicts arise, on the same principle as the opinion that foreign travel broadens

---

[40] T. Skutnaab-Kangas: "Bilingualism or Not?" in *The Education of Minorities*, Clevedon, 1984. See also Chapter 6 in L. Arnberg: *Raising Children Bilingually. The Pre-school Years*, Clevedon, 1987.
[41] B. Bullivant, *The Pluralist Dilemma in Education*, Sydney, 1981.

the mind, which is not always true. The counter-argument may be that to pinpoint differences such as in matters of religious belief or sexual equality is to provoke hostility rather than tolerance. The assumption was also that the immigrant pupil would achieve more in school if he were more self-assured regarding his origins. Such growth in self-esteem would help to equalize opportunity.

This meant a broadening of the curriculum, and, for example, caused in Britain, where at the time school programmes were not centrally determined, a rash of subjects such as "Asian studies", "Caribbean studies", even "European studies" [sic], began to appear on the timetable. Some of the curriculum "packages" that were devised had an element of folklore about them. The religious assembly each morning was used as the occasion to speak of faiths other than Christianity. This aroused dissatisfaction among many parents. One parent – incidentally, of a child partly of immigrant origin – requested that her daughter be transferred to a different school, because she had been taught to recite nursery rhymes in Urdu rather than English! [42] (Despite cries of 'racism' the transfer was allowed, since the law stipulated that children must be educated in accordance with the wishes of their parents.) Parents complained that since time in school was limited, subjects more germane to the daily life of their children should be chosen.

Multicultural programmes can have a number of aims: assimilationist; remedial, so as to put an end to racial discrimination; or egalitarian, to ensure that minority cultures are granted equal status. However, how far non-immigrants should be taught about immigrant cultures remains a vexed question. It is argued that it is uneconomic both in terms of time and finance to attempt to do so. Some hold that, because one of the political aims of schooling since public education began is to promote national and cultural unity, the enterprise is counter-productive. Such multiculturalism might entail school reorganization to ensure that a mix of immigrants from different cultures were present in each class to act as a resource for the others. Otherwise the exercise would lack reality. Even so, what aspects of such a variety of cultures, some richer than others, should be taught? On what criteria should the selection of content be made? This in turn leads into a host of pedagogical questions. In two or three hours a week – more time would probably be ruled out – what can be taught that is worthwhile? Should the teacher, incidentally, insist upon differences or similarities? The latter would seem preferable if we are thinking in terms of 'one world'. Might not immigrant parents object to their child being used as a 'resource', and want him to use the time learning more about the majority culture? Parents of the majority of

---

[42] Quoted in *The Times* in a leading article: "Culture and the Classroom", 24 April 1990.

children would possibly not view benevolently such a use of school time. The questions are many, and all urgently require to be addressed.

At present, however, if an OECD report is to be believed, [43] the discussion is academic, for "everywhere multicultural education is practically non-existent, or [in the European countries] is in an embryonic phase". We are, it seems, far from forming for all children "a new open cultural identity which is neither Eurocentric nor ethnocentric".

Some xenophobia was perhaps to be expected in a Community that comprises 14 million migrants, a "Thirteenth State" within the EC, mostly from the Third World, which have seen the barriers progressively go up against them in every member country, thus excluding them from the "People's Europe". [44] The discussion has in fact become bound up with the perennial problem of racism, from which the school is not exempt. This sensitive question is perhaps best considered in the English context. The existence of 'colour prejudice' is seen as a latter-day parallel to turn of the century anti-semitism, which it is the task of the school to dispel, aided by a number of race relations laws that began with the Race Relations Act of 1968. The politician, Roy [now Lord] Jenkins, a one-time head of the EC Commission, has advocated not "a flattering process of uniformity" but one of "cultural diversity" and mutual tolerance, as a desirable characteristic of modern society. [45] It is therefore ominous that such a multiculturalistic stance is viewed with suspicion in some quarters as a strategy for [white] 'ethnic hegemony', or as mere 'trendy rhetoric'. Moreover, the historical position of the three main immigrant groups in Britain must be considered: Irish, Asians and Afro-Caribbeans were all once ruled from London. [46] Bearing in mind that the status of minority cultures is determined by the status of its users, and by past and present power relations between the majority culture and the countries from which the immigrants originate, it would therefore seem salutary that the richness of the immigrant cultures should be appreciated by the majority no less than by the immigrants themselves. The schools must be provided with the means to do so.

The counterbalance to this is the need for immigrants to learn as much as possible about life in the host country: a knowledge of social mores and prevailing economic conditions may prevent a retreat into a cultural

---

[43] OECD, *One School, Many Cultures*, Paris, 1989, p. 65 and p. 8.
[44] Cf. Chapter 1.
[45] Quoted in Rex and Tomlinson, *op. cit.*, p. 170.
[46] Irish in Britain number just under 700,000 and form 1.3 per cent of the total population, a figure that includes merely those born in the Irish Republic and not first- or second-generation dependants born in the United Kingdom and possessing dual nationality, which probably makes them the largest minority of all in the United Kingdom.

'ghetto' that inevitably gives rise to feelings of resentment and alienation. Here, however, difficulties have arisen. In Britain the Swann Report [47] claimed that 'ethnocentrism' – here is meant the European view of the world – should be eliminated from the teaching of history, and that immigration should be depicted as a legacy of empire and slavery. [48] This somewhat extreme view understandably found little support among the majority community, particularly when the Report also advocated that *all* schools should provide courses in the Asian languages and in West Indian Creole, and some instruction be given *in* these languages. Nevertheless, some critics have argued that the teaching of British history, which deals in part with the decline and fall of empire, continues to depict one-time colonial peoples as in some way inferior, and that there is inherent in the educational system an element of racism, a charge that has been fiercely rebutted.

That problems of a cultural nature (here a blanket term used to include religion, institutions, the state, politics, the law, technology, philosophy and knowledge) exist is undeniable. The greater the 'cultural distance' that separates the immigrant from the indigenous culture the more impact this will have upon education. Thus between Germans and Turks, or British and Pakistanis the cultural differences are greater than between, for example, Germans and Swedes, or British and Danes. Yet within Europe it has taken centuries, and a continual history of conflict, to overcome even these latter differences. Present divergent notions of national and migrant cultures need reconciliation and cannot be resolved on such a long time scale. In this process it is generally admitted that education has a key role to play. Integration requires the elimination of tensions and abstention from derision of the other culture, the removal of aggression, and intermingling, particularly in the school, as much as possible.

One indicator of successful integration is in fact intermarriage; for this to happen, intermingling of the sexes at school is desirable. Where cultural distance is greatest, the chance of this occurring are lessened. Thus in Germany in 1982 only 1.2 per cent of Turkish migrants had a German spouse, as compared with 6 per cent of Spanish and 5.3 per cent of Italian migrants. [49]

---

[47] Department of Education and Science (DES), *Education for All* ("The Swann Report"), Her Majesty's Stationery Office, London, 1985.
[48] Cf. R. Lewis, *Anti-racism. A Mania Exposed,* London, 1988, p. 142.
[49] Cf. Note 32.

9. Religions in European Society and Schools

The greatest stumbling block may well be religion. A European society that still esteems Christian values – even although some may no longer accept Christian beliefs – finds it difficult to come to terms with the values of Islam. The reverse applies with equal force. Moreover, where immigrant children bring home from school norms that their parents do not feel able to share, this leads to a conflict of generations, and perhaps causes parents to retreat even more into a cultural ghetto from which the father may only emerge for his job. Even in the workplace the 1.6 million Turks now living in Germany – often called in Turkey the '68th province' (there are 67 in Turkey proper) – may have only Turkish workmates. In Turkey itself the question is often raised as to whether their fellow-countrymen should live any differently in Germany than they do at home. The Koran schools teach national pride and reject accommodations with German life styles. This has aroused a backlash among certain German groups. Graffiti on school walls such as "Türken raus" (Turks out) are not rare. Nor are teachers free from such hostile sentiments. In 1981 fifteen professors published what became known as the Heidelberger Manifest, protesting against what it called the *Unterwanderung* of the German majority. Similar groups exist in Britain, even among educationists.

Both in Pakistan and in Turkish Anatolia, from which areas the bulk of Asian immigrants spring, the Moslem faith is very strong. What many of the majority population fail to realize is that Islam is not only a religion but also a legal and customary system that prescribes in some detail the social, political and private life of the believer. Two researchers into conditions in Birmingham sum up their findings on the matter: Asian children have "a home culture based upon a distinct system of morals, language and religion set within a kinship structure". [50] Legal mores and ethics, even down to the details of clothing and hygiene, are determined by religious belief. The role of the family, and of the institution of marriage, is central. The respective parts to be played by men and women, and sons and daughters, are patriarchally ascribed. The choice of a life partner is determined by the extended family. Women are highly respected, but are assigned to a subordinate role: the Koran lays down that the husband can exact unconditional obedience from his wife, and her duty of compliance even extends to the father or brother or other male relations.

One can perceive immediately how this clashes with the principles underlying the educational systems of European countries. Educationists everywhere now preach the equality of the sexes, and whole school

---

[50] Rex and Tomlinson, *op. cit.*, pp. 28 ff.

programmes have been devised to combat what is termed sexism. Everywhere in Europe co-education is now the rule rather than the exception. In Bradford, where one-third of all pupils are from Asian Moslem families, much thought has gone into attempting to accommodate their legitimate religious demands. The city council pays for nearly 70 "supplementary schools" where children may learn the Koran after normal school hours. However the subsidizing of exclusively single-sex Moslem schools, in the same way as the State subsidizes extensively Catholic, Anglican, Methodists and Jewish schools – the so-called "voluntary-aided" schools – has not yet happened, but since each application for voluntary-aided status is considered on its merits the first all-Moslem State-subsidized school cannot be far off.

A further vexatious issue besetting European schools concerns religious education. Recent legislation in Britain has maintained compulsory religious instruction – unless parents withdraw their children from it – and the practice of a morning assembly, and specified that these should be predominantly of a *Christian* nature. Unfortunately, the mere act of withdrawal from a religious education lesson sets those who absent themselves from it on grounds of conscience apart from their classmates. Some local authorities have arranged special religious instruction for Moslems in the Koran and for Sikhs in Sikhism. In less significant matters such as Moslem dietary laws special school meals are sometimes arranged. As regards clothing Sikh boys are allowed to wear their turbans, and Moslem girls their *shalwars*, rather than the conventional school uniform. The *affaire des foulards* in France had repercussions in Britain, where some girls also came to school in their head scarves, and, unlike in France, were allowed to continue wearing them. More could perhaps be done to achieve harmony in such matters, both great and small.

However, the opponents of separate state-aided Moslem schools in Britain maintain that it would be politically divisive to set them up: [51] so long as one of the aims of a national education remains the promotion of national cohesion, and this has been the case in every European country since at least the French Revolution, all measures that isolate one section of the national community must be considered dangerous in this view.

10. Culture Gaps and School Policy

If Asians in Britain possess a secure home culture, with many distinctive elements, this is not true of Afro-Caribbeans. It has been held that West Indian immigrants started with great disadvantages: they came

---

[51] For a discussion of this problem, cf. P. Ely and D. Denney, *Social Work in a Multiracial Society*, London, 1987, p. 51.

from "an inferior position within a variant of English culture", speaking a variety of dialects of English, or even patois, mainly lacking a culture, except music, that they could specifically call their own. Educationally it was assumed that West Indian children spoke English as their mother tongue – although in fact Creole, which is spoken in some parts of the Caribbean, differs considerably from standard English – and it was wrongly assumed they required no special assistance in the area of language. What was insufficiently realized was that their need for initiation into English culture was as great as that of Asians. In short, they suffered from a lack of *identity*, which led to an inferiority complex both in relation to whites and Asians. Some Afro-Caribbeans have proposed that Black Studies should be offered in schools, as they have been in American universities, as a counterweight to both English and Asian culture.

Education authorities have, however, been far less responsive to demands for action, although some have leaned over backwards to avoid anything that might be held to be racist. Textbooks and other books have been bowdlerized: the Inner London Education Authority issued a list of books that should not find their way into schools. Thus *Dr. Doolittle* was banned because in one section it depicted a black man as a buffoon. Another London borough, Lambeth, banned the American classic *Uncle Tom's Cabin*, on the ground that it was both patronizing and paternalistic. In a way, however, all such efforts to influence the content of education to promote a positive image of West Indians may be counter-productive, because they may be tantamount to a refusal of the majority society – just as are, to a certain degree, the demands of Moslems. Migrant children have, so to speak, to live culturally "between two worlds", that of the home and the host community, as represented by the school. Customs and usages that were once taken for granted are called into question by the new social environment. Parental influence may become less strong; families may feel their identity threatened. In the long run the cultural problems may be more difficult to solve educationally than those raised by language. It is possible that this has not yet been taken on board by national educational systems. Most host countries are still far from the "internationalization" of schooling advocated by the Swedes and, more recently, by the Japanese.

Although some immigrants prefer to maintain a 'social distance' from the host society – one thinks of the Greeks in Germany or even Greenlanders in Denmark – efforts are being made to forge links between them and the local community. The Frais-Vallon Project started in 1976 at Marseille, the Zürich project, for Italians, and the Coventry Community project are examples of how attempts have been made to bridge the gulf. In the Coventry scheme inner city areas whose schools contained between half and 80 per cent of immigrant children were particularly targeted: five community centres, aimed at drawing together the school and parents, especially mothers; community schools, open to all for activities from 9 a.m. to 10.30 p.m.; and a teacher training centre – all these have been started. But other ways of knitting the host and immigrant communities closely together are needed.

11. Immigrants and School Performance

The degree of achievement attained by immigrant children is of course crucial. Failure at school, leading inevitably to truancy and early drop-out may have more dire consequences than usual for their future prospects. There is some evidence that certain immigrant groups – West Indians in Britain, Turks in Germany, for example, – constitute an undue proportion of what has been termed the "new underclass" in Chapter 4, consisting of some 10 to 20 per cent of school leavers in the industrial nations who have effectively "contracted out" of the society in which they live, "no hopers" who have given up trying, sometimes even by the age of 14. [52]

Many research studies have shown that the greatest factor – for all children – in school performance is the home background, the social and cultural capital that they bring to school with them, with linguistic competence as a key variable also. [53] The concept of a "cultural deficit" was used in the 1960s to explain the under-achievement of Blacks in the United States. That this does not apply to children from Asian homes has also been demonstrated. [54] In countries such as the United States, Australia and Britain Asian pupils more than hold their own against other minority groups and not infrequently against those drawn from the majority groups. The Rampton Report, [55] an interim report to the voluminous Swann Report previously mentioned, demonstrated statistically that Asian and White pupils in Britain performed two to six times better than West Indian pupils at the 16+ examination and six times better at A-Level, a leaving examination usually taken at 18 years

---

[52] T. Husén, "Schools for the 1990s", *Scandinavian Journal of Educational Research*, Vol. 33 (1), 1989, p. 6 ff.

[53] The role of home background factors in student achievement is discussed in Chapters 3 and 4. See also, e.g., G. Verma and B. Ashworth, *Ethnicity and Education: Achievement in British Schools,* Basingstoke, 1986.

[54] Asian students are performing very well in the United States school system. There are now proportionally more Asian students in 'top of the bill' American universities, such as Stanford University and the Massachusetts Institute of Technology, than Americans with a "European" background.

[55] Cf. Note 47.

of age. The Report explained the plight of West Indians by racism, negative teacher attitudes and an unsuitable curriculum – all factors that can be rectified. However, as one hostile critic pointed out, it failed to mention two other possible explanations: the fact that far fewer Asian mothers went out to work than did West Indian mothers, and that the latter group had 13 times more single parents. [56] The explanations given of this phenomenon, the higher motivation of Asian pupils, their disciplined home life and the desire of their parents to see them succeed, have not been universally accepted. Another study [57] also showed that Asians fared better than West Indians, allegedly because of the different family climate and lack of basic skills of the West Indians, and even underprivileged white children outshone West Indian pupils. The former principal of a school in England composed largely of immigrants, who was accused of 'racism' in a *cause célèbre,* ascribed the comparative failure of West Indians to their loose family structure and different values, and to the publicity given to the efforts of radical teachers to stir up trouble. [58] Others have held that Afro-Caribbeans are more economically deprived, and this bears down on achievement. On the other hand a report to the Runnymede Trust [59] showed Afro-Caribbean pupils took school more seriously than white pupils, and their mothers had a great interest in their children's schooling and future career. One explanation, which was advanced by Jensen in the 1970s in the United States, that Blacks were genetically not so intelligent, was rightly categorically rejected. On this side of the Atlantic, Hans Eysenck, also favouring genetic explanations, did dispel the myth of IQ as an explanation: in 1971 he averred that classifying "Blacks" and "Whites" on that measure alone would be tantamount to tossing a coin. Rex and Tomlinson [60] declare that tests have shown there is no proof of "hereditary differences in intellectual capacity between ethnic and racial groups".

This brief review of some of the research shows that social factors have remained predominant as an explanation for underachievement in the case of West Indians, but the nature of these has still to be clarified. On the other hand an OECD report [61] declares that school failure is a

---

[56] Lewis, *op. cit.*
[57] A. Little, "The Background of Under-achievement in Immigrant Children in London", reported in: G. Verma and C. Bagley, *Race and Education Across Cultures,* London, 1975.
[58] Cf. B. Troyna and J. Williams, *Racism Education and the State,* 1986, p. 96. This is a seminal work for the study of the relationship between racism and education.
[59] Cf. P. Gordon, "The Rise of Racism in Europe", In: *Fortress Europe? The Meaning of 1992,* The Runnymede Trust, London, 1989.
[60] Rex and Tomlinson, *op. cit.*, p. 166.
[61] OECD/CERI, *One School, Many Cultures,* Paris, 1989, p. 60.

psychological trait ascribable to lack of confidence and that the school has to exercise a therapeutic function. British research findings upset Caribbean parents greatly, particularly when it was revealed that a disproportionate number of West Indian pupils were located in schools for the educationally sub-normal – four times as many in ESN schools as the proportion of them in schools would appear to warrant. Laudable efforts were made privately to start up 'Saturday schools' to remedy this state of affairs, with what success is not yet known. These private initiatives were in some respects a counterblast to the Saturday 'Koranic schools' set up by the Moslem community. The 'family' explanation for Asians cannot be discounted. In a study of a Pakistani community in Oxford, [62] it was shown that upwardly mobile Asians who enter higher education retain close social and financial links with their families; those with less education keep up stronger links still through the mosque, the events that mark stages in family life, and their work, mainly in factories; contact with native-born colleagues is not carried over to their leisure time.

In any case, it is likely that explanations cannot be generalized on a European scale, as the case of others, such as the Turks in Germany, makes plain. The Turks tend to be less middle class than Asians in Britain. Their educational aspirations are therefore not so high, as can be inferred from Table 7.8, which shows a break-down of the schools attended by their children in Germany in 1981.

Table 7.8: *Turkish Children in German Schools*
(1981, per cent of all Turkish children in Germany) [63]

Level and type of school attendance	Attendance %
Grundschule and Hauptschule	85.4
Special schools	4.8
Realschule	4.4
Gymnasium	3.8

Notes: (a) Grundschule = primary school. (b) Hauptschule = secondary modern. (c) Realschule = intermediate school. (d) Gymnasium = grammar school. (e) It should, moreover, be mentioned that combining "Grundschule" and "Hauptschule" may be misleading, because the former is compulsory for all children.

---

[62] A. Shaw: *A Pakistani Community in Britain*, Oxford, 1988.
[63] Source: Zentrale für Politische Bildung, Berlin.

The low proportion attending the *Gymnasium*, which remains the royal route to higher education, is symptomatic perhaps of a lower value placed by parents on education. A caveat is that the *Gymnasium* is highly selective, so that it is not only the aspiration of the parents but primarily a formal testing system that causes these allocations. Another factor may be that some of the Turkish students are urged at this stage in their lives to earn their living. It is alleged that the more uneducated immigrants are, the more likely they are to remove their children from school at the earliest opportunity. However, other surveys have shown that parents with a poor education nourish high expectations for their children once they become aware of the importance of the educational "ladder". Thus the evidence here is clearly contradictory.

We may perhaps be on firmer ground in seeking the causes of under-achievement in teachers' expectations of how pupils will perform. Research has shown that children tend to perform in direct proportion to these, and that those who teach immigrants do not have high expectations of their charges. [64] Some teachers resist the concept of multiculturalism as their main task. Others believe that all pupils should be treated equally – any form of positive discrimination is thus ruled out. They are more dedicated to an assimilationist policy.

Germany has seemingly grasped the nettle of positive discrimination more thoroughly than other countries. The federal nature of the educational system makes generalization difficult, but it would appear that the following kinds of experimental programmes have been mounted in the country: [65]

- support courses in German
- teaching in two languages of instruction at once
- special help with activities outside the schools, such as homework and the organization of leisure pursuits
- the use of special textbooks
- additional vocational training

12. Schooling and Employment

Up to now the interface of schooling and employment has not been discussed. Germans see the problem as one in which the solution lies in persuading immigrant youngsters to continue their schooling for as long

---

[64] M. Abdallah-Pretceille: "Multicultural Policies and the Consequences for Teacher Education" (OECD working document). See also M. Rey: *Training Teachers in Intercultural Education,* Council of Europe, Strasburg, 1986.
[65] Cf. J.J. Smolicz, *op. cit.*

as possible. At present only some 60 per cent of immigrants leave the *Hauptschule* with the requisite leaving certificate, a qualification that is practically indispensable for gaining an apprenticeship. Only one in five receives any vocational training. Even the compulsory part-time vocational school, the *Berufsschule* for youth in the group 15 to 18 years of age, in 1981, was attended by only 54 per cent of immigrants. Thus they have not been trained in what the Germans call the necessary *'Industrietugenden'*, defined as the virtues of discipline, attentiveness, the ability to concentrate and the capacity to take the long view. In an age of credentialism such qualifications are vital.

One German vocational programme for immigrants, however, deserves a brief mention here, the *Massnahmen zur Berufsvorbereitung und sozialen Eingliederung* (MBSE: Measures for preparation for an occupation and for social integration.) In Hesse this is a work-based course in which language teachers, vocational trainers and social workers cooperate. Between 1980 and 1984 over 40,000 young immigrants followed it, although girls were under-represented. [66] Nevertheless, some 35 per cent of migrants obtained employment at the conclusion of the programme, and 15 per cent obtained apprenticeships. Sadly, however, the fact that migrants are over-represented generally in vocational programmes of this nature implies that they are less successful in obtaining admission to courses that lead to more remunerative employment.

13. A Note on Teacher Education

Most teachers, with the exception of those dealing with languages, have not been trained to work with immigrants, and lack understanding of what multicultural and intercultural education entail. However, it has been estimated that in Britain courses on how to teach immigrants, if these were given in initial training, would be of value to only one in six teachers in their first appointments. In-service courses have been mounted, particularly in the late 1960s, but these have been poorly attended – participation was mainly voluntary – save for those teaching English as a second language. It is true that some 800 black teachers have been trained on special Access courses, but this only scratches the surface of the problem. The extreme view has even been argued, that migrants should teach migrants, since they alone have a grasp of the culture and language, but this might well defeat the purpose of integration and lead almost to a situation of educational apartheid.

---

[66] Source: Zentrale für Politische Bildung, Berlin.

Various proposals have been made as to how teachers could be better trained. It has, for example, been proposed that they should be given an understanding of the socio-psychological and anthropological mechanisms involved in the relations between majority and minority groups. According to an OECD source, [67] an ideal programme would consist of additional training in philosophy, cultural anthropology, linguistics, educational sociology and the psychology of learning. This would seem a very ambitious enterprise.

14. Concluding Remarks

This chapter has identified the problems associated with minorities and immigrants obliquely, partly because some, at least, of the solutions must be contextually specific. Others are capable of generalization.

The age-old problems associated with minorities within national boundaries, because they are so well-known, may today be more easily resolved. This is not the case with the problems following in the wake of more recent mass migration, the central question in which seems to be how to teach the children of the minority the skills and knowledge required to achieve success in a majority culture whilst at the same time preserving their distinctive linguistic and cultural identities. What has been set out above is a mere juxtaposition of the issues – with huge gaps in our knowledge of the general European context. Since this chapter has concentrated largely on France, Germany and the United Kingdom it has been thought useful to provide, in an annex to this chapter, information regarding the Netherlands, a country with a relatively small population but one that has nevertheless tackled the problem of immigrant education energetically. Not only the background of the Dutch position is summarized but a series of recommendations made by the Dutch Advisory Council, whose report has already been mentioned (cf. note 29), is reproduced. It is indeed through more detailed knowledge of how European countries are coping with the education of minorities in general, and immigrants in particular, that progress towards the solution of problems can be achieved.

---

[67] OECD/CERI, *One School, Many Cultures*, 1989, p. 75.
[68] ARBO *Een boog van woorden tot woorden, Utrecht, 1986*

ANNEX

Recommendations of the Dutch Advisory Council for Education with respect to the education of minorities [68]:

I   *Generalities* (pp. 20-21):
1. Recognition of equal rights for immigrants, who are here to stay.
2. Steps must be taken to avoid undue numbers of immigrant pupils being placed in special education.
3. Complete mastery of Dutch is essential for school success.
4. Immigrant educational policy is not free-standing: it must be worked out in conjunction with overall policy.
5. The goals set for immigrant education must be concrete.
6. Continuous assessment of progress must be made.

II   *Dutch as a "Second" Language*:
1. Policy must be flexible, varying according to situations.
2. A 'total immersion' model must be adopted.
3. But a distinction must be drawn between those under seven years old, or who were born in the Netherlands, and those arriving later.
4. For younger children playrooms (where Dutch can be learnt by play) should be made available.
5. An inventory of textbooks and materials for learning Dutch should be drawn up.
6. More statistical information should be made available.

III   *Instruction in "Own Language" and Culture*:
1. The aim is to support the identity of the ethnic group.
2. Statistics are required regarding present teaching of the immigrants' languages and cultures.
3. Some teaching also should be undertaken out of school hours.
4. Teachers, if immigrants themselves, should know Dutch well.
5. Cultural programmes should be worked out in detail, fitted into the overall school programme, and adequate teaching material made available.

IV   *Intercultural Education*:
1. The aims are: to get pupils to appreciate similarities and differences between cultures – to impart knowledge of the immigrants' background, circumstances and culture, and for

immigrants to acquire knowledge of the indigenous culture — learning to live together — the dispelling of racism, prejudice and discrimination.
2. Intercultural education should be backed up by activities outside the school.
3. Pupils must learn how to live in a multicultural community.
4. Regular reports by schools on their progress should be made.

V *"Concentratie-scholen"* (schools where either "black" or "white" predominate):
1. The need is to maintain parental choice, to foster the idea of neighbourhood schools, and provide immigrants with information about schools.
2. Where "white" and "black" schools exist side by side local authorities must take steps to rectify the situation.
3. Neighbourhood schools should be an accurate reflection of both the "white" and "black" population.

Chapter 8

School Leaving Examinations and the Role of the School in Preparing for Higher Education

1. Introduction

European countries are at present faced with the difficult task of managing change in the relationship between upper secondary and higher education. Change is reflected in two key development factors: enrolment growth and a simultaneous diversification of programmes. The traditionally close link between the academic secondary school and the university has as a consequence become weakened. Moreover, upper secondary education is now far less linked to higher education in general, as the upper secondary school is becoming a free-standing institution, with most students leaving school at 18 years of age, either postponing entry into higher education until later, or entering, on a part- or full-time basis, another form of tertiary education, or even going straight into employment. A number of policy problems and challenges of a technical nature arise as a consequence of the dual change involving increased student intake and a widening of study options. It has become imperative, for example, that the goals and aims of secondary education be reconsidered, so that the organization of upper secondary schools be adapted to meet new conditions.

Among the policy problems dealt with in the present chapter are, firstly, the setting of the goals and aims of upper secondary education, and secondly, the managing of increased social demand for both upper secondary and higher education. Furthermore, the policy dilemmas inherent in the strategy of expanding upper secondary education while at the same time attempting to channel a large proportion of upper secondary students into appropriate courses to meet labour market demands, and to achieve equity and accountability, are also considered. Technical problems discussed in this chapter relate to the development of an appropriate curriculum and the use of various criteria for testifying to the mastery of skills and abilities, such as different school leaving examination techniques employed for the allocation of places in higher education.

2. Goals of Upper Secondary Schooling

As mentioned previously in Chapters 2 and 3, secondary education has expanded dramatically during the 1970s and 1980s. From a policymaker's viewpoint, the overriding objective may well have been to raise the overall level of education in the population. From the standpoint of parents, however, the expansion was called for because they realized that attendance beyond lower secondary school was necessary in order for their children to qualify for jobs leading on to careers.

The structure of secondary education in Europe differs from country to country (see supplementary Table S.4), as does the breakdown of the school population by grade and type of education (Table S.5). In Austria and Germany, for example, secondary education usually starts with the education intended primarily for the 10 year-olds, compared with 13 year-olds in Denmark and Finland. The duration of secondary education is also variable, ranging from five years in Greece and Ireland to nine years in the case of the *Gesamtschulen* and *Gymnasien* in Germany. Secondary education is commonly divided into two stages, of which only the first tends to be compulsory for all children. Since the statutory school-leaving age varies from, for example, age 14 in Italy and Spain to age 16 in France, Belgium and the Netherlands, the duration of upper secondary schooling varies as well.

The most striking difference between the European countries does not concern the duration of secondary education, however, but the way the educational systems are organized. After completing primary education the children are sorted into groups attending different types of school in countries such as Belgium, Germany and the Netherlands. The full impact of early differentiation is moderated in these countries because the children who attend the first, or the first two, grades of secondary school – the *cycle d'observation* (Belgium), *Orientierungsstufe* (Germany) or *basisvorming* (Netherlands) – have, in principle, the right to transfer from one school type to another, provided that their performance warrants such a move. In other educational systems all children attend a common lower secondary school which, as in the cases of the Finnish *Peruskoulut* and the Swedish *Grundskola*, is considered to be part of first-level schooling.

Upper secondary education tends to be part of the post-compulsory system. This explains, at least to some extent, why the range of schools offering education at this level may be considered bewildering in a pan-European perspective. Yet, although the *concept* obviously varies across school systems, a high degree of communality appears to exist in the *objectives* and functions of upper secondary education in Europe. This can be established on the basis of legislative acts and official documents, such as the "Reports on the Development of Education" which most European countries occasionally prepare at the request of the International Bureau of Education in Geneva. Leaving aside rhetoric, the objectives of upper secondary education in Europe can be formulated as follows:

1. It should allow for differentiated provision, offering vocational training as a preparation for work and general education for personal development, taking into account the student's intellectual capacity and his educational and occupational interests;
2. It should equip all the students with the skills they need for effective communication in their own and foreign languages;
3. It should develop in all the students the ability and understanding they require for scientific literacy;
4. It should confer study skills as a basis for lifelong learning; and
5. It should offer orientation in the humanities and in the natural and social sciences as a preparation for university and other forms of higher education.

The latter goal, preparing for university, is by tradition the principal objective of the academically oriented upper secondary schools in Europe: *Oberstufenrealgymnasium* (Austria); *Cycle de détermination général* (Belgium); *Lycée* (France); *Gesamtschule* and *Gymnasium* (Germany); *Licei* (Italy); *Gymnasium* and *Atheneum* (Netherlands); and *Maturitätsschulen* (Switzerland). The point of departure for a discussion of the general eligibility qualifications for university entrance lies with the goals for undergraduate education. Study programmes at the initial level of higher education have, in many countries in Europe, become more vocationally oriented. Consequently, a distinction is now often made between *general* and *specific* eligibility criteria for access to courses in higher education.

Because of increased student mobility and international cooperation in higher education in the European region generally, it is desirable to discuss the elements of a common core of skills and abilities prospective students in European countries should have acquired. Such a discussion might be focused on the following criteria: [1]

1. General study skills, such as reading comprehension, ability to take notes and summarize texts, and independence in information-gathering and processing;
2. knowledge of modern languages, first and foremost English, but also French and German;
3. knowledge of elementary mathematics and statistics;
4. basic grounding in the human sciences, particularly in civics and social studies, psychology, economic and cultural geography and, to some extent, the history of ideas.

As noted in Chapter 3, contradictions arise between these goals for secondary education. These are apparent not least in the eligibility criteria used to regulate admission to higher education. One such criterion is the school leaving examination. The various uses to which school leaving examinations are put in different countries and educational systems in Europe are now considered. [2]

3. The Uses of School Leaving Examinations

Teachers have, in the main, established examination pass rates in practically every European country in the past. This was a powerful weapon in their hands, as it gave them a means of controlling the flow of students. However, the question may be posed concerning the criteria used for this control. A major study, conducted in France by Laugier, Piéron and Darmois in 1936, on the scientific validity of the marks given to the papers of the candidates, showed wide variations among the examiners; this research gave rise to the sub-discipline called "Docimologie" by H. Piéron. In the past 50 years, the pass rate at the *baccalauréat* examination taken in the 11th and 12th grade has been almost constant (around 67 per cent for each.) When the first part of the *baccalauréat* was abolished in 1960, the pass rate in the 12th grade remained at about 67 per cent. This raised the chances of being admitted

---

[1] See, e.g. Torsten Husén (1990), *Education and the Global Concern*, Chapter 13, in which the goals of the undergraduate curriculum are discussed.
[2] Differences as to the forms and functions of secondary leaving examinations are reviewed for six countries by Eckstein and Noah (1988). See also Noah and Eckstein (1989).
[3] OECD, 1990, p. 71.
[4] The source is OECD, *Education in OECD Countries: A Compendium of Statistical Information*, Paris, 1990, p. 71 and p. 91.

by one-third. The exception "*qui confirme la règle*", as the French like to say, came in 1968, after the student upheavals when, since no oral examinations were held, record numbers passed, thus presenting the government with a problem in the autumn, when many more students came knocking on the doors of the universities.

As hinted at above, the school leaving diploma or certificate awarded at the end of upper secondary schooling has traditionally been intended as a qualification testifying to the mastery of a certain level of skills and abilities. Now that in most European countries the great majority of the students in the relevant age group continue their education beyond the compulsory period, the school leaving examination has lost some of its former importance. As will be seen, some countries have even abolished the examination altogether. Yet the certificate obtained at the end of upper secondary school has retained its importance in some countries, particularly in those where it is a necessary condition for admission to university type higher education. The data recorded in Table 8.1 illustrate some of the major differences between the European countries in the importance they attach to the school leaving certificate. They show the number of students per thousand in an age group obtaining a qualification allowing access to higher education. Moreover, the data also indicate the number of new entrants per thousand students in educational programmes leading to the award of a full first university degree or an equivalent higher education qualification. It can be seen that the proportion qualifying varies from 25 per cent in Austria to 49 per cent in Greece. The estimate of the number of students obtaining a school leaving certificate in the Netherlands cannot be compared directly, as a distinction is made in this country between education qualifying for university admission, "voortgezet wetenschappelijk onderwijs", and education leading to a diploma, "hoger algemeen voortgezet onderwijs", conferring general eligibility for admission to non-university type programmes of higher education.

Variable pass rates in upper secondary education militate against convergence in examinations in Europe. In France and the United Kingdom about one-third of all candidates fail; in Germany and Austria the failure rate is less than 10 per cent. The pass rate is low in most European countries compared with, for example, Japan and the United States, where, according to the latest data available, three-quarters or more of an age group obtain a secondary school leaving certificate providing access, at least in theory, to higher education. [3] Cross-national variation with respect to the value attached to certificates awarded at the end of general or comprehensive secondary education, on the one hand, and technical and vocational education on the other, must also be noted.

Table 8.1: *Proportion of an Age Group Obtaining a Second-level Qualification Allowing Access to Higher Education, and Proportion Entering Higher Education (1986-1987 data)* [4]

Country	Age group in years	General or comprehensive[a]	Vocational or technical[a]	Total qualifying[a]	Total entering[b]
Austria	18-19	134	111	245	184
Belgium	18-19	195	201	396	—
Finland	19-20	424	—	424	177
France	18-19	213	111	324	200
Germany	19-20	284	—	284	174
Greece	17-18	485	—	485	180
Netherlands	18-19	303	262	565[c]	168[d]
Spain	18-19	248	96	345	304
United Kingdom[e]	18-19	278	83	361	175

Notes: (a) Number of students qualifying per thousand persons in the corresponding age group. (b) New entrants per thousand persons in educational programmes leading to award of full first university degree or an equivalent qualification. (c) Including those awarded qualifications not allowing access to higher education. (d) Universities and distance university only. (e) School leavers with at least 5 A-C awards at GCE O-level or equivalent are considered eligible for higher education.

Moreover, subsequent experience is also variable. As can be seen from the data, [5] in France, about 20 per cent of the relevant age group go on to traditional university education. In Austria, Finland, France and Greece, the proportion is 18 per cent, compared with 30 per cent in Spain. Yet the end-result is the same in, for example, France and the United Kingdom. The two countries produce roughly the same number of graduates because in France the attrition process is high during the first year at the university. By contrast, in Germany the attrition process goes on throughout secondary education, with the major drop-out occurring at the end of the sixth year, although nevertheless 28 per cent eventually emerge with the *Abitur*. In all countries considered – and this is in striking contrast to the situation a generation ago – the final destination may not be higher education. In Britain the successful may enter some type of further education, for example in polytechnic institutions; in France and Germany they may embark on technical courses – in the case of the latter country, even recently on apprenticeship training. Thus

---

[5] *Ibid.*

although pass rates have remained fairly constant over at least twenty years, goals have been diversified, largely due to changes in government policy. [6]

Politicians are becoming increasingly aware of how examination pass rates can provide a check for accountability, namely, how well pupils, teachers and schools are faring. Areas can be compared with one another, as can types of school (e.g., public versus private, selective versus non-selective.) Decision-makers are also, however, well aware of the public demand for credentials, i.e., documents certifying success. English figures show that only 12 per cent of the age group leave school without an attestation of examination success – even though it may only be at a very low level and in only one subject, such as cookery. Some of the certificates awarded may, from the point of view of employment, be hardly worth the paper they are printed on. But "credentialism" is not solely an English disease. [7]

Examinations are also customarily cited as being instruments for the maintenance of standards over time. In point of fact this puts an impossible demand upon their validity. Syllabus content and methods of examining change over the years, so that fair comparisons are difficult, not to say impossible, to make. In Europe since Napoleonic times the ultimate control of examinations has largely remained in the hands of government. Britain has been an exception, although since 1918 the central authority has made its influence stronger. [8] Because British examinations have been so closely linked with university entrance, the universities exercised the dominant influence through the examining boards they mostly created. [9] Recent moves towards standardization and rationalization of the organization have in fact been aimed at reinforcing central control. Elsewhere, since the leaving examination confers legal rights, governments have been paramount. Even when school education is a non-Federal matter, as in Germany, the central administration has had to intervene in order to ensure parity between the federal states.

---

[6] The variable role of grade repeating in the four systems of education must be noted.
[7] This point is made very well by Ronald Dore in *The Diploma Disease*, 1976.
[8] Ireland may also be an exception to this rule. The extent to which the governments of European countries have assumed responsibility and control over the school leaving examination is one of the issues dealt with by E. Egger (1971) in a review of examination practices.
[9] In Britain, there is one board, the Associated Examination Board, which has no formal link with a university.

4. National Differences in Admission Procedures to Higher Education

Different criteria are used in the regulation of university access in Europe. The following approaches are prominent among those used in different countries in the assessment of the qualifications of students seeking university entrance.

Some countries employ a model in which one uniform, national examination provides general eligibility for admission to all programmes of study in higher education. This method of assessing the qualifications of prospective students is, in principle, employed in Belgium, France and Germany.

Another way is to rank students according to criteria such as marks obtained in the final grade of upper secondary school, weight of courses, and even work experience. This approach, for example, has been taken in the Swedish system, where upper secondary schooling has become comprehensive and almost universal. The method of assessing the qualifications of students, usually in relation to school marks, is also employed in educational systems principally relying on uniform admission examinations in instances where the number of applicants for a given programme exceeds the number of available study places. In the case of Sweden, ranking according to school marks is combined with other criteria, such as work experience. Furthermore, students seeking admission are put in different quota groups. [10]

In a diversified system of education, such as that of the United States, use may be made of general scholastic aptitude tests in evaluating students applying for admittance to higher education. This model is not generally employed in Europe, although some *Länder* in the Federal Republic of Germany have in the past contemplated doing so.

It is possible, but difficult and controversial, to use work or employment experience as an additional assessment criterion. In practice, this means that age becomes a factor of importance in the screening of prospective students. This system is used mainly in some Nordic countries, although a variant can be found in Germany and the Netherlands, where prospective students can enhance their chances of being admitted by spending several years in succession in the queue. In this way they can improve their chances of being admitted to oversubscribed faculties such as medicine or law, notwithstanding an average performance record.

---

[10] L. Boucher (1982), S. Marklund (1988), U. Teichler (1988b) and A. Tuijnman (1990) discuss aspects of the Swedish approach to university admission, for example the quota system. See V.D. Rust (1989) for a discussion of the Norwegian situation.

Positive discrimination with regard to social background was for some time used in former socialist countries such as Poland and Czechoslovakia, where working class students seeking admission to higher education were favoured over those coming from comparatively better-off families. However, other criteria are also used, for example a student's performance in entrance examinations. The successful completion of upper secondary schooling in these educational systems normally provides the student only with a general eligibility to apply for university admission.

In Eastern Europe, including the Soviet Union, the condition for university or college entrance is a successful extrance examination, organized centrally by the State. These examinations are extremely competitive in some areas of specialization (i.e. medicine, law and science) and only a minority of the applicants pass them. (During the Communist era, working class children were positively privileged by the condition that at least 25 per cent of all the admitted students had to be from the working classes. However, even these children are required to sit for the entrance examination.)

The examination system used in Central and Eastern European countries is generally seen as a guarantee against nepotism: the most able students have a guaranteed access to higher education. The disadvantage of the system is that the schools and teachers are put under high pressure, mostly by the parents of comparatively well-to-do families, to offer the skills and knowledge required in order to excel in the examination. This phenomenon can also be observed in regions in Asia where adherence to a strong meritocratic policy has made the examinations for entrance to higher education very competitive. [11]

A lottery system is being used in Netherlands when there are too many applicants to higher education with more or less identical educational background and school marks. The lottery system is sometimes also employed in combination with other screening methods, for example interviews. [12]

---

[11] This applies especially to regions such as Japan, Hong Kong and Korea. It may be of interest to note that in certain countries, also in Central and Eastern Europe, students who excel in the national student competition are offered free access to the faculty and university of their choice, independently of their school performance.

[12] A student normally becomes eligible for Dutch higher education upon passing the school-leaving examination taken at the completion of the Gymnasium or the Atheneum. Students lacking such a qualification are required to pass a viva voce (colloquium doctum) examination. A decision is taken each year at the central level about the number of study places per degree course that are open to first-year students. Students are selected for courses subject to a numerus clausus by means of a lottery weighted so that the school marks a student obtained in the leaving examination are taken into account (Netherlands Ministry of Education and Science, 1989, pp. 46-47).

It is possible to base university admission on a combination of criteria such as the level and choice of subjects a student has taken in the upper secondary school. Where this type of information is of importance for university placement it is commonly used together with other criteria, such as an applicant's marks in an entrance examination or an aptitude test, school testimonials and, sometimes, an interview.

In France, and similarly in Italy, everyone having passed the *baccalauréat* can automatically be admitted to a university, but not to the other types of higher education. As a consequence, drop-out rates are very high in the first two years of university courses (40 to 50 per cent, according to the disciplines.) Successive reforms of the 1st two-year cycle, which leads to the *diplome d'études universitaires générales*, have with some success tried to restructure the curriculum and teaching methods in order to facilitate the transition from secondary *lycées* to university. The Dutch government, faced with the impossible tasks of increasing the intake into higher education, which is asked for by the public, and a demand to cut down expenditure on higher education, which is part of a general strategy of reducing the scope of government intervention and finance, is at present studying the feasibility of implementing a qualification at the intermediate level such as the French *diplome d'études universitaires générales*.

In countries where university admissions are centrally controlled and the quality of educational institutions within a certain category is, in principle, considered to be equal, students can readily be allocated to a course in a university where places are available. However, in countries where quality differs markedly between institutions, those who have completed an upper secondary schooling with good results tend to apply for admission to several universities. In the Netherlands where public and "private" universities exist, the demand is being put forward that institutions of higher education be ranked in a prestige hierarchy, as is the case informally in Japan. The most prestigious universities in Japan and the United States may even require students to take entrance examinations, for which they can prepare themselves years in advance. The same principle of admitting students to prestigious forms of higher education on the basis of proven merit and excellence is used by the *grandes écoles* in France, where one or more preparatory years in school are required over and beyond the *baccalauréat*, and there is a competitive examination. Those who complete a course in a university enjoying the highest esteem, for example the universities of Tokyo and Kyoto in Japan, are in high demand by employers in different labour markets, public as well as private.

English universities have full freedom to select their students. As a rule the minimum demand is that applicants have passed the General

Certificate of Education at Advanced Level in three subjects. Over and above this qualification, assessments and references from head teachers and interviews are generally used in the screening process.

Sweden stands alone in Europe in the use of a quota system based on the type of formal education and employment experience of the applicants. By dividing applicants into different quotas, and by allowing free competition for places on the basis of merit and other criteria to occur only *within* these categories, the system seeks to guarantee places also to those applicants who would be at a relative disadvantage if the admission procedures were based exclusively on academic merit. [13] That Sweden provides an interesting case-study of the way in which examinations have developed in Europe in recent years is an argument implicit in the paragraphs that follow.

It can be concluded on the basis of the descriptions that there are marked differences across countries in the organization of both upper secondary and higher education. The role of the upper secondary school in preparing and selecting students for higher education varies also across European countries. It is clearly evident that no universally satisfying procedure for admitting applicants to higher education has yet been devised on a European scale. This leads one to wonder about the value of this objective, and whether it can be achieved in the future.

5. Examinations in Europe: The Case of Sweden

The traditional academic secondary schools in Europe prepared an intellectual and social élite for university education. Universities in most countries had the right to assess the knowledge of prospective students without any interference from the State. This was the situation in Sweden until the 1860s. A matriculation or "maturity" examination was introduced at that time, as it was in other educational systems inspired by German university traditions. This examination was prepared and monitored by central educational authorities. It consisted of uniform written tests in some subjects. Oral examinations were administered by the secondary schools themselves. So-called "censors", as a rule university professors, were in attendance when these oral examinations were conducted.

---

[13] However, the results of recent evaluation studies show that social bias in university admission and enrolment has not been reduced to the extent anticipated by advocates of the - much publicized - reform of Swedish higher education implemented in 1977 (Vogel *et al.*, 1988). Some explanations for this apparent paradox are provided in a critical study of higher education in Sweden conducted by Ulf af Trolle (1990).

This system of assessing the knowledge base of prospective university students with a uniform "matriculation" examination was abolished in 1968. Until then the purpose of the *Gymnasium* had principally been to prepare students for higher education, since the majority of those who passed the *studentexamen* continued either to universities or to colleges of higher education. The 1905 statutes of the *Gymnasium* stated explictly that the purpose of schooling at the upper secondary level was to "confer the basic scientific insights that are further developed at the university or college". [14] The 1933 statutes specified that the *Gymnasium* was to equip young people not only with the general education they needed for continuing with university studies but should also provide a vocationally oriented education of a somewhat briefer duration at a level below that taught in a traditional *Gymnasium*. This change was introduced because a minority of the students who passed the matriculation examination entered occupations for which preparation in institutions of higher learning was not required.

About 25 per cent of an age cohort were entering the upper secondary school at the time when the matriculation examination (*studentexamen*) was abolished in Sweden, compared with only four to five per cent at the beginning of the 1930s. Lower secondary education had become universal as a result of the comprehensive school reform of 1962. The reform created an "enrolment explosion" in the new upper secondary school which, in a very short time, was expanded so as to cater for the vast majority of children aged 16 to 18 years. 22 different programmes, of which most were vocationally oriented, were introduced into the upper secondary school. The different vocational schools were integrated with traditional academic schools at the upper secondary level into this new common *gymnasieskola*.

In Sweden, the recent decision about the prolongation of vocationally oriented studies from two to three years is intended not only to accommodate workplace education but also to broaden the curricula by increasing the proportion of general subjects, such as mathematics and languages. This Swedish development illustrates a general trend in European educational systems at present, in which the importance of general rather than vocational curriculum elements is emphasized. [15] These elements of study include work experience, information technology and personal and social education. The objective is to improve the communication skills of students. Another objective is to

---

[14] It is acknowledged that the use of the term "scientific" is not altogether unambiguous, as its meaning varies from country to country.
[15] According to UNESCO statistics (cf. statistical appendix), England may be an exception. Yet there is also a parallel development in this country, as there is a policy of incorporating "core skills" into the secondary curriculum.

enable students to be better informed of academic and career possibilities beyond basic schooling. As was the case in the United States, expansion of upper secondary education in Sweden has resulted in a rapid increase in demand for tertiary education. Reforms are introduced at the upper secondary level to give more options to students having reached 16 years of age. The principle that the educational system should offer meaningful educational alternatives to all compulsory school leavers is emphasized. As a result, an attempt is made to adapt the working methods, content and organization of the upper secondary school to the diversified needs of all young people continuing education after compulsory schooling.

If the Swedish precedent is anything to go by, then European countries experiencing an explosion in numbers as regards upper secondary education will also have to face a corresponding demand for higher education. Yet despite high demand, the data included in Table 8.2 show that the enrolment level of Swedish youth aged 18 to 20 years is low compared with Belgium, Denmark and Finland, France, Germany and the Netherlands. [16] Less than 10 per cent of the 20 year-olds in Sweden were enrolled in 1986/87, compared with 24 per cent in the United Kingdom and 30 per cent in Switzerland. Part of the explanation is that the Swedish government has, since the early 1970s, pursued policy to divert young people from transferring directly from upper secondary school to the university. [17] "Diversion" policy aimed at containing social demand for university education has been successfully implemented since, in 1986/87, 44 per cent of the student body in Swedish higher education was 24 years of age or younger. [18] However, the critical question is whether diversion policy can be considered beneficial, and if so, to whom. [19] The data in Table 8.2 show that the age-specific enrolment rates in Sweden are exceptionally low, particularly those of the group 18 to 22 years of age. Yet, as will be seen in a subsequent section, the problem of coping with demand for higher education is being experienced in all European countries at present.

---

[16] Source: OECD, *op. cit.*
[17] Diversion policy may be considered a translation of "avlänkningspolitik" ('shunting-off policy'), a concept introduced by Murray (1988) in his description of the elements and consequences of policy aimed at containing "excessive" demand for higher education by diverting young people from education to work.
[18] Source: Statistics Sweden (1990).
[19] This critical question is raised by Tuijnman (1990), who concludes that diversion policy may be in the interest, at least in the short term, of government administrators and educational planners. However, neither students nor higher education institutions would seem to benefit in the long run.

Table 8.2: *Enrolment Rates for Ages 16-24, Full-time and Part-time Students (1986-1987)* [20]

Country	16	17	18	19	20	22	24
Austria [a]	85.6	78.1	47.7	23.0	16.2	13.7	12.5
Belgium	92.7	86.4	68.2	53.0	46.0	22.5	12.0
Denmark	89.9	75.4	68.0	52.3	36.4	24.1	20.0
Finland [b]	92.1	83.3	64.8	26.4	22.4	21.6	17.6
France	87.9	79.7	60.1	43.0	28.1	13.6	7.3
Germany	100.0	99.7	83.9	57.2	36.6	28.3	15.4
Greece [a]	75.1	58.7	43.2	28.6	29.0	12.2	9.3
Ireland [a]	82.0	64.7	40.8	24.1	17.6	6.6	2.0
Netherlands	92.9	78.3	60.4	43.2	31.9	17.7	11.1
Norway	86.7	76.2	61.6	30.1	23.4	24.1	18.2
Spain [c]	62.3	53.1	28.6	15.0	—	—	—
Sweden [d]	81.0	83.0	43.3	21.7	—	12.4	11.4
Switzerland [a]	85.2	83.1	74.5	50.6	30.2	16.9	13.0
Turkey [a]	35.9	31.5	15.4	8.5	—	—	—
United Kingdom	68.9	49.3	32.9	—	23.8	—	—

Notes: (a) 1985/1986 data. (b) Includes only university students from age 19. (c) Excluding students in technical and vocational education. (d) Full-time students only.

6. Insufficiencies of the School Leaving Examination

We have witnessed a progressive restriction of the prerogatives that the leaving examination has conferred. For example, in Germany the leaver with an *Abitur* from the classical *Gymnasium* could once, despite having studied little science, embark upon medical studies or enter a technological university. The French *baccalauréat* in a similar way was the passport to *all* forms of higher education. Today this no longer holds good in either country. In some countries it was never the case. Spain and Belgium had their own university entrance qualification following upon the *bachillerato* and the *certificat de fin d'études secondaires (het getuigschrift van hoger middelbaar onderwijs.)*

In France, in spite of several university reforms (including the "Loi Savary" of 1984), and in spite of the debate on this principle, the right of the holder of the *baccalauréat* to a university place has never been

---

[20] Source of data: OECD *op. cit.*, 1990.

modified. However, all other types of higher education, including the short cycles of a vocational type, are selecting the applicants. This dualist system of admission, especially with respect to the *grandes écoles* and their "preparatory classes", which select the best 15 per cent of the *bacheliers*, has until now remained unchanged through the wave of post-War reforms. The increased mobility of European students through ERASMUS and similar programmes might lead to changes in the traditional cleavages in higher education, in France and possibly also in other countries.

Germany sorts students through a central registry system, allocating places according to norms and conditions of different kinds. Although France, Germany and Switzerland have preserved the constitutional right of the holder of a secondary leaving certificate to a university place, no guarantee is given of where, in what subject and even when, that right may be exercised. The original purpose of the qualification, to certify fitness for higher education, has thus been only nominally preserved.

As distinct from admission to higher education, the attitudes of employers regarding the leaving qualification are in general consistent with the convergence postulate described in Chapter 1. In France and Britain it would seem that potential employers of qualified school leavers do not enquire too closely into the standard reached in the examination but pay more attention to content and the formal level attained. In other words, the examination is used as a mere filtering mechanism. To be included on the list of candidates for a job, it is sufficient to hold the qualification – "*y nada más*". Indeed until recently, except in Sweden, where employers have taken an active part in designing the structures and curricula of upper secondary education, industry and business seem to have been much more interested in incommensurables, i.e., character traits such as persistence, consistency and thoroughness, or personality traits such as the ability to get on with others. [21]

---

[21] Urban Dahllöf's (1963) innovative analysis of the demands on the Swedish upper secondary curriculum on the part of universities and colleges, government administration, and the business sector, which was put to good use in the redesign of the curriculum for the new upper secondary school in Sweden, is still outstanding today, as it presents the most promising approach yet to solving the policy problem of balancing divergent demands on the upper secondary curriculum, particularly those of employers. This example illustrates that there may well be discrepancies between the needs of higher education and demands expressed by employers.

7. Leaving Examinations and the Syllabuses

Perhaps the greatest technical problem examinations have had to solve over the past twenty years concerns the rapid growth of new knowledge, elements of which one way or the other have to be covered in upper secondary and higher education. This has come about in at least three ways. Firstly, new subjects such as computer education, industrial design, economics, business studies, and even psychology, have been added to the curriculum. Secondly, there has been an "explosion" of knowledge, particularly in scientific subjects. Moreover, it is now deemed feasible to teach somewhat complicated aspects of science at school level, such as elementary atomic physics. Whereas thirty years ago, the teaching of calculus in European schools was comparatively rare, today it is commonplace. Thirdly, new subjects and new knowledge inevitably call for displacement of knowledge previously included. Twenty years ago topics like heat, light and sound were main topics in the physics syllabuses in Europe. Today they occupy a much more subordinate place. Some subjects traditionally taught have either faded away or are learnt by only a select few. First Greek, and then Latin have in many countries practically disappeared. History and geography were devalued – although there are now signs that the former is making a comeback. Humanities in general were merged into a blend of social studies – *maatschappijleer* in the Netherlands and *Gemeinschaftskunde* in Germany, for example. In spite of the drive towards European integration, Britain has reduced modern languages, probably because English for the time being has become a *lingua franca*.

The question of a suitable syllabus content became acute in the 1970s. The implicit solutions provided by the application of the curriculum theories developed by Bruner and Bloom, which crossed the Atlantic about this time, never took on in continental Europe, although for a while they were in vogue in Scandinavia, the Netherlands and the United Kingdom. [22] Theories of "general culture", of élite secondary education being for "life" rather than mere "living", such as Capelle's [23] idea of the "languages" of knowledge, or German ideas of paradigms –

---

[22] Instructional models of concept attainment are often derived from early studies conducted by J. S. Bruner (1960) and Bruner *et al.* (1967), in which the processes by which people recognize and group the essential features of an event or thing into a cohesive category are analyzed and described. Likewise, the taxonomy of educational objectives developed by B.S. Bloom *et al.* (1956, 1964) has been an influential source of inspiration for many educationalists on both sides of the Atlantic, particularly during the 1960s.

[23] J. Capelle (1964) developed a theory of learning based on the idea of "communicators" in his *L'École de demain reste à faire*, translated with an Introduction by W.D. Halls (1965): *Tomorrow's Education: The French Experience*, Oxford.

exemplification theory and "overlapping knowledge" – were ousted by the pragmatic or economic principles applied to school policy. If, for example, science syllabuses were too large, they should be divided and become specializations in their own right. Hence the formidable growth in the number of "sub-lines" within the Swedish upper secondary structure, which originally comprised six "lines" only, and in the number of options in the French technicians' baccalaureate. For students there is "*die Qual der Wahl*" – the agony of choice. Under such euphemisms as "education for working life", what is offered today in upper secondary schools is increasingly based upon the market place "principle". Educationists may propose, politicians dispose.

So far it may seem that tendencies in examination programmes favour convergence in educational policy in Continental Europe. However, this may be only superficial. One matter to be resolved is the lack of precision in stating what has to be taught. Whereas the Netherlands, and France to a lesser extent, prescribe in detail, countries such as Germany and Sweden, allow greater latitude by merely laying down general guidelines. A much more difficult problem, however, is the philosophy behind the compilation of the programmes.

Thus the well-known English "specialization in depth" means that the pupil will study three, more rarely four, subjects drawn from one of three groups, namely, the humanities, the sciences and the social sciences. Although choices are permitted across group boundaries, this is not the rule and runs counter to tendencies elsewhere in Europe. Thus a student in a sixth form in Britain choosing English, French and, exceptionally, mathematics at A-Level, follows the same specialized mathematics programme as the one who has chosen chemistry and physics as well as mathematics. The contrast with France or Germany is apparent. The underlying rationale of these two countries' programmes is that of the *dominante* or *Schwerpunkt*, – the emphasis or "bias" within a particular group of subjects. For instance, in France the future university mathematics specialist will have (Section C) mathematics as his "dominante", but will continue also to take subjects such as modern languages, and history and geography. Thus in his last year he may well attend nine hours of mathematics, and three hours each of the other two subjects (history and geography count as one subject.) He will of course also be studying "scientific" subjects. By contrast, a pupil studying for a technician's *baccalauréat*, with mechanics as his emphasis, will have twenty hours workshop practice, but only three and a half hours of mathematics, together with some history and geography, but no modern language. Moreover – and this contrasts strongly with English practice – his mathematics syllabus will be geared to his "dominante", mechanics, and differ considerably from that being followed by the future mathematics specialist just mentioned. Likewise his syllabus in history and geography will be different.

The continued expansion of knowledge – it is alleged that scientific knowledge is doubling every decade – renders very difficult not only the choice of the combination of subjects to be studied, but also the content of the syllabus for each subject. Perhaps the Swedish expedient, in which there is cooperation between teachers in higher and other post-school education, public and private employers of labour in the civil service, industry and commerce, trade unionists and other representatives, offers the most imaginative compromise to date in the devising of "relevant" courses. All this, however, would imply rethinking teaching and examining methods, to which we now turn our attention.

8. Pedagogical Problems and Examining Methods

Some practical obstacles to convergence in secondary education should be mentioned here. The first concerns the *length of secondary schooling,* which varies from six years (Belgium and Sweden) to nine years (Germany). It is clear that the examinations rounding off such studies vary according to the duration of the course. A related concern is that the number of hours of teaching provided in a particular subject also differs from country to country. *Syllabus diversity,* which depends on different historical, cultural and economic conditions, has already been touched upon. *Grading* also varies from one country to another: the British have a five-grade letter system, France marks from 1-20, whereas Germany and the Netherlands have a points system. Moreover, standards of marking may differ for very good reasons, as does the latitude allowed to individual markers. Some systems take the candidate's record into account: others do not. France produces the results of the *baccalauréat* quickly because much discretion is allowed to the individual jury; England, with its cumbersome structure of examining boards, may take up to three months to publish results.

Examining procedures in Europe continue to be rather conservative. The essay-type question is still the rule, despite its notorious disadvantages, such as "luck" in the choice of question and, above all, variability in scoring, which Hartog and Rhodes commented on as early as 1935 in their *An Examination of Examinations,* and which was confirmed in the French Carnegie Report of 1937. [24] At one stage

---

[24] R.J. Hartog and E.C. Rhodes (1935). See also [Carnegie Foundation] International Institute Examinations Inquiry, *Essays on Examinations*, London, 1936. The fact that examinations were being questioned as to their validity and reliability already half a century ago demonstrates that this is not a new problem.

Luxemburg even adopted triple marking in order to guard against such variability. On the other hand, the multiple-choice questions common in the United States, which are used to achieve objectivity in marking – although subjectivity on the part of the question setter is of course not eliminated – are viewed with suspicion, despite the fact that they may well allow a broader spread of the syllabus to be examined. Against the use of multiple choice questions it is argued that the items do not test powers of judgement or the ability to pursue a logical argument. There are also alleged "coaching" effects. Another disadvantage may be that multiple choice favours convergent thinking.

The tradition of appending an oral examination to written papers continues, although the multiplication of candidates for examinations has considerably curtailed this system of assessment in recent years. In the name of greater fairness, such as to counter the effect of a candidate going through a "bad patch" at the time of the written examination, forms of continuous assessment are not infrequent, in spite of the risk of teacher bias. In some German *Länder*, candidates are passed or failed on their work throughout the course, but if doubt arises, or the candidate thinks the teachers' judgement unfair, a formal examination can take place. The Swedes are reverting to a more traditional form of competence assessment, in part presumably because their previous evaluation process hung too much on the teachers' assessment. [25]

Moreover, considerable standardization in pass rates would also have to occur before a greater congruence between the national examinations could be a reality. [26] In any case present examining methods, it has been argued, do not discriminate sufficiently between candidates. In a high-level examination given to some 25 t0 40 per cent of the 18 year-olds, of which the majority may go on to higher education, 'fine tuning' is essential. Present methods are to be seen as blunt instruments rather than precision tools. In European countries where places in higher education are still limited, this is a grave disadvantage. In the United States and Japan, where many more places are commonly available, and supply can even exceed demand in some places, no great damage need arise from this lack of discrimination, although it may signify lower standards.

---

[25] The university admission procedures used in Sweden are being revised at present. Less emphasis is currently put on working life experience while more significance is attached to, for example, school marks. In future all applicants to higher education will be required to take an admission examination. There is also a discussion in Sweden as to the desirability of reintroducing traditional, strict ways of assessing student achievement in the upper secondary school.

[26] The pass rate depends of course to a high degree on the proportion of the age group sitting for the examination. Where a "filtering out" occurs at age 16, for example, the proportion of those continuing in school for a further two or three years who are successful in the final leaving examination can be expected to be greater.

9. Coping with the Rising Demand for Higher Education

Closely linked to the foregoing is the use of examinations for regulating the flow of students from the different "tracks" in the secondary school into institutions of higher education as well as certain occupations. This facet of policy has come increasingly to the fore in recent years. When entry to secondary education was the initial gateway to higher education, governments, by selection, could control numbers very effectively. The centre of gravity in the selection process has since moved to the school leaving examination, now that secondary education is open to all. However, the pass rate for this examination is one that, by and large, teachers, not governments, control. To limit access to higher education, governments have therefore had to resort to other expedients, for example by erecting post-examination barriers. In the Netherlands and Britain the recent cuts in the level of financial support for students may have the effect of restricting demand.

It can be concluded that in industrialized, technologically advanced societies, it is no longer possible to run a system where the sole criterion for admission to higher education is based upon a uniform certificate testifying to the successful completion of upper secondary schooling. Since upper secondary schooling now tends to become nearly universal in the most developed European countries, institutions of higher education cannot be expected to handle the large numbers of entrants that would result if completed upper secondary schooling were the only criterion employed in the screening of applicants. Secondly, courses differ in attractiveness and in the skills and qualifications they demand of students. Medicine provides probably the best example of this, as the countries in Europe that have applied the *numerus clausus* principle have tended to introduce this rule first in medical faculties. Thirdly, systems of higher education have become so diversified that the applicability of a general entrance qualification based on completed upper secondary schooling must be considered to hold limited value. However, experience has indicated that it is still advantageous if all students have acquired a common core of skills and competencies in upper secondary school. This is to guarantee that all university students adhere to a common intellectual framework of reference – regardless of the disciplinary and vocational orientation of the course they may actually embark upon.

The Americans have managed to cope with a comparatively high demand for tertiary education by allowing it to diversify in accordance with the requirements of the labour market. Thus many students are accommodated not in the traditional universities but in clearly vocationally oriented community colleges offering only two-year study programmes. Moreover, most mature students entering higher education in the United States after some years of work enrol in community colleges, so that the majority of the students are often between 35 and 40 years of age.

Selecting students for prestigious and attractive courses is relatively easy (even though difficult in an absolute sense) because it is not so hard to discover whether and to what extent an applicant has acquired the knowledge and posesses the intellectual ability that are prerequisites for success in higher education. The problem is rather to identify the criteria that should be used in predicting and validating which students are likely to be successful in the occupation for which they are trained. Furthermore, a sound selection of students means that a university can raise its internal efficiency, as more students can reach the end of the study programme within the appropriate time.

On the contrary, efficiency indicators relevant to success in working life are more difficult to establish, among other reasons because a substantial proportion of those who complete a course in higher education end up doing jobs they were not trained for. Not surprisingly, this applies especially to students in the humanities and social sciences. This was the case even in former socialist countries with a planned economy, where admission quotas were set centrally, in accordance with the demand for skilled labour in certain sectors of production as predicted by manpower planning models. Such demand models have as a rule not been very accurate. Even in democracies such as Germany and Sweden, one finds examples of the gross inadequacy of educational planning, for example in the notorious failure to match the intake of students into teacher training institutions with the expected demand for teachers a few years later. [27]

Examinations in European educational systems equivalent to the *Abitur* in Germany, the *baccalauréat* in France, *studentexamen* in Sweden, and the General Certificate of Education, A-level in England confer eligibility for applying to a university for admittance. As been noted, although Sweden has abolished its matriculation examination and has made an effort to decentralize control over the content of higher education courses, the examinations marking the end of upper secondary schooling are still in use in other European countries as a means of regulating the flow of students. Yet the leaving examination has, on the whole, become less of a sufficient condition for university admission in the great majority of European countries. This explains why the question of the role of the upper secondary school in preparing for higher education is being urgently asked in Europe today.

---

[27] The "classic" case illustrating the difficulty of forecasting accurately the demand for educated teachers on the basis of simple extrapolations of available demographic data and trends of enrolment can be found in *Die deutsche Bildungskatastrophe* by G. Picht (1964).

10. The Upsurge in Technical and Vocational Education

The growth of technical and vocational education may well tend to change the character of leaving examinations in the future. Their previous "academic" hallmark and general character will to an increasing extent be supplanted by vocationalism. The expansion of the range of *Fachreife* in Germany, and of technician baccalaureates in France, and the recent discussion in Britain to link a new certificate of vocational education with a revised form of the A-Level examination, are symptoms of this.

That the share of vocational and technical courses in total upper secondary provision has generally increased in Europe since the mid-1970s can be seen from Table 8.3. It is interesting to note that the trend in Japan seems to be reversed, as the share of vocational education has been decreasing gradually since 1975. The United Kingdom and Japan now have about a quarter of their students in vocational courses. Additional data on the share of vocational education in secondary and post-secondary education are presented in the statistical annex. Here it can be seen that this share is high in former socialist countries such as Albania (70 per cent), Bulgaria (60 per cent), Czechoslovakia (59 per cent), Hungary (74 per cent) and Romania (93 per cent). However, the share of vocational education at the secondary level is comparatively low in the Soviet Union (16 per cent). There are indications at present that the level of provision of vocational secondary education is dropping in Central and Eastern European countries. The expectation is that this tendency will continue in the coming decade, perhaps with the exception of the Soviet Union.

The proposals made by Pierre Chevènement, a recent minister of education in France, whose plans were endorsed in principle by his successor, René Monory, provide a striking example of the "rising tide of vocationalism". [29] Chevènement's timetable ran as follows: by 1987 there would be a French baccalaureate with eight different sections; by 1990 there would be a new and expanded technological baccalaureate offering twelve options in a two-year course, together with a practical vocational baccalaureate with 30 options. Moreover, up to 80 per cent of all pupils would be following one or other of these courses. [30] This is in contrast with Britain and, indeed, the Soviet Union, both of which are preoccupied with the right relationship between school and working life, where change is slow. Higher education institutions such as the German *Fachschulen*, the French IUTs, the Dutch institutions of *Hoger*

---

[29] Cf. W.N. Grubb (1985). See also UNESCO, *The Integration of General and Technical and Vocational Education*, 3 Vols, Paris, 1986.
[30] Cf. J.-P. Chevènement (1985).

beroepsonderwijs, and the English Polytechnics, which largely date from the 1960s, have at last begun to spread their influence downwards to the schools. Exponents of convergence theory in education can point to the technical and vocational sector in upper secondary education as the one where most international consensus as to the future seems to exist.

Table 8.3: *Students in Vocational Education at the Secondary and Post-secondary Level (as a percentage of total enrolment)* [28]

	1975 %	1980 %	1986 %
Austria	25	33	54
France	35	36	40
Germany	32	33	48
Japan	34	32	28
Netherlands	51	53	57
Sweden	45	45	50
United Kingdom	16	18	26

11. Links between Secondary School and Higher Education

As was noted in previous sections, the links between the upper secondary school and the university system were once very close, since the purpose of upper secondary schooling was mainly to prepare university students. During the 1950s the great majority of German *Abiturienten* continued on to higher education. The university was involved in discussions concerning the design of curricula for the upper secondary school, and in preparing examinations. It was even not uncommon to find teachers with doctorates in these schools. This has become much rarer, and is symptomatic of the weaker link between schools and the traditional university.

The expansion in lower secondary schooling in the 1950s and 1960s also led to that of upper secondary education. The internal growth dynamic was, in a number of European countries, so strong that upper secondary education in some countries became available for the majority in a relatively short time. Educational planners in some countries anticipated an enrolment in upper secondary school of some 80 to 90 per cent of an age cohort. As the upper secondary school would thus have to

---

[28] Source: UNESCO, Statistical Yearbook 1989.

accommodate a broad span of interests, abilities and learning potential, it was imperative that the provision of education at this level should become diversified.

Because one had to provide room for new school subjects, especially in the vocationally oriented upper secondary school, the core of common subjects traditionally taught in the academically oriented school could not always be maintained. The result has been that upper secondary education has tended to develop into a continuation of a basic type of schooling. The majority of upper secondary students today have no intention of applying for admission to universities or colleges.

The result is that the mutual interests between the upper secondary school and the university have weakened. Many structural links that used to exist between them have disappeared altogether over the years. This development is apparent also as regards the orientation, methods and the curricula of teacher training. It is common today to hear complaints among university staff about lack of academic preparation among students coming from upper secondary schools. This criticism has focused, first and foremost, on the inability of students to handle both their own and foreign languages. University teachers are, quite naturally, discipline-oriented in their choice of content and methods of instruction, whereas teaching in primary and secondary schools today often follows an integrated and cross-disciplinary approach.

12. Efforts towards Cross-national Convergence

During the past quarter of a century in Western Europe a number of attempts at "convergence" of examinations have been made, by substituting for national examinations supranational ones, such as the one-time NATO Diploma. Even though these efforts to arrive at a degree of harmonization in upper secondary education have all involved the setting up of *separate examinations,* they are well worth considering – particularly in view of the harmonization objective that may well be considered central to the ERASMUS programme set up by the EC Commission with the aim of increasing international student mobility. Table 8.4 shows that the foreign enrolment ratio was still low in EC member countries in 1986-1987 (with the exception of Luxemburg, which is a very special case, as well as Ireland, Portugal and Greece.) In numbers, however, those enrolled abroad were a large group.

Table 8.4: Higher Education in the EC Countries (1986-1987) [31]

Country	Number of domestic higher education students	Higher education students per 100,000 population	Number of students enrolled abroad	Foreign enrolment ratio (%) [a]	Distribution of higher education students			Distribution of higher education students by field of study	
					University and equivalent (%)	Other tertiary education (%)		Social sciences [b] (%)	Applied sciences [c] (%)
France	1,327,771	2,395	13,305	1,0	75	25		X	X
Germany	1,579,085	2,592	24,867	1,6	87	13		49	51
United Kingdom	1,068,386	1,880	15,354	1,4	34	66		49	51
Netherlands	399,786	2,749	6,980	1,7	43	57		62	38
Belgium	254,329	2,566	3,699	1,5	41	59		57	43
Ireland	70,301	1,979	2,835	4,0	56	44		45	55
Luxemburg	843	232	3,709	440,0	0	100		54	46
Denmark	118,641	2,314	1,879	1,6	79	21		56	44
Italy	1,141,127	1,995	18,478	1,6	99	1		55	45
Spain	954,005	2,542	10,310	1,1	94	6		70	30
Portugal	103,585	1,020	4,711	4,5	68	32		59	41
Greece	197,808	1,987	34,049	17,2	59	41		48	52

Notes: (a) Col. 4 = (Col.3 divided by Col. 1) times 100. (b) Includes education, humanities, arts, law, social science, business, mass communication, home economics and related fields. (c) Includes natural science, maths, computer, medical and health science, engineering, architecture, commerce, industrial, transport, agriculture and related fields.

---

[31] Source: UNESCO, *op. cit.* Data for 1985-1986 are presented by G. Psacharopoulos in an article dealing with higher education in the EC countries, *European Journal of Education*, Vol. 25 (2), 1990.

At least two examples of "forced" convergence can be cited, arising from what has been described as "transaction theory", i.e. an educational need arising from mixed national groups thrown together for a common purpose. Thus in the early days of NATO, when France was still a member, a secondary school-leaving examination entitled the NATO Diploma mentioned above was created, which entitled children of NATO personnel who passed it to enter higher education in any member country. How widespread it was is not known, although it was taken at such schools as the then NATO school at St-Germain-en-Laye, in France. Another example is that of the European Baccalaureate, which is still taken by children of EC officials in the European Schools, and which confers similar privileges. [32] Both examinations were to a large extent modelled on the French *baccalauréat*.

One should in this context cite a case of "natural convergence"; the creation by private groups of the International Baccalaureate, which from being initially a European phenomenon, has become worldwide. [33] In certain private schools, such as the *École Internationale* in Geneva, or Atlantic College in South Wales, the first of the United World Colleges, in order to avoid pupils at a crucial stage being forced to go their separate ways and take their own national examinations for entry to higher education, the International Baccalaureate serves a useful purpose. To set up the examination involved the development of a genuinely international syllabus, the recruitment of an international team of examiners, and, above all, the achievement of recognition by governments and universities.

Another promising experiment, however, failed to reach fruition. In the mid-1960s circles within the Council of Europe thought that it might just be possible to set up a Council of Europe baccalaureate for use in its member countries. It was argued that this would facilitate student movement in Europe. The postulate was that *natural* convergence was occuring. At the upper secondary level, it was argued, syllabuses, and

---

[32] It is appropriate in this connection to reflect on the results obtained from a series of comparative studies of the upper secondary syllabuses in European countries that were carried out, under the auspices of the Council of Europe, in the late 1960s and early 1970s. The findings show that there were great disparities in the syllabuses, so that substantive harmonization would have been difficult to achieve at that time. The studies were published as nine separate subject monographs in the series, *European Curriculum Studies,* (edited by W.D. Halls), Council for Cultural Cooperation, Council of Europe, in French and English, 1965-1972. See also *International Study of University Admissions. Access to Higher Education,* edited by W.D. Halls, 1963. Paris: UNESCO and the International Association of Universities.

[33] Cf. A.D. Peterson: *The International Baccalaureate: An Experiment in Education,* 1972; and G. Renaud: *The Experimental Period of the International Baccalaureate: Objectives and Results,* UNESCO, 1974.

methods of teaching and examining tend to strive to be similar. This was, however, not the case. The content of what was deemed necessary for entry to higher education was very diverse among the various countries – this may well still be true, even in scientific subjects. Even the style of posing questions was different. Moreover, the political will at the time to achieve some kind of harmonization was lacking. Recent moves in Brussels, however, seem to mean that unification of examinations or harmonization of qualifications is no longer being sought. Within the EC, qualifications at a similar level will be automatically recognized. What will be the effects of this one can only speculate.

13. Concluding Observations

While it would seem that industrial and trade imperatives may favour some convergence of curricula, it would seem unlikely that nation-states will so readily abandon their own national examinations. Indeed, among the major powers examinations have, to some extent, been a form of "cultural imperialism". Belgium, Britain, France, Spain and the Netherlands have imparted models to their former colonies that still in part subsist. English and French examinations linger on stubbornly in Africa and elsewhere. The relative similarity between educational systems in Latin America results from most of those countries adhering, in the past, largely to the Spanish model and more recently, the American model. The countries of Central and Eastern Europe in some respects followed Soviet educational trends, although reluctantly. Other examples could also be cited. Education, both internally and internationally, is a political football. Nations whose cultural legacy, particularly in the field of assessment, dominate the world, are likely to be reluctant to give up the influence they can still exert. This is perhaps the most convincing argument, since education today is above all a political matter, against any swift translation of convergence theory into reality.

It would seem that, at present, we are a long way off from achieving any "harmonization" in upper secondary education at a European level. The disparities between European countries in the levels of average student achievement, as measured in participating educational systems by IEA in science and mathematics, might be taken as tangible evidence of a lack in coherence. [34]

---

[34] That the average achievement level in science of 14 year-old students varies substantially between countries is shown previously in Chapters 2 and 5. See also figures 1-3 in the Statistical Annex.

Chapter 9

The Recruitment and Training of Teachers

1. Introduction

The importance of the recruitment and training of teachers has recently come to the fore, and it is likely to loom ever larger as 1992 approaches.[1] The latest official document on the subject dates from 1987, when the European Ministers of Education, meeting in Helsinki to discuss "New Challenges for Teachers and their Education", passed a number of resolutions. In brief, these sought to improve the "image" of teaching, wanted entrants to the profession to have a broad general education and recommended studies be undertaken to discover optimal recruitment criteria. Where shortages existed, candidates needed to be attracted from other walks of life. Since teachers acted as role models, steps should be taken to redress the balance between the sexes, now biased towards females. They set out a list of points relating to initial training but did not neglect the need for permanent in-service training. Importance should be given in training to teaching practice periods, to knowledge of the school system and to the development of the future teacher's own personal and social skills, here defined as communication, adaptability, creativity, self-confidence and empathy. It went without saying that the recruit should not only know his subject but also how to teach it, acquire some knowledge of research methods, guidance and counselling, education for intercultural understanding, new technologies, special education, human rights and democratic citizenship, "European and global awareness", and health and safety education. He must also be capable of engaging in "philosophical reflection about values and their transmission to young people", know how to appraise his own teaching performance and provide remedies for his shortcomings. Lastly, the training of teacher trainers, head teachers, inspectors and educational administrators required to be reviewed. [2]

This imposing catalogue of desiderata pinpoints where educational authorities themselves perceive weakness in teacher recruitment and training. That twenty-three educationally very disparate countries could find such common ground somewhat confirms what has been termed the "convergence theory": simply, this states that educational concerns are increasingly international rather than national. This may likewise be applied to the smaller Europe of the Twelve. After 1992, Vonk has asserted, the EC will also be faced with "new challenges in education", and already "problems [are occurring] and developments are more and more taking place along similar lines in all European countries". [3]

The imminence of the date envokes the most general practical question of all. After 1992 there will be greater mobility of teachers, who will see career prospects extending beyond narrow national bounds. How best may such movement be achieved, and what intellectual baggage will these new itinerant scholars need to carry with them? Such questions relate to recruitment and training, and may eventually even require resolution in terms of a wider conception of Europe. Within the EC alone, in 1986 there were 3.5 million teachers, perhaps the largest concentration of "cultural capital", in Bourdieu's terms, centred round any single intellectual activity. The teacher's task, which is essentially concerned with the passing on of knowledge and know-how, and willy nilly, with the instilling of values, cannot therefore be treated lightly. This last facet, the imparting of values, as one aspect of the teacher's mission, and one perhaps not sufficiently brought to the attention of intending teachers, raises of course philosophical problems of the utmost

---

[1] We would like especially to acknowledge our debt to Dr Alan Wagner, who supplied various OECD sources, and in particular a pre-print copy of the final report of the CERI project on teacher education: *The Teacher Today*, Paris, 1990, which incidentally lists all the papers and meetings preparatory to this report (unless otherwise stated, all tables are drawn from this recommendable study.) Also acknowledged is an important unpublished monograph by Guy Neave (1988), *The Status and Employment of the Teaching Profession in the European Community* (A Report to the EC Commission.) Chapter 4, "Quality, Control and Recruitment" of this study overlaps with the present paper. It has been so frequently used that it is otiose to cite every reference made to it.

[2] *New Challenges for Teachers and Their Education.* Report of the Fifteenth Session of the Standing Conference of European Ministers of Education, Helsinki, May 5-7, 1987, Document M-ED-15-9, Council of Europe, Strasburg, n.d. See also Document M-ED-15-4, prepared for the Conference, *National Reports on Teacher Education,* which contains an introduction by G. Neave: "Challenges Met: Trends in Teacher Education, 1975-1985".

[3] J.H.C. Vonk: "The Education and Training of Teachers in a Changing Europe" in: J.T. Vorbach *et al.*, *Teacher Education: Research and Developments on (sic) Teacher Education for the Netherlands,* Vol. 5, The Hague, 1989.

importance: which values – national, "European", international or universal – should be inculcated? The Ministers of Education, wisely or unwisely, only touched upon this in passing. *A fortiori*, such ultimate questions, which educators such as Hartmut von Hentig nevertheless affirm should take priority over lesser matters of *Praxis*, are dealt with in Chapter 3.

This paper will therefore, as regards recruitment, concentrate on practical problems of a managerial nature, such as shortages and surpluses of teachers, which relate to demographical considerations, and the associated financial questions; the control of the profession; and the social and psychological problems influencing the choice of teaching as a career. [4] As regards training, it will treat recent developments, looking particularly at the types of institutions involved. Training models and those appointed to carry them out have to be considered. Finally, the diversity of views on such programmes is examined.

2. Shortages and Surpluses: Introduction

Already, large-scale concerns are recruiting internationally within Europe for jobs in industry, commerce and the law. Once formal barriers are down, and many countries still impose nationality and other restrictions on the teachers they employ, teachers may be expected to flock to where the market suits them best. The reasons may well relate to salary and status – in Germany, for example, the status of the *Gymnasium* teacher dealing mainly with the *Oberstufe* is still very high compared with his British colleague teaching an academic Sixth Form. Although the language hurdle may well prove difficult to surmount in many subjects, the incentive to overcome it in order to achieve better conditions will be compelling. One consequence could be that recruitment bottlenecks will arise, and this in turn would necessitate coordination ("harmonization"), if not standardization, of recruitment procedures, and later, of training, of which it is a synthesis.

---

[4] This essay lays no claim to completeness. This is partly due to lack of information on specific topics. Thus, for example, the problems raised by the influence of teachers' unions on recruitment and training have not been studied, nor, save incidentally, the growing power of parents, either grouped as associations within schools or in national associations – "consumer preference" is becoming increasingly important in education; nor the influence of other pressure groups, from the churches to child-action associations; also omitted, through lack of comparative data, save for an incidental reference, is any detailed investigation of selection processes, although the subject is raised in the chapter concerned with the role of upper secondary school in preparing for higher education.

3. Demographic Factors Influencing Recruitment

Problems of teacher recruitment hinge initially upon demography – the number of young people requiring education. In the mid-1960s the birth rate peaked in France, Germany, Italy and the United Kingdom. By the mid-1970s, in a dramatic turnabout, the number of live births was not sufficient to replace the population in these countries (with the exception of Italy) nor in Belgium, Denmark, Luxemburg and the Netherlands. For the period 1985-1995 the average annual "growth" rate for five to 19 year-olds in Europe turns out to be negative, except for the former German Democratic Republic, Iceland, Ireland, Poland and Turkey, where small gains are shown. Table 9.1 below is culled from various OECD sources.

Table 9.1: *Population Estimates, 5 to 19 Year-olds (average annual growth rate 1985-1995)*[5]

EC Countries		Other Countries	
Belgium	- 0.8	Czechoslovakia	- 0.1
Denmark	- 2.0	Finland	- 0.3
France	- 0.6	Hungary	- 0.9
Germany	- 2.0	Iceland	+ 0.2
Greece	- 0.3	Poland	+ 1.1
Ireland	+ 0.6	Romania	- 0.2
Italy	- 2.2	Sweden	- 1.7
Luxemburg	- 1.2	Switzerland	- 1.6
Netherlands	- 1.7	Turkey	+ 1.0
Portugal	- 0.2	Yugoslavia	- 0.2
Spain	- 1.0		
United Kingdom	- 0.9		

Since the trend up to the middle of the decade is largely negative, the decline in pupil numbers may lead to a slackening in the demand for teachers. A further illustration may be given of what may happen. A medium-term prognosis from Table 9.2 bears specifically upon future secondary and post-secondary pupil numbers as compared with 1975.

---

[5] The table is adapted from EUROSTAT, *Demographic Statistics,* Theme 3, Series 3, 1987.

Table 9.2: *Population Size and Change, 15 to 19 Year-olds (index 1975 = 100)* [6]

Country	1995-1996	2000-2001	Difference
Belgium	76.1	72.0	- 4.1
Denmark	87.0	72.1	- 14.9
France	87.4	91.3	+ 13.9
Germany	64.9	67.3	+ 2.4
Greece	100.0	101.7	+ 1.7
Italy	91.9	73.7	- 18.2
Luxemburg	77.1	83.0	+ 5.9
Netherlands	77.0	75.7	- 1.3
Portugal	96.9	92.4	+ 4.5
Spain	106.1	85.8	- 20.3
United Kingdom	81.6	86.3	+ 4.7

If one assumes that those aged 15 to 19 years in 1995-1996 will be just old enough to replenish the pool of the pool of teachers required for the same age group in the year 2000 or thereabouts, *other things being equal*, it looks as if France particularly will have great difficulty in meeting its needs by the new millenium; other countries, with the exception of Denmark, Italy and Spain, which can meet the target with comparative ease, will be struggling to cover their needs, particularly Germany, Luxemburg, the United Kingdom, Greece and Portugal.

4. Problems of Supply and Demand

Population statistics, however, yield only very crude figures as regards the number of teachers required. The reverse of the medal is the numbers of teachers retiring or leaving the profession, as distinct from those entering it. By the end of the century the massive intake of those that began teaching in the 1960s and early 1970s will be nearing the end of their career. No firm statistics are available, but governments such as the French are already undertaking advertising campaigns to recruit teachers to deal with this contingency. This is because in France the incidence of

---

[6] EUROSTAT, 1987, *op. cit.*

the retirement "bulge" falls somewhat earlier than elsewhere, with the result that 34 per cent of all secondary teachers will be gone by the year 2000. [7] Moreover, a substantial increase in the number of secondary teachers will be needed to cater for the doubling in the number of pupils expected to complete the full "baccalaureat" course. Because of the retirement "bulge" a shortage can also be anticipated in the United Kingdom. On the other hand recruitment needs in Greece, Ireland and Spain will be comparatively modest. However, without wishing to repeat the (fortunately!) inaccurate predictions of Georg Picht for an earlier era, given in his *Die deutsche Bildungskatastrophe*, when he asserted that *all Abuturienten* would have to enter teaching if German schools were not to be severely understaffed, [8] it would be wise if the EC faced up now to the general problem of shortages and surpluses.

5. The Impact of Vocational Education on Recruitment

One feature in Europe in recent years has been the growth in vocational education, as Table 9.3 shows. If the trend indicated in the data is extrapolated, then it looks as if a third of all pupils will be in vocational education before the century is out. This will pose the problem of recruiting good technical teachers, at a time when education agencies will be competing with occupations financially more rewarding for scarce skills. Since the salary of a skilled craftsman is rather low (see, e.g. Table 9.4), the blandishments of teaching are unlikely to attract recruits of quality. At the same time the demand for general education teachers will correspondingly decline. A strategy needs to be worked out to meet these complementary tendencies, which will also impact upon higher education and third level education in general. In passing, it should be noted from the data recorded in Table 9.3 how little training of teachers now takes place at secondary level.

---

[7] J. Lesourne, *Éducation et société. Les défis de l'an 2000*, Paris, 1988.
[8] Georg Picht: *Die deutsche Bildungskatastrophe*, 1964. See also the vols published from the project "Educating Man for the XXIst Century", which was conducted in the early 1970s under the auspices of the European Cultural Foundation. One report bears out the difficulty of forecasting the demand for teachers: R. Poignant: *Education in the Industrialized Countries*, The Hague, 1973.

Table 9.3: *Enrolments at Secondary Level by Type of Education (per cent; all European countries)* [9]

Year	General Education	Teacher Training	Vocational Education
1975	79.2	0.7	20.0
1980	75.9	0.8	23.4
1985	73.0	0.8	26.2
1986	72.7	0.8	26.5
1987	72.5	0.8	26.7

Note: Averages across the four European regions (Northern, Western, Eastern, and Southern) as defined by UNESCO (see supplementary Tabel S.18 for an explanation.)

6. Competition from Other Occupations

The competition from other occupations is indeed liable to grow fiercer, as two studies illustrate. Sweden anticipates during the 1990s the need to recruit more teachers: although demand declined from a base index of 100 in 1980-1981 to just above 70 in mid-1987, it is now rising again. But the demand for higher education graduates in the rest of the work force will also rise. This is against a background in which over the next decade the numbers of young people in Sweden aged 19 to 24 years will decrease by some 20 per cent. Places in higher education have been restricted over the last decade. The conclusion is therefore that these should be increased. [10] Likewise a recent British study [11] shows that by the year 2000 in the United Kingdom alone the number of jobs demanding above A-Level standard will rise by 1.75 million, of which one million will be occupied by graduates, and the rest by those possessing other higher education diplomas, including postgraduate qualifications. In Britain particular areas of shortage will be in engineering, computing, business studies, economics, law and accountancy. Significantly this list includes a number of specialities that are now being introduced into the secondary curriculum as vocational subjects and for which teachers have to be found. The report furthermore agrees that other European countries will be suffering from similar

---

[9] The figures are taken from UNESCO, *Statistical Yearbook* 1989, Table 2.9.
[10] Cf. D. Andersson: Higher Education in Sweden on the Eve of the Year 2000, *UHÄ Report No. 17,* Swedish National Board of Universities and Colleges, 1988.
[11] *Projecting the Labour Market for the Highly Qualified,* Institute for Employment Research, University of Warwick, 1990.

shortages and concludes, "the likelihood that they might be an alternative source of graduates is further reduced by their longer periods of investment in higher education and by their relatively high salary levels for graduates". In brief, the trend in the European economy seems to be for ever more qualified manpower, against which teaching must compete.

7. The Present Position: Oversupply?

For the moment, however, with the exception of the United Kingdom, teacher shortages are not general. Indeed, the opposite appears to be the case. In the arts subjects and the social sciences undoubted surpluses exist in Europe. The Ministers at Helsinki admitted as much. What is apparent is that planning departments of ministries of education are not, to say the least, very competent in their projections, unless they are deliberately aiming to create a reserve pool of teachers. At every stage the government has a duty to make firm decisions, planning at least for five years ahead, and allowing for drop-outs from the course and the profession. Planning, however, requires better techniques, as is self-evident. According to the national Reports prepared for the Helsinki Conference, in Austria, Belgium, Denmark, Germany, Netherlands, and Switzerland numbers of newly qualified teachers are unable to obtain an appointment. In Sweden a 1987 survey showed that 25 per cent of those qualified had not found a post. In the Netherlands in 1985 teacher unemployment was reckoned to be running at 15 per cent. In Belgium less teachers, 10 per cent less in primary and six per cent less in secondary schools, will be required over the decade. In Poland, by contrast, over the next quinquennium there will be an estimated shortfall of 33,340 teachers, of which 13,500 posts will be for teachers of English. France at present employs 40,000 "auxiliaries", who are not fully qualified, but who may have been only "near failures" in the competitive examinations to enter the profession. "Hidden reserves" also exist elsewhere. In England, despite an alleged shortage in some subjects and some parts of the country, there is a pool of 400,000 qualified teachers who have abandoned teaching as a career. [12] In general, therefore, at

---

[12] In England it is estimated that over half of all teachers leave the profession within five years of qualifying. Nor is this necessarily a matter of low salaries. A secondary headteacher can earn £35,000 per annum, and over the decade 1979-1989 teachers' salaries rose in real terms by 30 per cent – although from a very low base. Explanations for this anomaly point also to the deteriorating physical condition of buildings, inadequate staff facilities, lack of equipment and textbooks, and the retention of outmoded technologies. Comparisons in these respects with other countries would be interesting.

present such shortages as exist in Europe are confined to certain countries, levels and subjects.

At present the problem is thus essentially one of over-supply, although the perceived need for additional teachers can fluctuate rapidly. For example, in Germany since 1979 the proportion of posts available for newly-qualified teachers has diminished by over 70 per cent, whereas the number of qualified applicants has been allowed to increase by one-third. On the other hand in Ireland between 1978-1987 the proportion of applicants accepted for training has dropped by 17 per cent. Meanwhile governments have tried to deal with shortages and surpluses by not offering permanent, tenured appointments, but taking on fully qualified teachers on a temporary basis. This situation, which arouses the wrath of teachers' unions, is hardly liable to tempt the well-qualified applicant. Even where surpluses exist, there may be shortfalls in various subjects – in Germany, for example, in religious education and music. In England and Wales science, modern languages and craft, design and technology (CDT) teachers are insufficient in number – the latter subject may not even be offered. In Inner London in 1988-1989 schools were unable to provide teachers for all pupils. There may also be shortages in various types of schooling, such as vocational education and the teaching of the handicapped, or at particular levels – Denmark, for example, which otherwise has a surplus, has been unable to fill certain posts in secondary education. A country such as Norway has difficulty in providing teachers for remote rural areas, just as heavily industrialized nations fail to attract teachers to the inner city areas.

However, European comparisons of "shortages" and "surpluses" are obscured because of the lack of a common yardstick. Thus countries are content with different standards as regards the teacher:pupil ratio. Whereas the ratio in primary education for Luxemburg primary schools (1986) was 15.3, in Greek primary schools (1987) it was 22.8. In Belgian secondary schools (1987) it was 8.8, whereas in Greece it was (1987) 17.1. [13] For reasons discussed elsewhere, [14] the 'interpretation' of calculated teacher:pupil ratios in almost any context is problematic, however.

In *The Teacher Today*, the OECD states that some countries in Europe are finding it increasingly difficult to attract and retain good teachers. There would appear to be a lack of interest in entering teaching, although

---

[13] Calculated teacher:pupil ratios are given for the different European countries in Table S.5 in the statistical supplement. The data, which refer to first-level education only, are taken from an OECD source: *Education in OECD countries, 1990 Special Edition*.
[14] An excellent discussion of the limitations of estimated teacher:pupil ratios, and especially of the difficulties besetting their interpretation, can be found in *The Teacher Today*, a document summarizing the recent work of the OECD in this area.

the case of Norway may be atypical: in 1983 one quarter of all post-secondary female students expected to become teachers, by 1988 this had dropped to about 10 per cent. Incidentally, this may be accounted for by the fact that other occupations are increasingly being opened up to women, as well as alternative work styles, such as working from home, or sharing jobs, which appeal to married women with families as much as teaching. Moreover, the picture on the supply side is complicated by the fact that there are differential rates at which teachers move out of teaching each year, either because of retirement, or for other reasons. Of the secondary teachers leaving the profession in the United Kingdom in the 1986-1987 school year, only a quarter were doing so because they had reached retirement age, and over a half − presumably mostly graduates − were seeking careers outside the educational field. [15]

8. Enlarging the Pool of Recruitment

There is some evidence that, apart from entrants to teaching immediately upon completion of their training, and in England only one third of those selected for initial training enters the profession immediately, the pool may be enlarged by attracting fully qualified teachers who have left the schools, or young people who have qualified but have not initially taken up teaching, but gone into alternative employment, or by tempting other adults into the schools. In the Netherlands, for example, in 1985, Dutch secondary schools were staffed by only 40 per cent of graduates who had entered teaching immediately upon qualification; of new recruits in the United Kingdom, staff consisted of about 13 to 22 per cent of "delayed" entrants, and 25 to 34 per cent of "re-entrants". [16] However, if the long-term aim is to attract more and better-quality teachers into the profession manpower resources will have to be found elsewhere, perhaps at a lower level of ability or qualification. Where some of these have been tapped in the past, however, through the recruitment of unqualified auxiliaries (as in France) or likewise unqualified "classroom helpers" (as in Britain) strong resistance has been encountered from teachers' unions and, in some cases, parents.

In Britain, as has been seen, other expedients are being tried. "Mature" graduates are being tempted from other sectors of employment to enter

---

[15] Data supplied to OECD by the British authorities. For a general discussion of the factors influencing the turnover of teachers, see *The Teacher Today* (OECD, 1990) and *The Training of Teachers* (Education Document, OECD/CERI, September 1990). It can be noted that the Secretariat of OECD is publishing a new volume, *Teacher Supply and Demand*, in which teacher turnover rates for different countries are compared.
[16] Discussed in *The Teacher Today, op. cit.*

schools without training and learn "on the job". This has been seen by some as almost a reversion to the pupil-teacher system common up to the late 1920s and also reminiscent of the time when any university graduate, if he secured a teaching post and passed muster – the "testing" of his competence that took place after one year was very superficial – acquired qualified teacher status. Under the new scheme graduates are employed on a probationary basis initially for two years, and then become "licensed teachers". Again, their present and future salaries are by no means competitive. Moreover, to stimulate recruitment in shortage subjects, young science graduates are being enticed into undertaking initial training by offering an additional grant. But the sum is so derisory as compared with the lifelong increment in earnings that can be earned in alternative employment that it is a consideration unlikely to weigh greatly with the very best applicants. The pious hope expressed by the Ministers in Helsinki of "promoting excellence" in the future is indeed highly unlikely to be fulfilled, at least as regards scientific subjects. Moreover, these alternative sources of recruitment may or may not enhance the quality of the profession as a whole – the arguments are finely balanced.

However, three other options have also been considered. One pool from which good quality teachers may well be fished is the migrant population, who could particularly be enlisted to deal with the near-enclaves of migrants in the big cities of Western Europe. The largest alternative source, however, is, as in the 1960s, qualified married women teachers in their early 40s, who, having brought up their families, might be tempted back to the classroom, if other possibilities were exhausted. Again, the problem is to discover on what terms this might take place. Another British expedient to deal with the immediate shortage in the inner cities, particularly the poorer areas of London, has been to recruit teachers from other Common Market states, who would otherwise be unemployed. Although the experiment is not complete, the chances of success are small. A common axiom in British education – and this is most likely true in other countries for the native language – is that, regardless of the subject he teaches, "every teacher is a teacher of English". No matter how proficient a foreigner is in what is not his mother tongue, it is highly unlikely that he will prove to be the most suitable linguistic or national cultural resource for immigrant children, or for that matter, the general run of pupils.

One other expedient to deal with teacher shortages could be to substitute educational technology so far as possible for teachers. One recalls the optimistic note coming from the United States and the Nordic countries in the early 1970s regarding the use of teaching machines. One prediction was that by the year 2000, 70 per cent of teaching would take place by the use of such technical gadgetry and 30 per cent by the judicial use of the teacher, who, being released from routine chores, would be able to "individualize his instruction". These hopes were not fulfilled; the teacher cannot be substituted. With the great strides made in software since then perhaps "programmed learning", as applied to the school rather than higher education, might be intensified. Alternatively, the teacher:pupil ratio might be allowed to rise. Teaching involves interaction between the adult and the child. Therefore, the teacher cannot be completely replaced by educational technology – not even by the use of 'modern' media in individualized learning, in which self-instructional teaching materials play a prominent role.

9. The Limitations of Educational Planning

Planning the provision of teachers has often failed because of the unanticipated problems besetting the cycle of policy formulation and implementation, which substantially alter the balance between teacher demand and supply. [17] Other factors may be insufficient data, faulty or inadequately measured indicators, and errors made in the specification of variables in models predicting demand.

Since the "oil crises" of the 1970s, educational budgets have been cut, school enrolments have, on the whole, been declining, and rationalization by closures of schools has taken place. In this climate of decline these stringencies have nevertheless been accompanied by a prolongation of schooling by large numbers on a compulsory or voluntary basis, although the inevitable step, the raising of the compulsory school leaving age to eighteen, has still to be realised. One questions whether, if conditions were transformed, teacher planning could respond adequately to the situation. The fact that surpluses exist betrays a certain lack of flexibility that is not only detrimental to the European economy but the cause of much personal unhappiness. The problem here is how planning procedures can be improved.

If, as is suspected, there is a pool of "inactive" teachers in most European countries, since teaching is eminently a practical task, should not the opportunity be vouchsafed to as many as possible to take on part-time teaching so as to maintain some "hands-on" experience, to use the computer metaphor? At the Helsinki meeting, the Ministers of Education also suggested job-sharing and short-term teaching contracts. Although such arrangements are a feature of the modern economy in general, the evidence suggests they may not to be to the advantage of the

---

[17] This argument is elaborated in a recent OECD working document on teacher supply and demand.

beneficiaries of teaching, the pupils themselves. Furthermore, planning up to the present has dealt quantitatively with recruitment — with unfortunate results. Is it not time that quality should be taken more seriously? What concrete measures can be taken? In a time of teacher surplus it is perhaps impolitic to mention — which the Ministers do not — the one sure means of improving quality, as has been found in most other occupations: better salary scales.

However, summing up, it would appear that the EC will in theory be able to meet, and is meeting quantitatively, teacher requirements for the decade 1985-1995. Immediately afterwards, however, the "replacement pool" for teachers of the 15 to 19 year-olds, will be smaller and difficulties are likely to ensue. Looking farther ahead, in primary education the position will have eased by 2010 in most member states. So much for quantity, but quality is another story.

10. Quality Recruitment and Teachers' Salaries

General financial considerations are dealt with in the next chapter, but a brief note on how these influence teacher recruitment seems appropriate here.

The days when education was "the priority of priorities" in Western European countries are long gone, and consequently so has the expansion of financial provision. Governments have been reluctant to commit an ever-increasing proportion of the national budget to education. Educational expenditure as a percentage of the gross national product has progressively declined since the economic crisis of the mid-1970s, although this has not been true for Italy. The realization that "the party was over" hit educational institutions about this time. The powers that be seemed suspicious that in the previous decade and a half money invested in education had been squandered. Thus today the cry is for "value for money" and talk is of "cost-effectiveness", even for penny-pinching and making do without essentials such as textbooks. In fact, the demand from governments for "high-quality education" has not been proportionately backed by extra cash. Priorities have changed and in many countries education now ranks below expenditure on defence, health and other social services. There seems no likelihood of this contingency altering in the near future, although an improved situation as regards defence and a brighter outlook for employment — since payments to the unemployed represent a big drain upon resources — would induce a faint optimism. It is nevertheless against a background of financial stringency that payments to teachers, which relate closely to recruitment, must be viewed.

For recruits to teaching salaries loom large. A sentence from a Greek report to OECD [18] typifies the situation in most EC countries, where salaries fall behind those of other comparable graduate occupations:

> "This fact induces many graduates to view teaching as a 'last resort' occupation; they seek employment in the education sector only after they have failed to secure a position in some other sector of the economy".

Moreover, since the skills acquired by many, if not all, graduates — employers maintain, for example, that classics graduates make good computer analysts and programmers — are highly marketable elsewhere, particularly those in mathematics, science, technology and information science, the Greek statement should cause no surprise. The recruit will from the outset feel a certain financial malaise as compared with his non-teaching peers.

Before dealing with this disparity it may be of relevance to compare the wide range of differences that occur within the EC as regards salaries. That countries have different salary structures with different incentives for prospective and current teachers is a major issue to be discussed in anticipation of the 'Single Market', because it reflects divergent thinking in countries about teachers' careers, work conditions, and salaries. A Dutch study conducted for the EC presents data showing the position in 1988, [19] which officials at the Belgian ministry of education for the Flemish Community have converted into annual average salaries expressed in Belgian francs. The data are presented in Table 9.4.

Such a comparative analysis of data does not of course reflect the differential costs of living nor the rate at which promotion occurs. What does emerge is that teachers in Luxemburg, the Netherlands, Belgium and Germany appear much better off than, for example, their Italian, Scottish and Irish colleagues. Might this mean in 1992 a rush of Celtic teachers to the Benelux countries? Does it mean that in Britain the average secondary teacher does better without a degree? [20] Incidentally, in 1989, the average salary for all teachers, including heads and deputy heads, in England and Wales was slightly above the average earnings of

---

[18] *Greece on Educational Indicators.* A Paper for the OECD Washington Conference, Athens, 1987 (November), p. 7, quoted in Neave, *op. cit.*
[19] The study, *The Conditions of Employment of Teachers in the EC,* (Leiden, 1988) was conducted by the 'Stichting Research voor Beleid' for the Commission of the EC and the Netherlands Ministry of Education and Sciences.
[20] The ambiguity is due to the categorizations made, expressed in Dutch (Flemish) as "Onderwijzer" (primary teacher), "Regent" (secondary teacher), "Licentiaat" (teacher with a degree). Any attempt to equate with English terminology is dubious, and goes to show how difficult detailed comparisons are.

non-manual workers in administration, sales and commerce. One might add that the teaching force in Japan is paid very well in comparison with the majority of European countries. Teacher dedication may well be an important explanatory factor in the high overall performance of Japanese students. It is of interest to refer the reader back to Figure 2.4 in Chapter 2, which shows that Dutch students also perform rather well in science – almost on a par with their Japanese fellows – while children in a country such as Britain, where teachers are generally paid poorly, have an average score in mathematics and a science. The data recorded in Table 9.5 illustrate the argument with respect to the mathematics achievement of 14 year-olds. [21]

Table 9.4: *Comparative Income Position by Teacher Category (1988 data, average life career earnings expressed in 1000 Belgian francs)* [22]

	Average yearly salary			Indexed international salary		
Country	Primary teachers	Secondary teachers	Upper sec. teachers	Primary teachers	Secondary teachers	Upper sec. teachers
Belgium	828.0	895.4	1138.1	100	101	114
Denmark	796.8	796.8	958.1	96	90	96
England	772.7	905.1	905.2	93	102	91
France	666.7	705.6	883.6	81	80	89
Germany	952.5	1007.1	1092.7	115	114	110
Ireland	830.5	832.8	832.8	100	94	84
Italy	613.3	647.1	651.8	74	73	65
Luxemburg	1300.3	1415.4	1590.4	157	160	160
Netherlands	834.5	941.1	1220.2	101	106	122
Scotland	682.3	692.6	692.6	82	78	70
Average	827.7	883.9	996.6	100	100	100

[21] The rank order of countries is determined by the average score of the student population (10 year-olds) of a given country on a mathematics test administered by IEA in 1980.
[22] The data are from *The Conditions of Employment of Teachers in the EC*, and are quoted in Coens, *op. cit.*, Table 40, p. 207.

Table 9.5: *Student Achievement in Mathematics in Relation to Instruction Time and Teachers' Salaries (early 1980s)* [23]

Mathematics achievement [a]	Country	Hours of mathematics per year	Days per school year	Teacher salary index [b]
Higher				
↕	Japan	101	243	2.03
Math	Netherlands	112	200	2.33
learning	Britain	130	195	1.69
↕	Finland	84	190	--
	United Kingdom	144	180	1.44
	Sweden	96	180	1.37
Lower				

Notes: (a) IEA 1980 data. (b) Index is based on the ratio of average secondary teacher salary to per capita GNP.

The disparity of salaries between teaching and other occupations, has also been well documented. Guy Neave [24] cites the case of primary teachers, and adduces figures to show that by age 46 the manual worker in ten EC countries is earning half as much again as the top primary teacher. Even at age 32, a time when many teachers will have acquired heavy family responsibilities, the difference is already marked. The same comparison holds good for secondary teachers (1985 figures). At age 46 the manual industrial worker in Germany earns twice as much as the *Gymnasien-Lehrer*. In eleven countries no teacher fares better than the skilled blue-collar worker (Neave, p. 144).

The prudent recruit will doubtless note carefully the number of years, on an incremental scale, before a teacher reaches his maximum salary (assuming that he receives no promotion and remains an ordinary classroom teacher.) Table 9.6 sets out the position for three levels of school education in selected EC countries. The lengthy period of time, no less than the lack of standardization, undoubtedly poses a problem.

---

[23] Source: Heidenheimer *et al.*, *Comparative Public Policy: The Politics of Social Choice in America, Europe, and Japan*, Third Edition, New York, 1990 (Table 2.3, p. 42). See also: C.C. McKnight *et al.*, *The Underachieving Curriculum: Assessing U.S. School Mathematics from an International Perspective*, Champaign, Illinois (pp. 52-53).
[24] G. Neave, *The Status and Employment of the Teaching Profession in the European Community*. A report to the Task Force for Human Resources of the EC, Brussels, 1988.

Table 9.6: *Years of Teaching Service Required to Reach Maximum Seniority (EC countries, 1985 data)* [25]

Country	Primary	Lower Secondary	Upper Secondary	
Belgium	27	27		25
Denmark	20	18/20		14/16
Germany	28	28		28
England & Wales	20	25		25
France	18/28	20/30	Certifié:	19/30
			PLP:	20/30
			Agrégé:	20/30
Italy	40	40		40
Ireland	26	26		26
Luxemburg	23	23		21
Netherlands	26	23/21/22		21
Scotland	12	10		10

Wide differences are clearly noticeable in Table 9.6. Thus in Scotland it takes only 12 years for a primary teacher to attain maximum earnings, as compared with 40 years in neighbouring Ireland. The Irish secondary teacher also comes off worst, with a 40-year time-scale, as compared with a possible 14 years in Denmark. Data supplied by a Dutch source present the picture as regards the actual salary gradient. The data relating to teachers in upper secondary education are reproduced here (Table 9.7).

From the data given so far that deal with primary and lower secondary education it may be concluded that the primary teacher at the end of his career has only doubled his starting salary in Luxemburg and (barely) done so in the Netherlands. In lower secondary education this is not the case. In upper secondary education, where practically all teachers possess a university degree, as Table 9.7 shows, only in the Netherlands and in England and Wales may teachers succeed in doubling their salary before retirement, although in England and Wales the upper secondary teacher starts from a very low base as compared with his counterparts elsewhere in the EC. It would be a poor business man, engineer or scientist, university-trained like his one-time teacher fellow-student, who had not succeeded in earning twice as much at age 65 as he did in his mid-20s. The salary outlook for the teacher is hardly attractive,

---

[25] G. Neave, *op. cit.* Another useful source is the chapter on Teachers' Salaries in E. Cohn and T.G. Geske: *The Economics of Education*, 3rd Edition, Oxford, pp. 236-275.

particularly as security of tenure, which was one of the great perquisites of teaching, and liability to redundancy are now possible. Moreover, the tendency is to engage younger teachers in preference to older ones, since the former are paid less. The prospect, however distant, of this happening must deter a would-be recruit who may well have become so specialized that, if dismissed, he would experience considerable difficulty in finding alternative employment.

Table 9.7: *Indexed Comparison of Teachers' Salaries in the EC Countries (upper secondary education, index the Netherlands = 100)* [26]

Country	Job title	Minimum salary	Salary at age 32	Maximum salary
Belgium	GHSO	112	138	200
Denmark	magister	104	127	155
England and Wales	(senior) teacher	70	95	142
France	professeur agrégé	87	--	216
Germany	Oberstudienrat	125	143	185
Ireland	teacher	82	116	144
Italy	docent	79	89	117
Luxemburg	professeur	155	211	275
Netherlands	Grade 1 teacher	100	158	214
Scotland	teacher	73	109	109

The same recruit may also look carefully at future pay differentials in relation to promotion. Depending upon the country, there are considerable differences between the maximum salary a non-promoted teacher can earn as compared with a head teacher, ranging from only three per cent in France to 40 per cent in Italy.

If the aim is quality recruitment then the salaries of teachers may well require to be raised. However, since teacher renumeration is a factor influencing supply, the raising of teachers' salaries may not have universal validity in all European countries and at all levels of the educational system. Teacher effectiveness or retention may be improved if the salary gradient could be made comparable to that in other

---

[26] *The Conditions of Employment of Teachers in the EC*, Leiden, 1988.

professions, or extended beyond the peak that currently exists. Problems of undersupply and low motivation of teachers can also be overcome by other means than increasing salaries, however. One could introduce changes in the work organization of schools, for example, or iron out anomalies within the teaching profession itself – all steps calling for government intervention.

11. Factors Influencing the Decision to Become a Teacher

A number of considerations not yet dealt with arise for the recruit to the teaching profession. Factors that may well influence the decision to become a teacher are now discussed. Attention is given to the following: status, feminization, the teachers' role and workload, holidays, the teacher:pupil ratio, career prospects and job satisfaction.

(i) Status

Status has already been touched upon in relation to salaries. But the European Ministers of Education were also intent upon enhancing the teaching function in the interests of quality. However, quality is difficult to define: the intellectual and personal characteristics of recruits; proficiency in practice; adequacy of provision; the raising of standards; or perhaps all of these? The quality of teachers and teaching depends, among other things, upon status, of which the main determinant in the market-led, meritocratic society is financial reward. Certainly where salaries are higher, status is higher also: thus in the Netherlands upper secondary teachers are still rated among the top one-fifth of occupations. However, presumably the argument of the Ministers was that, if the market is to be believed, remuneration was adequate, because, apart from some geographical areas and certain subjects, *quantitatively* there were few shortages. But crude market forces and quality, which they fervently desired, do not necessarily go together.

Certainly over the last quarter of a century the average real salary of teachers has increased. The question is whether in the salaries' race other occupations have leapt ahead. The evidence on this is inconclusive. Some studies show that teachers' salaries are not "low" in an absolute sense but it must be remembered that since teachers are mainly women, the comparison has been made with women's salaries in other occupations, which tend noticeably to lag behind those of men. [27]

---

[27] As Alan Wagner rightly points out in an OECD working document, *"Social and Economic Aspects of Teaching: The Attractiveness of the Profession"* (OECD, 1987), the central question is not whether females are paid as much as males (of the same age and employed in comparable, private-sector occupations); this is a broad economic and social

It appears that teachers at the outset of their career earn above the average industrial wage. This is disputed by teachers' unions in Britain, where earnings have been the subject of much bitterness. Thus it has been calculated that in Britain, using 1985 prices, teachers' salaries only rose in real terms from £7,300 p.a in 1960 to nearly £11,000 in 1986 – just over 50 per cent, and far short of what could be earned with comparable qualifications elsewhere. One effect of this was that whereas in 1960 34 per cent of all male graduates entered teaching, by 1980 this had fallen to 14 per cent; the comparable figures for women were: 1960 – 61 per cent; 1980 – 35 per cent. [28]

A decline in status has come about also through the demise of teaching as a vocation, using the term almost in a quasi-religious sense. The young idealist was in the past attracted in this way into the profession. However, in a materialistic age, where teachers talk bluntly of "pay" and not "salaries", of "industrial action" and "withholding their labour", suave euphemisms for striking, rather than of supervising extra-curricular activities, giving their services free, of unionism rather than professionalism, the concept of vocationalism has flown out of the window. Responding to a "vocation" was agreeable when teachers were accorded a status in the community that did not depend necessarily on remuneration and which arose partly from a mystique: teachers were then a band apart, because they *knew* more than the majority of people, and few were aware of what their salary was. In other more privileged walks of life, such as medicine or the law, this mystique, sedulously fostered by professional associations, survives. Today, when everybody has been educated at least to secondary level, and when teachers' salary scales are a subject for public debate, the mystery has been dissipated. The Emperor has been found to have no clothes, or to be clad only in rags! Perhaps one of the most formidable problems facing recruiting agencies is the revival of idealism and enthusiasm. It may be one reliable way of attracting quality.

The teacher's self-image depends upon the esteem in which he is held by the public at large. This, it must be said, is at present not great. He is perhaps further off being considered a member of a profession than he was a generation ago. He has lost much scope for autonomous action; his work is more subject to intervention from local or central authorities – perhaps an inevitable consequence of the politicization of education. For teacher quality to be improved steps require to be taken deliberately to

---

question, which bears no immediate relationship with the attractiveness of teaching to females. The latter depends on the wages actually available to women in teaching as opposed to other, comparable professions – in which case teaching may hold a "salary advantage" for females.

[28] P. Molton, "Closing the Teacher Gap". Article in *The Times*, April 18, 1990.

improve the public image and the self-image of the teacher. It may well be that if specialization of the teacher's task continues apace, this decline in status will be reversed. Governments are certainly anxious to make teachers professional rather than the contrary.

Whether a generalization of civil service status for teachers would contribute to producing the desired effect is debatable. Certainly in Germany, where the secondary teacher has *Beamter* status, his ranking among the desirable occupations remains high. The assimilation to a specific grade of the public service might indeed assist status.

(ii) Femininization

At the end of the 20th century it is perhaps saddening to have to write that the femininization of the teaching profession has done little to enhance the status of teaching. To be fair, it must also be stated that there is no evidence supporting the claim that school standards have declined as a consequence of feminization. Increasingly one public view of teaching has been that it is "an occupation for girls until they get married". Although this is far from the truth, it is the public perception that is decisive as regards status afforded to teachers. [29]

Since 1975 the number of women in teaching has been steadily increasing. [30] For Europe as a whole, in 1975 already 71 per cent of primary teachers were female; by 1987 this had grown to 75 per cent. Extrapolating from these figures, by the year 2000 this proportion may well rise to over 80 per cent. It is true that in secondary education this femininization is not so apparent: in 1975 it was of the order of 55 per cent, and in 1987 only 57 per cent, giving perhaps 60 per cent by the end of the century. More detailed OECD estimates give the position by the level of education as in Table 9.8.

What conclusions may be inferred? It is likely that men will become even more reluctant, although they still occupy the top teaching posts for the time being, to enter a profession where women may well eventually predominate. The increasing femininization at primary level may not give rise to much concern, because the conventional wisdom, rightly or wrongly, is that women handle young children better. On the other hand, it could be argued that, particularly with the increase in one-parent families, as noted in Chapter 4 usually the mother, it would be beneficial if young children of both sexes could come more into contact with males. Moreover, the present shortfall in girls in secondary education, - in 11 European countries they are substantially fewer than boys, - will nevertheless mean that the number of them coming forward for teacher

---

[29] The reader is reminded of the argument presented under footnote (27).
[30] Cf. UNESCO, *Statistical Yearbook* 1989, Tables 2.5 and 3.7.

training by the end of the century will inevitably diminish, particularly if other more attractive careers are beginning to open up for women. Not so convincingly, it has been argued that since the majority of women teachers will be married, they will have less time at their disposal than men for lesson preparation, correcting homework or for the multiplicity of extraneous activities that the teacher is called upon to fulfil outside the classroom. At any rate, the case for maintaining a judicial balance between the sexes is a cogent one.

Table 9.8: *Females in Teaching by Level of Education (1986-1987)* [31]

Pre-primary:
Almost all women, with the exception of Ireland, where men constitute 25 per cent of the whole.

Primary:
Women over 70 per cent : Austria, Belgium, France, Ireland, Germany, Spain, United Kingdom, Yugoslavia
50-70 per cent : Denmark, Netherlands, Norway, Sweden
40-50 per cent : Greece, Turkey, Luxemburg

Lower secondary:
Women over 50 per cent : Austria, Denmark, France, Greece, Spain, Yugoslavia
30-50 per cent : Greece, Turkey

Upper secondary:
Women over 50 per cent : Austria, Belgium, Spain, United Kingdom [All only just over 50 %]
45-50 per cent : France, Greece, Turkey, Yugoslavia
under 45 per cent : Germany(30.6 %), Netherlands (27.5 %), Norway (34.1 %)

(iii) The Teacher's Role

Viewed from the individual angle, one factor that will influence a young person's decision to become a teacher is how he, or rather she, conceives the teacher's role in modern society, which surely requires re-definition, and perhaps limitation. A French educationist has remarked, although in a different context and with exaggeration, that "The teacher is the donkey of the Arab proverb. He prepares and teaches his class in

---

[31] The data are from *The Teacher Today* (OECD, 1990), Table 4.

his spare time, so numerous are the many functions he is called upon to fulfil". According to the Council of Europe, the work of teachers in lower secondary schools in France includes the following (non-exclusive) activities: [32]

1. Teaching, whether lessons or practical work;
2. Evaluating the results of teaching, whether for formative evaluation techniques incorporated into teaching activities or by summative evaluation procedures such as tests and examinations;
3. Preparing lessons and instruments of evaluation – both conceptual and material, on an individual or team basis;
4. The correction of evaluation tests on an individual or team basis;
5. Intellectual and personal support to pupils (tutoring) including relations with parents, the school administration, counsellors, and other teachers;
6. Contributing to the guidance of pupils on an individual basis (through tutoring) or collectively (in school councils);
7. Permanent in-service training in its various aspects, either within the framework of the school (in teaching teams), or outside in teacher training centres;
8. On a voluntary, optional basis: running clubs and societies within the school, maintaining links with cultural and other outside bodies collaborating with the school, and contributing to the provision of adult education in the community.

In the English view, school has traditionally been as much about the inculcation of personality and character as about academic learning. Hence, for example the greater emphasis on sport and extra-curricular activities as elements in personal development and a corresponding diversity of the teacher's tasks. This non-intellectual, even anti-intellectual approach, contrasts with that of the Latin countries, where the teacher has above all been charged with *instruction*, and the family with the other facets of education. However, it is now almost universally acknowledged that he has also a pastoral role that extends beyond the cognitive domain. In France, for example, there has been a trial experiment of a tutorial system. Practically everywhere in Europe nowadays the teacher has to deal with sex education, and drug and alcohol abuse – even "traffic education"! His functions as a kind of maid of all work (or Jack-of-all-trades) are all-embracing: he is the "manager" of learning, the dispenser of knowledge and skills, the preparer of pupils for the labour market, a child-minder, an administrator, counsellor and organizer outside the classroom proper, the implementer of reforms, the disciplinarian, the surrogate parent, even the technician: the list is long

---

[32] Council of Europe, "Pour un collège démocratique". In: *Faits Nouveaux*, 1983, No. 4 (pp. 8-21). Strasburg: Documentation Centre for Education in Europe.

and daunting, but still incomplete. One may take leave to wonder whether the raw recruit to teaching, who is often more enthusiastic about the subject that he is to teach than about all else, is fully aware of the burden he may have to shoulder. Moreover, initial training notoriously falls short in preparing him for such tasks, and in-service training on such duties is patchy and ill-conceived.

(iv) The Teaching Load

The number of hours per year that the teaching actually stands before the class has been computed for a number of EC member states. The findings are recorded in Table 9.9.

Table 9.9: *Number of Teacher Contact Hours per Year (standard 60 minutes, 1988 data)* [33]

	Number of contact hours		
Country	Primary	Lower Secondary	Upper Secondary
Belgium	758	698	637
Denmark	810	810	810
England	832	832	832
France	945	805	805
Germany	1,008	972	864
Netherlands	1,040	967	967
Ireland	920	810	810
Italy	1,032	774	774
Luxemburg	864	630	630
Scotland	950	887	887
Average	916	818	802

In the actual time that they spend every week in the classroom, at the "coal face", so to speak, teachers do considerably better than other workers, although even here there is great variation between countries: for primary education the range is from 16.5 to 28 hours; for secondary education, 12 to 32.5 hours. To these figures must be added out of class activities, which average 13 hours for preparation and marking, and another 8 hours for other pupil-related activities such as counselling and

---

[33] *Final Report on the Conditions of Employment within the EC*, Stichting Research voor Beleid, Leiden, 1987, quoted in Neave, *op. cit.*, p. 118. Similar figures are recorded in an OECD working document: *Teacher Demand and Supply* (Paris, September 1990).

administration. The weekly teaching load varies according to teachers' qualifications and status. According to the Council of Europe, [34] the position in the mid-1980s in France was as follows:

- 15 hours for agrégés (with higher degrees);
- 18 hours for certifiés (graduate teachers);
- 21 hours for professeurs d'enseignement général de collège (trained in colleges of education);
- 22 hours for teachers of physical education;
- 20 hours for arts teachers;
- 24 hours for primary school teachers.

Thus the average for all countries quoted in a recent OECD report [35] for the total time devoted to work is 41.5 hours in primary education and 40.5 hours in secondary education, although in Norway this can rise to 52 hours a week. The time spent on out-of-classroom chores seems to have increased, and is still increasing, whereas the average employee – as distinct from the professional or executive person, with whom the potential or current teacher often identifies – now usually enjoys a week of under 40 work hours. This is another consideration that the recruit to teaching must take into account.

(v) Holidays

One of the undoubted attractions of school teaching in the past has been the length of holidays, and for young mothers who were teachers the fact that their holidays coincided with those of their children at school. The prospect of extended periods of leisure, even if some time has to be spent in preparing for the coming school term, has been a factor of importance in recruiting. In recent years, however, holidays have gradually been whittled down by such necessities as in-service training, as has the comparative advantage, since the tendency in all occupations has been to shorten the working year. Today outside teaching the number of working days per year varies between 235 and 250, whereas for teachers it is between 180 and 220 days.

It can be assumed that, in countries where teachers now have longer work years than those in "comparable" occupations, teaching will have become less attractive compared with before, when teachers were generously provided with holidays and vacations.

---

[34] Council of Europe, *op. cit.*
[35] OECD, Working document CERI/AW/gk/600, Paris, September 1990), *op. cit.*

(vi) Teacher:Pupil Ratio

There is a clear evidence that the teacher:pupil ratio is not a significant factor in the effectiveness of teaching, despite the protestations to the contrary made by teachers' unions. [36] Where the professionals may be right is in the amount – unquantifiable – of stress generated in teaching a large class. What therefore is the trend? Reliable figures for the ratio are available [37] for primary education. (Estimates of the teacher:pupil ratio in primary education are given for all European countries in the statistical annex.) The data generally show an improvement since 1975, although there are marked variations between countries. Thus out of 22 European countries in 1988, 14 had a teacher:pupil ratio of under 1:20, although there are extremes: in Denmark and Austria a low figure of 1:11 is reported, whereas in Ireland it is 1:27 and in Spain 1:26. The recruit seeking an unstressful working life, and stress is today one of the main causes of drop-out from the teaching profession, may well be deterred if the ratio is deemed too high, although the question must be asked; What is the *optimum*?

(vii) Career Prospects

The far-seeing recruit to teaching will be looking at the prospects of promotion during his career: what percentage of classroom teachers reach senior posts, defined as head or deputy head teacher, within a school? Again, there is considerable variation within the EC. In some countries progress up the hierarchy is slow because of the number of levels through which the teacher has to pass in order to attain his goal. Data are incomplete, but Neave shows that in Belgium only 11 per cent reach the top rung of the ladder, as compared with 40 per cent in nearby Luxemburg. [38] In the secondary sector the situation is exactly reversed: Belgium heads the table of eight countries cited, with 35 per cent of secondary teachers obtaining senior posts, as compared with only 1 per cent in Luxemburg. Of course such figures are relative, depending upon population density and school size, which also is a determinant of salary. Thus promotion for the budding teacher looks just as much a lottery as it

---

[36] See, *e.g.*, Anderson, L.W. *et al.*, *The Classroom Environment Study: Teaching for Learning*, Oxford, 1988. One of the earliest meta-analyses on the subject is contributed by G.V. Glass and M.L. Smith, "Meta-analysis of Research on Class Size and Achievement", *Educational Evaluation and Policy Analysis*, Vol. 1, 1979, pp. 2-16. See also G.V. Glass, L.S. Cohen, M.L. Smith and N. Filby, *School Class Size: Research and Policy*, London, 1982.
[37] OECD, *The Teacher Today*, Paris, 1990. See also the data on teacher:pupil ratios provided in the statistical supplement (Table S.5), which are mainly derived from UNESCO, *Statistical Yearbook*, 1989, Table 3.4.
[38] G. Neave, *op. cit.* (Table 3, p. 124).

does in any other occupation. Moreover, just as any downturn in economic activity diminishes promotion opportunities, so do any falls in the school population or rationalization of schools, both factors which the newcomer to teaching may not be in a position to assess.

(viii) Job Satisfaction

What kind of feeling of well-being does a teacher derive from the exercise of his occupation? The incidence of stress has already been mentioned: in Denmark it is reportedly the cause in primary and lower secondary education of 12 per cent of teachers leaving the profession. A survey conducted in the United Kingdom showed that nine per cent of the teachers took early retirement because of ill-health. [39] Absenteeism has almost reached industrial levels. Yet the general frustration felt today by many teachers is a symptom of the malaise of modern times that education authorities alone cannot remedy. The main problem in teaching today is how to motivate pupils. Pupils perceive that school as an institution enjoys little esteem in society, and their teachers even less. The present rash of youth unemployment in Europe gives young people little faith that schooling can prepare them for working life. They are a prey to competing interests, superficially more attractive, from television to pop music. All this makes the teaching task more difficult. The potential teacher, whose own schooldays are not so far distant, may well ponder upon this and decide against teaching as a career.

To summarize then, these are the main factors in the problem of recruitment for teaching. The authorities have to wrestle with the control of supply and demand, which is cyclical. They have to cope with the swings and roundabouts of shortages and surpluses, sometimes by covertly lowering or raising the standards of qualification, as well as looking to the public purse. For his part, the individual will be intent on reviewing the conditions of employment, both present and future, in teaching. The task of reconciling the interests of the two main actors in the educational enterprise – to say nothing of parents, other interest groups, or the electorate in general, is daunting, to say the least.

12. Elements of a European Model for Teacher Training

Market forces cannot operate freely in the supply and demand for teachers, for, apart from the private sector, the central government has the whip hand on both sides of the equation. Where a private sector exists, its influence on the number of trainees will depend on its size – five per cent in Germany, seven in the United Kingdom, but as high as

---

[39] *Ibid.*

40 per cent in Spain. Since, however, many countries are now insisting that those who teach in non-State schools should have the same qualifications as their State counterparts, even the number of trainees destined for private schools can be regulated.

A general European model may be envisioned as comprising three stages. These are, firstly, selection for training. The second is actual teacher training and the third in-service and on the job training subsequent to appointment to a teaching post.

Selection for training. In some countries – and for some types of teacher – this occurs at the entry to higher education. The general hurdle for admission remains the secondary school leaving qualification, which is being acquired by an ever increasing proportion of the age group – for example, some 30 per cent in France, which eventually envisages a figure of 80 per cent. Standards for selection are being raised in some countries, such as Greece and Portugal. Other countries put up specific hurdles: in the United Kingdom future teachers must be qualified in mathematics and English; in Sweden and the Netherlands, in mathematics. It is also now common to involve practising teachers in the selection as well as the training process.

After selection, two types of training are usual. An academic course may precede a shorter training course proper – the so-called *consecutive* alternative; or both academic and professional studies and practice may occur together – the *concurrent* alternative. A third alternative is now emerging, whereby suitable candidates are admitted for training without formal requirements, and follow school-based courses, as with the "articled" and "licensed" teacher systems just introduced in the United Kingdom. Training courses may last *in toto* between three, now more usually, four or five years and end in qualified teacher status. In general, the shorter the training course the more attractive it is to mature entrants. If recruits are sent into the schools for a practice period early on in their course, it allows those manifestly unsuitable to be weeded out early on, to the mutual advantage of both sides.

Appointment to a teaching post or placement on a list of eligibility, both subject to satisfactory completion of a probationary period.

The model described above gives rise to a number of criticisms. The potential recruit may well prefer the consecutive alternative, because if he drops out after the academic course his options as regards a different form of employment are still open. Governments may also prefer it, because it gives more flexibility in numbers. In the past, moreover, the concurrent alternative has often been seen, as in France and the United Kingdom, as being of a lower intellectual level, and less academic in content.

The completion of the second stage, i.e. teacher training "proper", is crucial; here some wastage occurs, either through lack of success or because the newly-qualified holder of a teaching certificate decides that he does not wish to enter the profession after all. Money from the public purse has been spent in vain. How can such a situation be obviated? In France the acceptance of public funds to train for teaching entails a binding commitment to teach for at least five years in a state school; in England there was once the same obligation. Whether this solves the problem is arguable, however.

Other comments may be made regarding the model outlined above. The central body can intervene at any of the three stages. It controls generally the numbers entering higher education, the *sine qua non* for teacher training, although in countries such as France, Germany, Italy, Spain and Switzerland a secondary school leaving qualification confers an automatic right of entry. As regards the content of the course, the government usually refrains from direct interference and is content to leave it to the professionals, although it is in a position to intervene, particularly as regards the professional component. It may well be that it should be able also to tailor the academic course more to suit what future teachers will be required to teach. In the matter of the first appointment national practices diverge: in France, having surmounted intensively competitive examinations, the successful recruit is almost guaranteed an appointment; in Germany some *Länder* are not obliged to offer a post to all those successful; in countries such as the United Kingdom and Ireland the search for an appointment is up to the applicant. Such divergent practices within the framework of the general model seem to have grown up by tradition rather than derived from national considerations. A probationary period after the first appointment is obviously salutary, although its length is variable: for example, one to two years in Italy, England and Belgium, but as much as two to five years in Denmark, Portugal and Greece. The longer the probation, the greater flexibility the government possesses, although the potential recruit may view the matter differently. How long should this period be before the teacher receives a final seal of approval? Opinions are now being voiced that a license to teach should be given for only a few years, and only renewed subject to satisfactory performance and evidence that the teacher has updated his knowledge.

The level at which teacher education is located on the educational ladder is an important factor in recruitment. As mentioned, until the 1970s certain types of teacher were still trained in secondary education (France, Austria). Now, in the hope of enhancing status, content and quality practically all teacher education is at the level of higher education, it has been argued that, like all other professions, teacher education does not, as it has in the past, require a lengthy period of segregation in what has tended in some countries to resemble a religious seminary. Thus the 1989 French *Loi d'Orientation* proposes the creation of *Instituts de Formation universitaire* eventually to replace existing institutions. In Germany the *Pädagogische Hochschulen* are assimilated to the universities, sometimes within the framework of a "comprehensive university"; in England the colleges of education have links with the local university. In Ireland they have been so linked since 1975. Since 1986 in Portugal all primary teachers have been trained in the non-university sector of higher education. In Sweden all teacher education takes place within a common higher education sector. However, teacher training institutions or departments may well remain a poor relation within the higher education system, a factor that diminishes their esteem in the eyes of potential applicants. The problem of upgrading them, not only nominally, but in reality, remains. It is unlikely that universities, or even the less-esteemed polytechnic-type institutions will change their sometimes suspicious attitude until advances have been made in what might be termed a science of pedagogy. Those who regard the development as impossible should perhaps recall that Durkheim (who, somewhat surprisingly, held a dual chair in sociology and education), did not rule out this possibility. And, after all, nowadays universities admit the study of all kinds of preparation for non-liberal, even mundane, occupations, from brewing to cement technology, from business studies to criminology to publicity and advertising.

The more gruelling selection process now being adopted, in which national inspectors, local authorities, university representatives, trainers (in whose hands alone selection of trainees often rested), and teachers, may increase the prestige of teacher education. In France a certain guarantee of quality is provided by the competitive examination entry system, which the French insist is also needed to maintain equity. The stipulation of the European Ministers of Education, that student teachers should be capable of developing personal and social skills of a high order, might also require some kind of psychological aptitude test, as is now frequent in selection for other occupations.

13. The Trainers of Teachers

Increasingly the competence of the teacher trainers themselves is being called into question, as was hinted at in the final resolution of the Helsinki Conference. Some trainers of future subject teachers have been more than 20 years away from the classroom; a few may never even have faced a class. In the ivory tower of a college of education they may well have missed all the turmoil of reforms that has characterized the 1970s and 1980s. What then, it is urged, can they contribute to the training of the new teacher? Britain has recently decreed that all teacher trainers should have had "recent and relevant experience" in a school. All agree that such an updating of experience is essential. It has been said, somewhat mischievously, that it should not perhaps be confined to teacher trainers, but extended to inspectors and even administrators, who themselves have most frequently been recruited from the ranks of teachers. But how such a policy is to be implemented presents difficulties. A teacher trainer, who has in his time been presumably among the most successful of teachers, if he were to return to the classroom would require a long period of acclimatization before he regained his former competence and was able to pass on to future teachers an up-to-date version of his expertise. In any case, he may never find again his former peak of performance, and yet may know what has to be done although incapable of doing it any longer himself. The analogy may be with refereeing: the referee knows what is good play when he sees it, just as the spectator sees most of the game. After all, the teacher trainer, it is argued, is helped in his evaluation of the trainee by the supervisor in the school, and all the staff that come into contact with him on his teaching practice.

Nevertheless, all such arguments seem to indicate special pleading. To resolve the problem it might be better to abandon an institution-based form of training in favour of a school-based one. Trainers would become teachers in the school with a lighter teaching load, who could supervise trainees on the spot. This idea has much to commend it, although after a while the trainer might well become rather parochial in his outlook, and fail to keep up with the flood of professional literature and reports that would prevent this happening. In crude terms, the problem is: institution-based or school-based training: which is nearer reality? [40]

14. In-service Training of Teachers

However, in many ways the burden of training is being shifted from the pre-service period to later. As new developments follow in swift succession it is realized that the task of introducing them can no longer fall upon the young teacher. The reasoning once was that new practices would emerge in the schools as new blood was fed into the teaching profession. But the pace of change is such that it is now recognized that it is the older teacher with experience who can best implement innovations. It is one of the main functions of in-service training to

---

[40] Cf. S. Lawlor, *Teachers Mistaught* (A Report of the Centre for Policy Studies). London, 1990.

ensure that this happens. After a few years of experience the beginner is in a better position to apply new techniques and skills. The question is: how can such further training be systematized? Lifelong education is now a concept recognized by the universities, and they are increasingly organizing themselves to cater for it; the specific case of lifelong education for teachers must also be met.

In-service training can therefore either be *supplementary*, remedying gaps in initial training, or complementary, adding new know-how and updating knowledge. Ideally this might be done by granting teachers a sabbatical term once every five years for study in universities. Meanwhile valuable work has been done through the provision of local teacher centres, as in the United Kingdom, "centros de professores" in Spain since 1982, "Missions académiques" in France since 1981, and "Istituti regionali" in Italy. [41] In addition to general courses relating to new skills such as computer education, more specific training must also be catered for. Incidentally, as shown in Table 9.10, by 1988 many European countries had mounted large-scale courses in this subject, notably in Belgium and France, Germany and Hungary, Switzerland and the Netherlands. Teachers have to be trained for non-teaching tasks such as vocational guidance and counselling. A few will need further professional knowledge in a specific field, such as the psychology of teaching backward or gifted pupils

---

[41] Cf. J.L. Garcia-Garrido (Ed.), Primary Education on the Treshold of the Twenty-first Century, *International Yearbook of Education, Vol. XXXVIII,* Geneva, 1986.

Recruitment and Training of Teachers 281

Table 9.10: *Provision of Staff Development for Teachers Using Computers in Upper Secondary Schools, and Agencies Giving Support with Teacher Training* [42]
(1989 data, in per cent of all upper secondary schools)

Country/educational system	Provision of introductory course in computer use	Provision of course on the use of application programmes	Provision of computer science course	Provision of course in micro-electronics	Provision of courses for specific school subjects	Teacher training organized by a ministry or other central level authority	Teacher training organized by universities and higher education institutions	Teacher training organized by the business sector
Belgium, Flemish	90	94	81	19	55	19	11	22
Belgium, French	63	65	59	19	45	24	24	25
France	80	90	66	29	70	72	16	35
Germany	92	84	94	49	80	83	24	11
Greece	67	59	74	--	25	89	22	27
Hungary	94	99	80	42	63	13	28	--
Japan	89	75	35	25	51	*	*	*
Netherlands	90	92	66	21	56	36	13	22
Poland	85	57	87	19	41	9	22	--
Portugal	54	53	19	--	33	23	26	--
Switzerland	95	97	89	44	65	47	21	22
United States	84	85	76	55	69	28	43	39

Notes: (a) *: Data not collected. (b) --: Estimates cannot be calculated because of insufficient valid cases.

[42] The data were collected in 1989 for 21 countries in the IEA study *The Uses of Computers in Education Worldwide*, edited by H. Pelgrum and Tj. Plomp, Oxford, 1991.

Table 9.10 shows that the majority of teachers who *are using computers* in upper secondary schools had previously taken part in an introductory course on the subject. In France (72 per cent) and Germany (83 per cent) such courses are mostly organized by the Ministry of Education or some other central-level educational authority. Higher education institutions are particularly active in organizing teacher training courses on the use of the computer for education and instruction in the United States. An interesting observation emerging from the data is that, in Hungary and Poland, the responsibility for the upgrading of teachers and the implementation of a major innovation such as computer education seems to rest in the main with the individual school, since the central government, universities and colleges, and even the business sector, are not providing staff development programmes on computer education for secondary teachers.

The localization of in-service training has been a feature of the last decade. Sweden, for example, in conformity with a general tendency to decentralization, now merely lays down the general outline of a national curriculum; it is then up to the teachers on the spot to work out the details in accordance with specific local needs or their own classroom requirements. This can be done by local groups of schools or by individual schools. Many countries now set aside a number of working days a year for local workshops and seminars. Thus the distinction between development and training is becoming blurred. Whether research from the centre will now take place also on the periphery, which appears to be a logical consequence, remains to be seen.

15. Teacher Training Programmes

What should be the content of what the future teacher learns in order to equip himself for his task? Clearly both theory and practice are necessary. Most countries insist that at least three months should be spent in the school environment before the beginner becomes an autonomous teacher. Some demand even longer periods of *Prakticum*. The trainee not only sees how others teach and, under supervision, starts to teach a class himself but is also expected to learn about school organization, curricula and examinations, etc. On this much there is agreement. In Britain and Greece the development of closer links with the schools is now a notable feature of teacher training. Nevertheless, about the other professional studies the future teacher is expected to assimilate there is still great controversy. However, there is a consensus that he should learn something of what might be described as "general teaching methods", – perhaps to include such American innovations as "mastery learning" and "cross-age tutoring" – and more importantly, the methodology of teaching his one or several subjects. In recent years countries such as Greece and Portugal have been emphasizing these elements more. Elsewhere with future upper secondary teachers, whose courses, as in France and Spain, had little strictly pedagogical content, methodology has recently become more important. Incidentally, teacher trainers, in advocating certain methods, have been accused of "trendiness", putting forward the particular ones that are fashionable at a particular time. One thinks for example of the use of the language laboratory, stimulus-response techniques and pattern-learning, and the shunning of translation for foreign language learning that held sway in the 1960s and are now rejected by many language teachers. On the other hand, until the beginner has found his feet and discovered the kind of method that suits his pupils and his teaching style best perhaps a certain dogmatism is unavoidable. If, to use a French expression, "il faut sécuriser les enfants", the trainee himself needs assurance no less. This does not mean that he should not at least be familiar with all the arsenal of methods available, say, in teaching reading, from the traditional "look and say" and phonic methods, down to the controversial "contextual reading" approach.

More use requires to be made of research findings that have relevance for teaching. Thus, for example, IEA studies show that deeper knowledge of subject-matter is associated with an increase in pupil achievement. This would imply that trainees should be very deeply grounded in their subject – far beyond the knowledge required to teach it, and also be able to "transform their knowledge into forms that pupils are most likely to understand". [43]

There is a growing realization that the use of new technology may have far-reaching effects on the teaching of all subjects. Trainees will have to have an adequate knowledge of computer science and techniques such as word processing, data base construction and graphic display, in order to use microcomputers in teaching their subjects. The new methods are only in the process of being elaborated and are therefore likely to loom large in in-service courses. [44]

What theory, however, should be learnt apart from methodology? In Britain, following a Transatlantic example, in the 1960s students on a one-year course were often given a smattering of what were

---

[43] L.W. Anderson and T.N. Postlethwaite, "What IEA Studies Say about Teachers and Teaching", in: A.C. Purves, Ed., *International Comparisons and Educational Reform*, Alexandria, Virginia, n.d., pp. 75-76.
[44] M. Vickers: *Microcomputers and Secondary Teaching. Implications for Teacher Education.* Report on an International Seminar arranged by the Scottish Education Department in cooperation with the OECD, Glasgow, October 12-15, 1987. See also: H.M. Levin, "Economic Trends Shaping the Future of Teacher Education", *Educational Policy*, Vol. 4, 1990, pp. 1-16.

euphemistically called the "educational disciplines": psychology, sociology, and philosophy, with some history of education – often going back to the ancient Greeks – and other odd fragments of knowledge allegedly relevant for the future classroom teacher thrown in for good measure. Each of these disciplines in turn held pre-eminence. It was hoped in this way that the teacher would be more readily recognized as a professional. In point of fact such courses aroused derision, particularly in the academic world. Of recent years the fashion has been for curriculum theory and practice to be taught. Even more recently class management techniques, sometimes a circumlocution for learning how to keep order and maintain discipline, have been introduced. Although a case can probably be made out for each and every one of these, decisions have to be made as to what elements are essential and whether they are meaningful without a basic knowledge of the discipline from which they are drawn. A little knowledge can be as dangerous as too much.

Up to the late 1950s, when it was supplanted by sociology, in Britain at least, psychology was reckoned to be the one field closest to education. It was toppled from its pinnacle, some say, because the theories of learning that were taught seemed more relevant to the behaviour of monkeys than of schoolchildren! Perhaps the time has come to review once more the theoretical content of courses, particularly in psychology, for teachers, If some knowledge, say, of developmental psychology is held to be essential this of course throws open the door to other disciplines. It could be argued, for example, that sociology has a claim, in that a knowledge of the pupil's social environment is vital. Nor must the long enumeration laid down by the Ministers at Helsinki, [45] from familiarity with research findings to health education – ten topics are mentioned in all – be forgotten. [46] This obliquely raises the question of the balance of the components for the whole course, both academic, theoretical and practical . It may well be that the proportions, and the

---

[45] The Ministers specifically mentioned: "a minimum degree of familiarity" with: educational research findings and methods – guidance and counselling – education for intercultural understanding – new technologies – special education – human rights and democratic citizenship – European and global awareness – health education – safety education. This is a formidable list to be added to: the development of personal skills – teaching practice and knowledge of the school system – mastery of subject and methodology for teaching it – "philosophical reflection" about values.

[46] Soviet educators have compiled a similar list. In addition to teaching practice secondary teachers should study all or some of the following: theory or philosophy of education; history of education; general didactics; child psychology; specific teaching methods; use of teaching aids, including computer technology. Cf. Z.A. Malkova and B. Vulfson, editors, "Secondary Education in the World Today", *International Yearbook of Education*, Vol. 39, Geneva, 1987.

content, will vary according to the level at which the teacher is called upon to teach.

16. Specialization

Durkheim's characterization of modern society as one requiring an increasing division of labour — in effect, specialization — is relevant. Just as medicine has two main branches, institutional health care and care in the community, each with a large number of specialities, it may well be that teaching has now started down the same road. In teaching, specialization can be by level of teaching, or in many other ways. In most European countries the former is already well-established practice, although in Britain as late as the 1960s it was possible to obtain a polyvalent teaching certificate, and the qualification awarded is restricted to primary or secondary education. Transferability between levels is becoming more and more difficult, although governments faced with a shortage at one time of primary teachers, and six years later with one of secondary teachers, might on occasion wish it were otherwise. The career structure that enabled the village primary teacher in France, for example, by dint of hard work, to rise to the rank of university teacher — or even, in one case, to become minister of education! — is almost outmoded today. Other specializations also occur — it is alleged that the Netherlands has over a hundred different types of teaching certificate. In some cases, of course, specialisms are essential. There is a broad distinction to be drawn between general and vocational education, and within these two branches between subjects, although the secondary school teacher, or even the primary teacher, capable of teaching one subject to a high level and two or three others to a lower level is a dying breed. In countries such as France, Italy and the Netherlands upper secondary teachers now normally teach only one subject, although in Sweden and Switzerland two (occasionally even three) are possible. The teacher in the Danish comprehensive school can theoretically teach seven subjects up to seventh grade (Danish, mathematics, religious education, three practical and artistic subjects, and civics.) How far this versatility occurs in practice is not known with certainty. Other specialisms naturally occur in the field of physical and aesthetic education, or in the education of the physically handicapped or mentally retarded, although both groups are increasingly being taught now with their more fortunate peers. Even within the physically handicapped group there may be sub-specializations: the education of the blind, the deaf, the paraplegic, etc. There are the beginnings of a specialization in the teaching of what is termed the "slow learner" — those termed a future "educational underclass" in Chapter 4, who will leave school with no formal qualifications. Some would argue for specialization in the education also of the academic high flyer – the "gifted" pupil.

One may question as to how far down this road of specialization one should go. It does not always work to the teacher's advantage. The future teacher of a modern language, for example, may find the prospect of teaching practically nothing but the rudiments of the language to lower secondary pupils for the next thirty years somewhat daunting. Indeed there was much to be said for the old class-teacher system, particularly at primary level, where one teacher remained with his class almost the whole school day and was able to teach most subjects fairly competently. Of course, the training of the polyvalent generalist, as contrasted with that of the specialist, would demand a very different approach. Indeed, is it feasible in training of specialist teachers to find common ground between all specialities? On the analogy of courses in other subjects with many specialisms, such as engineering or accountancy this should be possible. Yet even within a discipline there may be a case for different academic and professional training. The general science teacher, whose task is to give a general background in science to the 11 to 16 year-olds may well require a different type of knowledge and know-how to his more specialized colleague teaching a specific science to upper secondary pupils. Are we, in fact, on the verge of a fragmentation in teacher training, because of the increasing complexity of the modern world?

17. The Educational Career of the Would-be Recruit

The traditional sequence of education for the future teacher: school – higher education – teacher training – return to school – is increasingly being called into question. From infancy to old age the teacher who has known nothing outside the world of education may well be an anachronism. Lack of experience of the "real world" is reckoned to be counter-productive. There is a demand that at some stage before embarking on teaching, and at regular intervals throughout his teaching career, he should be faced with other kinds of work experience, preferably in industry or commerce, in the world of work which his pupils will eventually enter. The problem of the sort of occupational experience that would be most valuable is therefore posed, and is one that has also an organizational facet.

18. Summary and Concluding Remarks

The reform of teacher education so that training will in future be located in universities or other institutions of higher education is a benchmark in the training of teachers. Whether, as in the reforms proposed in France, the content of courses will be raised to a higher level, is problematic. It can be argued, for example, that more advanced courses in child development and the psychology of young children would be more useful for nursery and primary teachers than raising their general academic standards. On the other hand, in secondary education the quality of graduates entering teaching, as evidenced by the lack of entrants with higher degrees, would be enhanced by more rigorous courses in the subjects they are to teach. Such an upgrading would moreover lead to an enhancement of the status and respect accorded teachers by the general public, which has slipped badly in recent years – although in some countries it could be argued that an improvement in salaries would do as much.

There is a general recognition in all sectors of employment today that updating and further training is imperative. This applies perhaps with even greater force to teachers, who tend to rely on a store of knowledge learnt in their youth and less on what they learnt in their initial training than on what they have drawn from experience. The different experience, however, of say, one term in five years spent away from the classroom in the company of their peers, discussing the problems of teaching generally, and of their subject(s) in particular, and learning new knowledge and methods, would do much to raise both morale and professionalism, and, ultimately the quality of the education dispensed in the schools. At present a large number of uncoordinated efforts to retrain are in progress. One can, however, envisage refresher courses eventually on a European basis. Such courses will also be needed if the weakening of the sharp division between primary and secondary education, which can be observed in many European countries at present, will continue in future.

Enhancing the quality of teachers and teaching has been a central objective of educational reforms implemented in European countries during the 1980s. This new interest in teachers and their professionalism has been fuelled by concern with international trade competition and by the promise of high teacher turnover in next decades. In the wake of these reform efforts there has been a significant growth in the number of research studies addressing the question of the relationships between policy, productive teaching, school improvement and student achievement. [47] At a general level, this research provides strong

---

[47] Over the past 25 years the body of research on teaching policy and the relationship between "productive" teaching and student achievement has grown by leaps and bounds. Several analyses of hundreds of studies are now available, see for example J. Brophy and T.L. Good (1986), B. Fraser *et al.* (1987) and H.J. Walberg (1989). M.S. Smith and J. O'Day (1988, p. 4) of Stanford University group the research into four categories roughly representing an historical progression in the literature. These categories, which are

arguments for the thesis that teachers ought to be well grounded in the subject matter of their discipline. Indeed, Marshall Smith and Jennifer O'Day [48] conclude on the basis of an extensive review of empirical research studies that "both depth and breadth of knowledge in a content area are critical for teachers entrusted with giving students the opportunity to engage in complex cognitive activities". The fostering of teachers' professionalism is therefore the key problem that national ministries throughout Europe face in the 1990s.

This paper has raised more problems than answers, but it has done so in the context largely of Western Europe. It is clear that the time is ripe for the review now under way of the questions raised by the recruitment and training of teachers. Whereas contacts among Western nations have been numerous for half a century, ever since 1939, educational developments in Central and Eastern Europe have remained almost a sealed book to those directly concerned with future teachers. Now the possibility of a fruitful exchange of views is possible for the first time. Divorced from all questions of ideology, it is certain that both East and West have much to learn from each other.

---

distinguished by the relative emphasis placed on content and context-specific approaches to teaching, are briefly: (1) research on teacher characteristics, (2) research on teaching strategies, (3) research on teaching special populations, and (4) teacher competency research.

[48] M.S. Smith and J. O'Day, "Teaching Policy and Research on Teaching", 1988, p. 24. This paper was originally prepared for an OECD study on the conditions of teaching, and an abridged version is presented in Chapter 4 of *The Teacher Today* (OECD, 1990).

Chapter 10

The Economic and Financial Aspects of Education: Prospects for the 1990s

1. Introduction

A crucial question for the development strategies of educational systems is the nature of relationships with the economy. A new conception of the role of education for the economy emerged in the post-War period, when growth became a major concern of economic development policies. Investment in physical capital was perceived as the main instrument of economic development; by analogy, education was identified as a factor of economic growth, and compared to an investment.

This conception was supported by empirical and theoretical research, and eagerly adopted in a simplified way by policymakers and supported by public opinion; the slogan "investment in education" replaced the traditional view of education as a social expenditure. The belief in the profitability of education was, however, progressively weakened in the late 1970s, as a consequence of the world economic crisis. More recently, in the more optimistic 1980s, new conceptions concerning the training requirements of the labour force have emerged.

Parallel to changes in the perception of the relationship between formal education and the economy, conceptions about educational expenditure, costs and financing have also varied.

2. Formal Education and the Economy

During the period of accelerated economic growth of the 1950s, 1960s and 1970s, the "thirty glorious years" as they have been called by Jean Fourastié, [1] the French economist, the conviction that education was a direct investment in the economy was one generally shared: education was considered as a productive investment, still more profitable than physical investments; the "theory of human capital" allowed this profitability to be measured. [2] Moreover, the application of planning methods to the "needs of qualified manpower" allowed the linkage to be made between economic objectives for each sector of economic activity and forecasts of needs in vocational qualifications, which were translated into training needs and forecasts of numbers to be educated in the different levels and branches of the educational system.[3]

The world economic crisis of the period 1975 to 1985 greatly contributed to these conceptions being called into question, sparking off a period of scepticism concerning the quality of education in general and its overall profitability. Since the mid-1980s, however, the economic revival in the industrial countries has once again laid stress upon the importance of education and training in a world of economic

---

[1] Jean Fourastié, *Les Trente glorieuses*, 1979.
[2] The OECD was influential in initiating research studies into the relationship between education and economic growth. The economist Denison's work was typical of the period. The concept of "human capital" has dominated the economics of education since the American economist Theodore Schultz (1961) analyzed educational expenditure as a form of investment, and others, among whom E. Denison (1962), Jacob Mincer (1962) and Gary Becker (1964), published seminal articles in leading periodicals in which the process of human capital formation was studied and the rates of return to investment in education and training were calculated. According to George Psacharopoulos (1981, p. 326), four consistent patterns have emerged on the basis of studies estimating the social and private rates of return to educational investment: (a) Both the social and private returns are the highest at the primary level compared to education at all other levels of the system; (b) the private rates of return are generally higher than the social returns; (c) most rates of return to investment in education are above the 10 per cent threshold level commonly used in assessing the opportunity cost of capital; and (d) the returns to education are generally higher in the developing than in the developed countries. Evidence supporting the human capital perspective can be found, for example, in Psacharopoulos (1984 and 1987), and in a chapter entitled "Educational attainment of the Labour Force", *OECD Employment Outlook, 1989* (July). Important criticisms of the human capital "paradigm" are contributed by Rosen (1979) and Blaug (1985).
[3] The "manpower planning" methodology was developed by OECD (H. Parnes, 1962) and applied to six European countries in the "Mediterranean Regional Project"; for a more sophisticated application of the methods, e.g. to Argentina, see OECD (1967); for a critical analysis of the forecasting methodology, see R. Hollister (1967). B. Ahamad and M. Blaug (1973) also present theoretical and empirical criticism worth noting.

competition.[4] The relationships, firstly, between initial and post-initial education and, secondly, between education and the economic competitiveness of the nation, have once again come to the fore. Today, the question of whether Europe is underinvesting in school education and in the development of human resources in a lifelong perspective, is insistently being asked. There is, of course, nothing new about this concern with further education and training. The basic ideas have been around at least since the 1960s, and appeared under labels such as recurrent, continuing, and lifelong education.[5]

However, the concerns are now different from those of the earlier period. For example, the extension of schooling to the end of the secondary level is increasingly recognized as a desirable objective for raising the general level of education in the labour force. Another novelty is that the policy context of recurrent or lifelong education has changed as a consequence of recent technological innovations, changes in the content of work and the organization of the workplace and increasing competitive pressures in a global market economy. Because the stakes are higher than before, [6] there is a chance that a political consensus about the desirability of developing an all-European strategy for investing in lifelong education will emerge this decade. Naturally, questions are being asked in the current policy debate about the role of school education in such a new approach. The restructuring of education in accordance with a lifelong model of human development necessarily has implications not only for the objectives of basic education but also for its financing. This is recognized in countries such as Sweden and the Netherlands, where questions concerning the design of a novel approach to educational finance are being debated at present (incidentally, J. Ritzen, the current Dutch Minister of Education, is a university professor in the economics of education by profession).

There is a new tendency to assign to the school system the objective of ensuring a minimum level of basic knowledge or competence for

---

[4] These views are well synthesized in: OECD, *High Quality Education and Training for All*. Report to the OECD Education Commitee and the Ministers of Education of the Member States, 1990.

[5] Cf. Chapter 3, Note (25).

[6] Gregory Wurzburg *et al.* (1991) present a comparative analysis of changes in the policy context of recurrent education and training in three OECD countries: Australia, Sweden, and the United States. They also discuss the nature of the "stakes" – social, economic, technological, – the different OECD countries have in the further education and training of their labour force. In all three countries, the current debate about what is referred to as "post-initial" education is mainly focused on economic issues. Prominent among these are questions concerning the implications of change in the interpretation of the relationship between initial and post-initial education for educational financing, and for the incentives useful in encouraging investment in education and human resources development programmes in particular.

everybody, and a level of general culture (which should include basic scientific and technical training): vocational training proper being given either alternately, or afterwards, as close as possible to employment. It is repeatedly being said that school syllabuses and diplomas must be validated through external assessment methods, either instead of internal examinations or parallel to them. Similarly, procedures for the certification of competencies (independent of the school system) should allow each individual to have taken into consideration his vocational qualifications in the employment market. Beyond compulsory schooling, educational expenditure should not come exclusively from national budgets, but from the increased participation of all concerned: local authorities, business enterprises, and families. Educational institutions should work in partnership with local authorities and local economic interests. Rather than making detailed manpower forecasts by occupational categories and levels of qualification, it appears desirable to improve the continuous flow of information concerning the functioning of the labour market by sectors of economic activity and occupations. One seeks "conjunctural" indicators concerning the supply of diploma-holders on the labour market, the distribution of wages and the way these evolve, and the structure of unemployment and its characteristics by sex, age, region, occupational and educational distribution, *et cetera*. [7]

The contribution of education is today no longer considered, to the same extent as it had been during the period of continuing economic growth, as a factor in the equalization of opportunity in society. Nor is it viewed any longer, as it was during the period of economic crisis from 1975-1985, as a component of public expenditure that had to be restrained. At present, education rather appears as a factor able to play an important role in the struggle against unemployment, and in international economic competition. [8]

Because educational attainment is likely to enhance a person's chances on the labour market, a higher level of educational attainment may make a person less vulnerable to long term unemployment compared with someone who has received a short and, from a job selection perspective, often a poorer education. The evidence available from several studies on the relationship between formal educational attainment and

---

[7] Educational attainment is increasingly being regarded as an economic variable, representing the labour force qualifications of an individual. The limitations of using estimates of the educational attainment of a population as an indirect measure of its qualification level are discussed in an important article by the OECD secretariat: "Educational attainment of the labour force", published in *OECD Employment Outlook* (Chapter 2, pp. 47-93), 1989 (July).

[8] Cf. OECD, *High Quality Education and Training for All*, Paris, 1990.

unemployment, [9] which have been conducted in different OECD countries, support this contention: with the exception of both young and older women, the risk of becoming and staying unemployed for a relatively long time period is lower the higher the level and quality of formal education received early in life. According to an OECD source, only the highest level of educational attainment is not consistently associated with the lowest unemployment rates. [10]

These remarks set out the general trends in the evolution of ideas concerning the relationship between education and the economy, as they can be observed in recent reforms or those in preparation, as well as in reports produced by OECD and the Council of Europe. Naturally such simplifications would have to be qualified, for there exists a strong diversity between European nations, although it is tending to diminish, through convergent trends of change and development, which might facilitate further harmonization of educational policy in Europe.

3. The Supply and Demand of Basic Education

The matching of supply and demand for education has been a major concern since the end of the Second World War. It raises new problems in a Europe that has set itself on the path to the free movement of people, including the recognition of qualifications, so that every citizen has the right and the possibility to carry out, in the country of his choice, the job that corresponds to the qualifications attested to on his diploma. In the long run this implies an increasingly greater harmonization of the educational systems of Europe.

The notions of supply and demand are employed by economists to analyse the functioning of markets for goods and services. By analogy, they can be used to describe the evolution in the demand for education, and the supply of educational services designed to satisfy that demand. It can also be used to describe the matching of the economic demand for qualified labour with the supply of those emerging from the training mechanism with a diploma. This analogy has some utility in the analysis of the imbalances of supply and demand of qualified workers on the labour market, and of the adjustment mechanisms as a function of salaries and numbers of qualified workers. It is less accurate when it is a matter of describing the supply of educational services provided by the

---

[9] This relationship between educational attainment and unemployment is well documented. Examples of studies in which the relationship is investigated are Bednarzik and Shiells (1989), OECD (1989) and van Leeuwen and Dronkers (1991).
[10] Cf. OECD, "Educational attainment of the labour force", *OECD Employment Outlook, 1989*, Paris, p. 79.

public authorities in the form of "public goods" which are distributed free, or quasi-free, to the population.

As regards the demand for educational services, two great tendencies will continue to exert an influence. Firstly, social demand for education will continue to grow this decade. Secondly, demographic fluctuations will have an effect on demand for education and, eventually, on educational provision. The conjunction of these two factors will continue to cause severe tensions in educational systems.

It might be thought that the abrupt slow-down in live births experienced in Europe from the 1960s onwards, which contrasts with the post-War "baby boom", would be translated into a progressive easing of the explosive expansion in school attendance at all levels. In reality it has been found that the tendency to prolong attendance at secondary and higher levels (as well as in pre-school education) has, almost everywhere, outweighed the effect of the demographic decline. Most European countries are still far below the levels of attendance achieved in the United States and Japan, but they are continually drawing closer to them. Even though this general tendency carries with it variations depending on the countries and their incidence over time, it may be anticipated that the 1990s will confirm the development that was begun towards the generalization of secondary education, and the access to higher education of an ever-growing proportion of the relevant age group.

As regards supply, it must be noted that the public authorities have attempted to satisfy this social demand by the expansion of educational systems, rather than to resist it. Countries that have indulged in planning based on forecasts of social demand have often underestimated the rates of growth of this demand. Countries which have sought to give priority to forecasts of needs for qualified manpower have not succeeded in controlling through voluntaristic measures the development of this increasing demand for courses leading on to higher education. As argued in Chapter 9, Sweden may be an exception.

This expansion of school systems in which social demand exceeded the supply of educational services by the public authorities has had a number of consequences. The first is an unprecedented increase in the level of public expenditure on education, the share of which has increased considerably in national budgets, as well as in national income. Secondly, the number of secondary schools and higher education institutions has increased substantially in order to meet the growing social demand. A third consequence has been the massive recruitment of new teachers at all levels of the educational system.

The world economic crisis, which lasted from about 1975 to 1985, slowed down the growth in public educational expenditure, with some delay and in very different ways, depending upon the country. But social demand has continued to grow, and this has brought with it tensions, and in several countries changes in attitude towards the public educational system. A growing dissatisfaction of public opinion regarding the quality of education can be noted. This, in turn, has led to an increased pressure on the part of the public authorities to demand accountability from the educational system, and the development of out-of-school educational provision. Moreover, the sudden and unexpected fluctuations that have often characterized the evolution of the social demand for education have brought about an increased interest on the part of decision-makers as well as researchers in the motivations (economic or otherwise) behind individuals' behaviour, influencing their choice of education and training. Parallel to this, families appear to attach ever greater importance to freedom of choice as regards educational institutions, as well as to information regarding the qualitative results achieved by them.

The pace of teacher recruitment has undergone abrupt variations because of demographic reasons: in order to provide for the "boom" generations after the War, massive recruitment of teachers was necessary from among the much less numerous pre-War generations, often through expedients that diminished the length of professional training. When the more numerous post-War generations arrived at higher education, the recruitment needs for teachers had been drastically reduced, since the number of young people undergoing compulsory schooling was decreasing because of the considerable decline in the birth rate, and at a time when the teachers recruited in the preceding generation had not yet arrived at retirement age. It is in the 1990s that a massive new recruitment of teachers will have to occur. In France, for example, it is thought that it will be necessary to train each year 25,000 primary and secondary teachers,[11] i.e. four times more than in preceding years, whereas the length of professional training for teachers has been extended by one year under the new reforms.

Will the free movement of workers that should come into force in the countries of the Community from 1993 lead to a migration of teachers from countries where there is unemployment among teachers (as in Germany and the Netherlands), to those countries where there is a shortage (as in France)? The prospect of such an adjustment shows the importance that the teaching of foreign languages will assume in the coming years, as well as the harmonization of professional training.

---

[11] Some of the reasons for this development of teacher supply in France are discussed in the preceding chapter.

4. Financial Issues

Financial questions include several distinct areas: the expenditure, costs and financing of education. The main questions that decision-makers and public opinion raise concerning them can be formulated as follows:

Firstly, is there a limit, or an optimum level, for national expenditure on education, and especially for its share in public expenditure as a whole? Do international comparisons, as well as the extrapolation of recent trends, provide forecasting criteria for the 1990s?

Secondly, is the long-established trend in the rise of education costs irreversible? Is there any prospect of controlling the way in which they evolve?

Thirdly, can one count upon an increase in sources of financing other than those coming from the public purse, by requiring families, business enterprises or local authorities to pay more?

For each of these issues, concepts have evolved, and changes can be observed from one period to another. These concepts are closely linked with views and theories concerning the economic function of education. Finance is just one aspect of the economics of education. It cannot be examined in isolation, for example as a mere problem of financial logistics. Although the main emphasis is concentrated on the present period and prospects for the next decade, it is helpful to recall the principal ideas on which researchers, decision-makers and public opinion have relied in the two preceding periods.

5. Educational Expenditures: What are the Limits?

There were three different views concerning the economic and financial issues of education in the post-War period of economic growth from about 1950-1975. Even though they were partly contradictory from a theoretical point of view, their policy implications nevertheless converged, and were favourable to a considerable increase in educational expenditure.

Firstly, most educational planners have engaged in forecasting manpower needs in order to determine the levels and types of vocational qualifications that correspond to the growth objectives of the different sectors of the economy, so as to translate them into training objectives, according to the level and type of qualification. From this viewpoint education was conceived of as a training system that had to be matched to the structure of employment.

The "integrated planning of human resources" was the model for this approach in the 1960s, one that was often termed "manpower needs"; it has been used in many industrial as well as Third World countries, either as a part of social and economic planning, or as simple forecasts. [12] All the applications of this forecasting model highlighted the considerable needs for "highly qualified manpower" (which were considered as "deficits" that had to be filled). This led to advocating the expansion of the higher levels of the education system, particularly in the scientific and technical fields, and consequently to recommending an increase in educational expenditure as a priority over other public expenditure.

By contrast, the concurrent model advocated by many neo-classical economists, referred to as the *human capital model*, measures in monetary terms the economic profitability of education, by interpreting differences in the wages and salaries of people with different levels of educational attainment as reflecting differences in labour productivity. In comparing the additional advantages for the life time period with the additional costs of education (direct and indirect, such as earnings foregone), the rate of return to education can be estimated. The cost-benefit analysis technique can thus be used for estimating the "rates of return" of the various levels and types of education. Calculations of the profitability of education show that the rates of return to education are higher for individuals than for society (since most of the direct educational costs are financed through public funds); both are generally greater than for other forms of investment. [13] This led to recommending increases in educational expenditure considered particularly profitable. "Education as investment" became a slogan during this period, during which the World Bank began to extend its loans to educational programmes. [14]

A third conception of education considered the *demand for education* from families as a social objective that the developmental policies of governments had to satisfy. Taking into account the increased social pressure for access to secondary and higher education in all countries from the 1950s onwards, all the forecasts for an increase in numbers based on the extrapolation of "social demand" led to an unprecedented growth in educational costs.

The increase in educational expenditure has been even faster than the exponential growth of numbers in education (in this respect the talk has been of "the enrolment explosion in the schools"). A comparison over a

---

[12] See, e.g. footnote (3) above.
[13] Psacharopoulos, *op. cit.*(1984). See also G. Psacharopoulos, Returns to Education: A Further International Update and Implications, *Journal of Human Resources,* Vol. 20 (4, Fall 1985), pp. 584-604.
[14] The involvement of the World Bank in the development of education in Third World countries since the early 1960s is described in the *Education Sector Working* [or Policy] *Papers,* which are published irregularly since September 1971.

15-year period carried out by OECD from 1950-1965 showed that the share of resources devoted to education had everywhere clearly grown much faster than the Gross Domestic Product, and than that of total public expenditure. [15] It was plain that this relative advance could not be sustained indefinitely, but signs of a slow-down only began to be noticeable at the end of the 1970s, several years after the beginning of the world economic crisis. The same comparative study has shown that governments were not succeeding in controlling this increase in expenditure. Governments have only recently recognized the need to monitor the evolution of educational expenditure and its components.

This historically unprecedented increase in educational expenditure does not seem to have been a consequence of research findings, planning methods or even deliberate policy measures. It may rather be attributed to the "social pressure" of public opinion, which was very favourable to this particular priority in public expenditure, setting aside differences between the conceptions as to how educational reforms should be carried out, or between the programmes of the various political parties.

Changes in views and attitudes have in the main been brought about by the economic crisis and its consequences for public expenditure. Yet we also find that research developed over this period, the outcome of which produced results that were more refined and diversified. Table 10.1 shows the development of public educational expenditure in the Netherlands from 1900 to 1988. It can be seen that public expenditure in Dutch florins has increased continuously since 1900. The share of the public budget in educational expenditure has also increased, from 42 per cent of the total in 1900 to 97 per cent in 1980. Expenditure measured in per cent of national income increased until 1975, when it reached 9.3 per cent. It can be seen that the world economic crisis of the 1970s had a dampening effect on the growth in educational expenditure in the Netherlands. Although expenditure has continued to increase in current prices, educational expenditure as a percentage of national income and expenditure measured in per cent of total public expenditure has decreased since about 1975.

The increasing use of computers in economic and social research has allowed the testing of conceptions and theories linked to the "manpower approach", which assumed a strict congruence between the structures of occupations, qualifications and formal training through the educational system. It was found that the occupational and educational structure of the labour force shows large (and unexplained) variations between countries, regions, and firms.

---

[15] Cf.M. Debeauvais, *Etude comparative sur les dépenses d'enseignement dans les pays de l'OCDE et leur évolution,* OECD, Paris, 1970.

Table 10.1: *Public Educational Expenditure in the Netherlands, 1900-1988* [16]

Year	Educational expenditure in millions of Dutch florins	Of which paid from the public budget (per cent)	Per cent of national income (net market prices)	Dutch florins per capita	Public educational expenture in % of total public expenditure
1900	26.9	42	1.5	5	—
1905	34.2	45	1.7	6	—
1910	47.2	64	2.0	8	—
1915	61.1	62	1.9	10	—
1920	180.4	71	2.9	26	—
1925	193.4	70	3.4	26	—
1930	255.6	58	4.1	32	—
1935	210.1	65	4.4	25	—
1939	199.6	68	3.5	23	—
1946	324.5	76	3.3	34	—
1950	557.4	66	3.3	55	7.3
1955	1,126.0	64	4.1	105	10.4
1960	1,994.3	88	5.2	174	19.8
1965	4,319.5	91	6.9	351	22.9
1970	8,535.1	91	8.1	655	26.2
1975	17,694.0	93	9.3	1,295	24.5
1980	26,001.0	97	8.6	1,845	19.7
1985	26,480.0	—	7.3	1,835	15.1
1988	29,241.0	—	7.6	1,989	15.9

On the other hand, the theory of "human capital" postulated infinite possibilities of substitution between occupations, qualifications and training, in accordance with variations in relative wages in the labour market. It was further assumed that all differences in salaries were due only to differences in education. When individual-level data were used instead of "age-earnings" profiles based on national averages, new variables were introduced in earnings functions for analysing the wage

---

[16] Source of data: The Netherlands Central Bureau of Statistics (CBS) and J.M.M. Ritzen, 1989.

structure, or through path analysis identifying the relative impact of variables over time. [17] These investigations have reduced the role attributed exclusively and a priori to education in the theory of human capital and have shown the importance of other structural variables in the distribution of salaries, especially the socio-economic characteristics of wage earners. These analyses of individual-level data have also emphasised the wide spread of variations in the relationship of education to employment: from one country to another, one region to another, one firm to another, one social group to another, *et cetera*. Moreover, the empirical observation of the evolution over time of school populations has shown that the "social demand" for education is subject to rapid, unexpected fluctuations and this has underscored the shortcomings in forecasting school numbers based only on the extrapolation of past trends.

The theoretical discussions between economists and planners on the respective merits of the three conceptions and the methods of analysis to be employed have naturally hardly penetrated beyond specialist circles. What decision-makers and public opinion have extracted from them was what was in line with their own expectations. To the extent that the conclusions of research legitimized for all three conceptions, as a priority, increases in educational costs, which in any case the decision-makers, under the pressure of public opinion had decided to make, these conclusions were favourably received and widely publicized.

In this respect we may note that a retrospective examination of the theoretical conceptions relating to the period of growth has not been systematically undertaken because from the end of the 1970s onwards the context of the crisis had led the decision-makers and public opinion to substitute for the question of the absolute priority given to education questions relating to: the reduction of public expenditure, the effectiveness of the utilization of resources, and the quality of education.

The different levels and sectors of the educational system have expanded at different rates in relation to developments in social demand and budgetary considerations. Table 10.2 shows the tendencies in the Netherlands during the "period of decline" from 1975 to 1984. The data mirror the priorities of the government with respect to the development of education. It can be seen that the intake capacity of pre-primary education did not increase over the period and that primary enrolment

---

[17] Examples of such studies are M. Boissiere, J.B. Knight and R. Sabot, Earnings, Schooling, Ability, and Cognitive Skills, *American Economic Review,* Vol. 75 (1985), pp. 1016-1030; and A. Kalleberg, Job Training in Norway: Organizational and Individual Differences, *International Journal of Educational Research,* in press. A follow-up study is presented in A. Tuijnman, Continuing Education, Training, and Life Chances: A Linear Structural Relations Analysis, *International Journal of Educational Research,* in press.

declined, as did expenditure. The indices show that priority was given to the development of special education, and lower and middle-level technical and vocational education. Whereas university enrolment increased by 30 per cent during the decade, expenditure was not increased to the same extent. Instead of additional investment in university education, the Dutch government increased expenditure on higher vocational education. The trend emerging from the data is that enrolment increased at all levels, with the exception of primary education, due to the birth rate decline. The development of educational expenditure shows that the emphasis was on the strengthening of vocational rather than general education.

Table 10.2: *Indices of Public Educational Expenditure in the Netherlands by Type of Provision (in 1975 prices) and Enrolment (number of students)* [18]

Type of provision	1975	1981	1984
Pre-primary education	100/100	103/77	85/77
Primary education	100/100	103/87	84/75
Special education	100/100	126/113	122/118
General secondary education	100/100	107/108	98/107
Lower and middle-level vocational education	100/100	106/93	113/124
Higher vocational education	100/100	119/121	125/132
University education	100/100	105/127	93/130

The period of world economic crisis that began in the mid-1970s modified from the early 1980s the context that had been favourable to educational expenditure, under the pressure of an imperative need to reduce public expenditure generally.

This change in prospects concerning expenditure on education is faithfully mirrored in the following quotation: [19]

---

[18] Source: J.M.M. Ritzen, *op. cit.*
[19] This quotation is taken from Hasso von Recum, "Financing the Future of the Educational System of West Germany", 1985.

> "It is more essential than ever that educational policy develop an awareness of the principles of economic efficiency. The arguments derived from the time of growth-oriented educational policy can no more provide suitable strategies for survival for this post-expansionary era than can hand-wringing self-pity, pedagogical extremism, or the simple denial of the existence of scarcity".

Although statistics concerning educational expenditure are scattered in many budgetary documents, and take a long time to become available, several comparative studies have shown a decrease in educational expenditure in most countries over the last ten years. [20]

The consequences of such reductions have been serious in the majority of European countries: a decrease in expenditure on maintenance; a slowing down or a halt in investment expenditure; in certain cases a relative or absolute decrease in teachers' salaries. One may wonder whether these tendencies will continue during the present period in which the economic crisis is generally considered to be over, and is replaced by a certain resumption of growth, even if this growth is weaker than that in the "Thirty Glorious Years", and qualitatively very different from the post-War period. In any case it appears unlikely that there will be a return to the period of the "fat years". Public opinion is apparently less unreservedly favourable to an indefinite and global increase in educational expenditure: there are other competing priorities (health, pensions, *et cetera*); the increase in the burden of public expenditure seems to have reached a ceiling; the attitude of families to public education is no longer as unconditionally favourable as before; the advances of "consumerism" now extend to the school.

However, one can also observe several factors that may possibly entail an increase in expenditure on education and training: the increasing pressure of international competition has led governments to consider education as an essential factor in economic competition.

Moreover, the priority recently given to the "quality of education", as well as the measures taken by many European countries favouring a decentralization that would bring the school closer to its "users", are liable to influence the amount of resources devoted to education. But it is not certain that this will be to the benefit of the public education sector, without compensating factors such as a fresh distribution of responsibilities and power over decision-making.

---

[20] Cf. J.-C. Eicher, *La crise financière dans les systèmes d'enseignement*, 1989. Eicher presents the available evidence on the trends in educational expenditure by world regions from 1960 to 1985, as measured by ratios to total public expenditure and to gross national product.

6. The Costs of European School Education

During the period of growth it was usually considered satisfactory to calculate the cost per pupil by dividing the total public expenditure relating to a level or type of education, for example general or technical education, by the number of pupils concerned. Certain advances have been made in the allocation of public expenditures on education so as to assign them more accurately to the various items, and in accounting according to types of expenditure (teaching and non-teaching staff, running costs, depending upon the nature of these and the way in which they are apportioned). Techniques of "programme budgeting", which allow the measurement of the cost of reforms or specific programmes, have been introduced.

Yet it may be said generally that these aggregate methods have not allowed an effective control of increases in expenditure nor any means of forecasting it correctly. Methods of cost analysis, systematically practised in the production sector of the economy, and increasingly in the public sector, are still used infrequently in the field of education. They require precise studies at the school level, for it is in this sphere that it is possible to study the effective use of financial and non-monetary resources, such as depreciation costs, real or imputed rents, the number of teaching hours provided in the different subjects using different methods. Moreover, it is in the individual school that educational services are supplied to the users, who are the pupils. Reforms seeking to increase the autonomy of schools should include a knowledge of school costs as well as of its methods and pedagogical outcomes.

Studies of costs undertaken in the schools of various countries [21] highlight a certain number of converging conclusions: although almost all countries use egalitarian criteria to finance their official schools, the costs per pupil present considerable variations, according to the schools. These differences, which only come to light through this type of study, arise from factors such as the number of pupils per class, the choices or the subjects offered, the qualifications and status of the different types of teaching and non-teaching staff, the teaching methods used, *et cetera*.

Such cost variations are to be interpreted with caution, since schools have tasks broader and more diversified than minimizing costs. Cost disparities should not be viewed only as evidence of inefficiency or wastage, but as a tool for analysing the educational process, taking into account its multidimensional objectives, pedagogical methods and outcomes, and relevant characteristics of schools and students, in the context of the utilisation of resources. Moreover, a knowledge of the real

---

[21] The International Institute for Educational Plannning (IIEP) in Paris carried out comparative studies of the costs of education in ten countries (Tibi, 1989).

costs is an important element in the efficient management of schools, particularly in the framework of a policy of school autonomy. Finally, for the authorities that oversee them, data on costs at the level of the institution are a pre-condition for a policy of cost control and for improving the ways in which resources are allocated.

Up to now cost studies have mainly been carried out for institutions of higher education. Their extension to primary, secondary and technical schools seems to be a task necessarily linked to policies regarding the autonomy of institutions and the control of the use of resources.

7. The Rising Costs of Schooling

The rising costs of schooling are a problem even in the richer countries of Europe, in that an increasing share of the public recurrent budget is allocated to the salaries of teachers and non-teaching personnel. In a situation in which the public budget is generally not allowed to increase, increases in the salary costs of teachers often result in cut-backs in other, quality-related areas of educational expenditure, such as spending on the maintenance and improvement of school buildings, new text books and other learning materials and special facilities for handicapped students.

The secular rising trend of educational costs has not been put to an end during the economic crisis by the cuts in public spending. A part of the explanation is that the relative costs of school administration have tended to increase. Few European countries have experienced a drastic reduction in teachers' salaries, even if salary readjustments in the public sector often tend to lag behind during periods of rapid inflation. Few studies have assessed long term trends of teachers' salaries, which would require careful consideration for structural changes in the various categories, such as age composition and fringe benefits. Most of the data available at present have been collected by teacher unions or governments for supporting their respective positions. However, there is some evidence to suggest that teachers' salaries follow (with some fluctuations) the overall trends of wages and salaries in European countries. It is also likely that the long term increase in their salaries has not been paralleled by a similar increase in their productivity. The issue of rising costs in education is therefore linked with the debate on teacher performance and the efficiency of the school system. The relationship between class size and student performance is an excellent example of this disputed issue.

Not surprisingly, teachers are known for expressing strong feelings about the optimal size of classes. It is generally claimed that an improved teacher:pupil ratio will have a substantial and beneficial effect on student learning outcomes. However, the empirical research so far available does not tend to support this position. The contrary is more accurate.[22] Because the debate is still open, policy implications should not be hastily drawn from the negative or inconclusive findings. Even if lowering the teacher:pupil ratio does not necessarily improve student achievement, a reduction in class size may have positive effects on other, mainly non-cognitive outcome variables. For example, as such a policy can be expected to contribute towards improving the attractiveness of the teaching profession, it may also positively influence the recruitment of good teachers.

It should be noted, incidentally, that a reduction in the teacher:pupil ratio does not necessarily imply that there also will be fewer students per teacher in the classroom. Data from Sweden can be used to illustrate this point. Although the system employs one teacher per nine children attending school, there still are 20 to 25 children per school class. This is explained as the result of a combination of factors. First, there has been a marked increase in the number of special teachers, for example those employed in order to provide language training for children with an immigrant background. Second, a larger proportion of the teachers is absent from work today than was the case some decades ago. This is due not only to the provision of and extensive participation in study programmes designed for the upgrading and retraining of teachers, but also to the introduction of generous rules for income compensation during sickness.

Thus, the problem of rising costs in public schooling is, at least to some extent, a problem of making choices as to priorities of spending. The following example may illustrate this. Since the mid-1970s a lot of money has been spent on the buying of computer equipment for schools. As the costs of manufacturing personal computers and small networks has gone down considerably, the purchase of new technology is not a major issue anymore. Instead, the problem is that schools are not likely to succeed in pushing effectively the argument that an investment in new equipment is required as long as they are packed with outdated units. The question is rather as to how the school can be guaranteed a continuous access to improved technological products and efficient instructional software.

In a comparative study on educational expenditures carried out under the auspices of OECD in the late 1960s[23] it was found that most European governments were unable to provide data, or even estimates,

---

22 The relationships between school and class size, on the one hand, and indicators of student learning outcomes and school effectiveness have been the subject of much empirical research during the last two decades. Examples of such research studies are Glass et al. (1982) and B. Fraser et al. (1987).
23 The work of CERI, for example, on the Mediterranean Project.

on the distribution of public expenditure on education (not to speak of private expenditure) by categories of current spending, such as salaries of teaching and non-teaching personnel and instructional materials, et cetera, for the main levels and types of schools. This was a sign that governments were not particularly interested in measuring and controlling educational costs, at a time when public expenditure on education had a steady and more than proportional increase in the national budgets. This lack of interest can be explained as a result of the fact that potentially controversial questions are implicitly involved in an analysis of the costs of education. For example, who pays and who benefits from education and who should bear the costs of education, the users or the taxpayers? Most studies have shown that the effects of free education are more regressive than progressive, contrary to the common belief. The upper middle class tends to receive more than their share of education, and the lower middle classes bear more than their share of the costs, since they benefit less from post-compulsory educational provision than their overall share in tax revenue warrants. The question that should be asked is whether equality of opportunity is best taken care of by providing free education or by operating cost-recovery mechanisms in educational systems. Behind the factor "who pays for education", there is the choice factor. This is influenced by public opinion rather than by governments.

The unit costs in education have gone up in real terms. The cost of educating a child in school is now two to three times higher compared with the situation prevailing about twenty years ago. However, it should be noted that there are problems even with the use of seemingly straightforward measures such as the unit cost in education. For example, the aggregate unit cost may mask extensive variations between regions and among educational institutions.

An interesting comparative study of the costs and quality of child care provision in the Nordic countries was carried out by the Swedish Agency for Administrative Development in the mid-1980s. [24] There were two main reasons for undertaking the survey. Firstly, there was a general concern with the high costs of child care provision. Secondly, there was a rapid rise in public demand for this provision. The fear was expressed that the expansion of child care, to which the Social Democratic Party had made a commitment in its, successful, election campaign in the early 1980s, would turn out to be a real burden on the national budget. There was an anxiety on the part of the politicians and the public to know whether Swedish day care was more expensive compared to other

---

[24] The Swedish Agency for Administrative Development (SAFAD): *Child Care in the Nordic Countries: Costs, Quality, Management*, 1988. The study was carried out by Sture Strömqvist.

countries and whether the costs could be reduced without harming quality (or, from the point of view of administration, whether the parents could be asked to increase their financial share). Table 10.3 shows some of the main findings.

Table 10.3: *Child Care in the Nordic Countries (supply and demand, costs and finance, and quality; 1985)* [25]

Criteria	Denmark	Finland	Norway	Sweden
Supply Percentage of children 0-6 years enrolled in child care institutions (1984-1985)	52	31	28	53
Demand Estimated demand for pre-school places, per cent of all 0-6 year-olds (1983)	70	55	52	73
Unit Cost Unit costs in pre-schools, index Sweden = 100 (1985)	54	48	43	100
Finance Financing of pre-schools. Per cent private expenditure of total public and private expenditure (1985)	21	14	25	10
Quality/Cost Indicator Number of children per full-time staff equivalent and hour open in pre-school institutions (1985)	6,2	5,5	5,1	4,4
Quality/Cost Indicator Square meter per child in pre-school institutions	6,7	8,9	6,4	12,2

About half of the children aged 0 to 6 years in Denmark and Sweden were enrolled in pre-school institutions in the mid-1980s, compared with about one-third in Finland and Norway. In all four countries demand outstripped supply. The results also show that the unit cost in Sweden

---

[25] *Ibid.* The data are culled from Tables 4, 5, 8, 9, 19 and 20.

was twice that of the other countries, and that the share of the Swedish parents in pre-school finance was the lowest. Given that the cost of Swedish child care is twice as high as that of other countries, one could ask whether its quality is also twice as high. The two indicators of quality in Table 10.3 show that the Swedish pre-schools had a lower ratio of staff to children, and offered more "living space". Whether quality improvements such as these warrant the high cost is an open question.

The above illustrates the fundamental economic question: are we getting value for money? During the growth period, the predominant thought was that the more we spent on education the better. Now we are no longer so naïve as to think this, as it has been demonstrated that increasing expenditure over and above a given threshold level may not have a significant effect on performance. Thus, the question as to how the optimum cost-benefit "equilibrium" can be identified concerns us now.

8. The Financing of European School Education

It is in those countries where a considerable number of schools belong to the private sector that questions of finance have been more deeply discussed, and where research has been carried out. Until recently in most European countries, the Ministry of Education financed official schools directly; the main concern was to ensure for them an adequate share of the national budget and to apply, in the distribution of funds between schools, simple, egalitarian criteria.

The economic crisis has stimulated a fresh examination of other possible sources of financing. It is conceivable that decentralization policies have frequently had as their aim to impel local or regional authorities to shoulder a greater share of the financing of schools, responsibility for which had been partially or entirely devolved to them.

On the other hand, adult training programmes which have undergone considerable development during the last two decades have diversified the sources of their finance: employers have been encouraged through various measures such as specific taxes, fiscal exemptions, and direct or indirect subsidies *et cetera*, to finance, wholly or in part, the further training of their employees.

These new sources of finance have partly benefited schools in the public sector, principally second-level technical and vocational schools, as well as higher technical education. Yet in the majority of countries employers have gained responsibility for decision-making either completely or shared with the unions and the public authorities, depending on the case, regarding arrangements for programmes of further education.

The increased participation of families in the financing of post-compulsory education, for a long while considered by political parties to be a taboo subject, is now a concern of most governments. It still goes against a long tradition of free education, linked to principles of democratization, in spite of studies concerning the "redistributive effects" of educational expenditure, which have shown that free education most benefits the upper classes.

Until now it has principally been concerning higher education that these questions have been examined. For second-level education the debate will probably continue to be limited to the resources that technical schools can obtain, through payment, from a growth in their training activities for adults outside working hours. Other sources of finance that have been considered in some countries consist of imposing upon families the cost of certain ancillary services provided by schools such as meals and textbooks, and in payment for the use of their sports or other installations by the outside public.

Financial questions have been a priority concern since the economic crisis. Teachers have stressed in the main the deterioration in their living conditions, and on disparities indicated by comparisons with other occupations requiring an equivalent level of studies. Research on this has, however, yielded mixed results, above all when non-monetary advantages or additional perquisites and promotions and salary rises by seniority are taken into account.

It seems important to undertake research studies and to develop surveys in order to add to our knowledge of teaching costs and teacher salaries. They should be carried out in consultation with teachers and educators; comparisons among European countries would also be of interest. However, the most important criterion concerns the recruitment of teachers. When there is a scarcity of qualified candidates it is a sign that there is an imbalance of supply and demand. A shortage of qualified teachers, either generally or in certain subjects, can have consequences which are all the more grave because the effects are only felt in the long-term. In this field, as in those of costs and financing, financial questions are linked to questions of educational policy, and are subordinated to it.

The extent to which the objectives of education are perceived as important, not only by the government and decision-makers but also by public opinion and individuals, governs the allocation of funds to the educational sector in preference to other allocations. Teachers have limited possibilities to influence these decisions; yet it is up to them to make public opinion, particularly parents and pupils, understand the importance of their educational task, and to accept their participation by favouring the opening up of the school to the community. Financial questions are indispensable to an educational policy, and a field in which research is necessary, especially as free or nearly free education and factors such as leisure during school years conceal the real costs and the real beneficiaries. But solutions are subordinate to an awareness of the priority that society accords to education, and to the way that it responds to the former's expectations.

9. Expenditure, Costs and Financing: Challenges and Prospects

If by challenges are to be understood the unsolved problems of the post-War period that will be posed once more in the future, the financial questions raised by the development of mass secondary and higher education force themselves upon our attention. Three questions sum up the financial aspects to which no satisfactory answer has yet been given, which will persist in the 1990s, and which now present themselves in a new light:

Firstly, can public expenditure on education rise once more in the same way as it did before the crisis, in the new period of revival of European economic growth?

Secondly, how can one control the continuing trend towards increases in educational costs and teachers' salaries without any apparent detriment in the quality of the educational services provided by society?

Thirdly, where and how can be found new sources of finance to deal with the growth in education and training needs?

The economic crisis in most countries entailed a decrease in public expenditure, which has had repercussions, more or less delayed, upon public expenditure on education. This development has taken on very varied forms according to the country, but it has been a general tendency since the beginning of the crisis. However, its effects have been scarcely apparent for several reasons. For a certain time one can delay or cut down on current capital requirements, for example by postponing the construction of new buildings and delaying maintenance and the renewal of equipment. One can also cut down on expenditure by increasing the number of students per class, limiting the recruitment of new teachers, and putting off adjustments to salaries in a period of high inflation, without the consequences becoming apparent in the short term.

The strong inflation of the crisis years has brought about an increasing disparity between the development of educational expenditure at current prices and at constant prices, which are more complex to appreciate. In cases where teachers' salaries have gone down in constant prices, which involves a lowering of the standard of living of teachers, it does not necessarily follow that the "quality" of teaching has likewise declined. Moreover, in cases where the quality of educational services may have in effect suffered, this deterioration in teaching may not have readily been noticed by the users of education, because of the difficulty in assessing the development of standards precisely.

Since coming out of the economic crisis education has begun once more to be considered by decision-makers as an essential element in economic competitiveness, as noted in Chapter 2. Two priority objectives of education in European countries at present are, firstly, improving the quality of training and, secondly, raising the level of knowledge and competence of the whole labour force. These factors contribute to an increase in educational expenditure and, hence, conflict with the effort to put a ceiling on public spending. This explains the trend in Europe to reduce costs, and to improve the efficiency of teaching and school management, as well as the pressure to find new sources for the financing of education.

The increase in the costs of education is a long-term trend observed in every country, without it being related to any corresponding improvement in "productivity" or the quality of the service. On the contrary, the view most widely held is one of continued deterioration (real or supposed, because on this point the signs are contradictory). This tendency to an increase in costs stems in the main from the rise in living standards of teaching and non-teaching personnel, which has followed the general increase in wages and salaries. Furthermore, the progressive introduction of new teaching techniques that followed the introduction of audio-visual equipment and, above all, computers in education, has had as its consequence an increase of costs per pupil rather than the expected effect of lowering them. Are the recent advances in electronics and software going at last to yield an increase in effectiveness at the same time as a decrease in costs? Many changes will have to take place in order to convince a public opinion rendered sceptical by the unproven prophecies of the last score of years. It is perhaps outside the school system that real changes will occur through new technology: through family computers, the potential market for which will stimulate the producers of learning software to develop attractive and efficient application products, at the same time as the "miniaturization" of materials, as well as their new capabilities, will continue to be accompanied by a reduction in cost.

Another possible way of reducing the costs of education lies in efforts to improve the management efficiency of the educational system from the top to the school level. In some European countries, modern management techniques tried out in the private sector are now progressively introduced in various parts of public administration, which give some hope for their extension to the public education sector.

Concerning the search for new sources of finance, a partially successful move has been made under the opposing pressures of the limitation of the public budget, and the persistent expansion of the social and economic demand for education. Efforts have been directed in three directions: regional or local collectives, business enterprises, and families.

Decentralization reforms, when they have effectively transferred a significant part of the powers hitherto wielded by the central government, have initially had as a consequence an overall increase in expenditure, but also additional funds for schools. This growth has until now, however, been insufficient in size to reduce noticeably the gap between social demand and the available public resources.

However, firms have perceptibly increased their financing of training costs in the numerous European countries that have set in motion national programmes for the vocational training of adults. The various reforms aiming at developing these kinds of training have been called, depending on the country, "continuing education", "permanent education" and "recurrent education". They are usually, as in France, characterized by a payroll tax usable by the employer for the training during employment of his workers. Table 10.4 shows that, in Sweden, private expenditure on job training, which comprises the financial contribution of both employers and employees, exceeds that of public expenditure on adult education and labour market training.

The data in Table 10.4 indicate that there are differences between the European countries in the approach taken to finance job training. Table 10.5 shows that employers carry about half of the burden in four EC countries investigated in a recent EC survey. [26] The differences between the countries concern the financial share of the government and the individual worker. In France, for example, the workers do not seem to have to contribute directly, although one can assume that they will pay in other ways, for example because the benefits of training in this country will probably mostly accrue to the employer and not to the individual.

Compared to the effort by the private sector, the financial contribution of families has remained rather small; most of the initiatives for raising fees and introducing user's charges have until now proven unsuccessful. The European tradition of free (or semi-free) education is not only linked to the belief that free education guarantees equality of opportunity for not only the period of compulsory schooling, but at all levels of the educational system, in spite of all the research that puts in doubt this correlation. It is also because the transition from a free, tax financed system of public services to the principle of payment by the users arouses considerable resistance, even when it is only intended to levy part of the real cost (*cost recovery*) and to top it up by aid to families. If

---

[26] The survey was conducted by the EC Commission, Task Force of Human Resources, in 1988 and 1989.

the political drawbacks in measures seeking to increase the financing of post-compulsory public education by fees appear to be too high, it could be argued that since the gap between the demand for training will exceed the possibilities of supplying it, this will favour the development of a private fee-paying sector. [27] This could increase the risk of worsening social inequality in education, as well as a risk of neglecting the maintenance and development of public educational institutions.

Table 10.4: *Expenditure on Adult Education and Job Training in EFTA Countries (1989 data, approximate estimates, in per cent of gross domestic product)* [28]

Country	Private investment in job training %	Public investment in adult education %	Public investment in labour market training %
Austria	0,6	0,6	0,01
Finland	1,0	0,5	0,27
Norway	—	0,2	0,07
Sweden	2,2	1,0	0,51
Switzerland	—	1,0	0,01

---

[27] The share of private spending in total national expenditure on education varies substantially from country to country. In Belgium, the share of private spending was 2.2 per cent from 1970 to 1974, and 1.9 per cent from 1975 to 1980. The comparable figures are 37.4 and 26.2 per cent for Greece, 25.0 and 21.6 per cent for the United Kingdom, and 52.7 and 51.6 per cent in Spain. By comparison, the share of private spending in total national expenditure on education in Japan was 61.2 per cent from 1970 to 1974, and 56.7 per cent from 1975 to 1980. Contrary to common belief, private expenditure in the United States over the same period was even slightly lower than that of the United Kingdom: 21.2 and 20.5 per cent (Source: United Nations, *Yearbook of National Accounts Statistics, 1981*).

[28] These estimates were provided by representatives of the governments of EFTA countries to a questionnaire distributed in 1990. The results are given in the paper "Education and Training of the Labour Force in the EFTA Countries: Summary and Comments", which was presented by L. Arvidson and K. Rubenson at the Conference: New Challenges in Education and Training of the European Labour Force, Stockholm, June 13-14, 1990.

Table 10.5: *Sources of Finance for Job Training in EC Countries (1988 data, in per cent)* [29]

Country	Government %	Employers %	Employees %
Germany	25	50	25
France	53	43	4
The Netherlands	39	50	11
United Kingdom	20	51	29

10. Conclusion

Education will play a crucial role in the economic development and social progress of Europe. Its relationships with economy are now perceived as much more complex and diversified than during the post-War period of growth, up to 1975. It is now recognized that the quality and efficiency of the education provided to youth and adults will be a decisive factor in the increased international economic competition.

However, financial problems posed by educational development prospects are not merely technical issues limited to costing the various components of educational needs and finding the ways for providing the required funds. In the end the solutions that will be applied to these financial problems will not depend on intangible limits fixed by economic norms, but on the degree of political priority given by our society to education vis-à-vis other economic and social needs, as well as on the distribution of power in the management of the educational system: those members of society who will be called upon to increase their share in the financing will want also to share in decisions in all domains of educational policy.

Education will be an essential component in the European cooperation and integration. Jean Monnet, the inspirer of the creation of the European Economic Community, is quoted as saying at the end of his life:

> "If it had to be redone, one should start Europe by culture and education".

---

[29] The data are taken from a newsletter published from the European Centre for the Development of Vocational Training, CEDEFOP, *Vocational Training News Flash,* No. 2, 1990.

Chapter 11

Challenges and Opportunities For School Education

1. The Task

Our main task has been to identify current issues and pervasive problems which beset European school education. We have also tried to spell out the ramifications of these issues, to identify the major challenges, and to present evidence in support of our analyses.

Rather than looking at problems within schools, the focus has been on the problematic relationship between the school as an institution and society at large. Readers of the report may with some justification ask us, why we as specialists in the field of education have limited ourselves to identification of problems and presentation of evidence. Why did we not try to come up with solutions?

The main argument is that the political, economic, social and religious conditions under which educational reforms are planned and implemented vary so much from country to country that effective solutions may be very different across countries. General solutions which do not consider these differences may do more harm than good. Another reply, which is not self-evident, is, that the specialist can at best produce diagnoses and technical means which are useful in efforts to implement solutions. The latter, however, always depend on political decisions. To be sure, solutions to problems dealt with in this report require rational elements in terms of hard facts and analyses. But rationally and scientifically valid evidence does not devise a unique solution to a given problem. Scientific rationality is a necessary, but not sufficient prerequisite for practical solutions but there are always alternatives, each with its own rationale. The alternative finally chosen depends upon the prevailing political will.

Some overriding problems of school education in the new European context have been dealt with at some length in the various chapters. Here we shall try to spell out the challenges for today and tomorrow which these problems pose. It is well worth repeating that the task of identifying and diagnosing pervasive problems, challenges and opportunities besetting the functioning of schools in the emerging new Europe is made even more complex by the change sweeping through Central and Eastern European countries. It is very difficult, if not impossible, at the moment of writing this report, to acquire an adequate overview of developments in education in the latter region. Given the uncertainties of integration of Western Europe the same problems apply to this region.

2. The European School Setting Today

One of the pervading changes in European school education this century was that the nation-state took over responsibility for the staffing, maintenance and curricula of schools from parents and other educative agents such as the church. Private schools were in many places brought within the realms of public access and inspection. However, the role of the state in governing and managing the school system has recently tended to become challenged and weakened.

As education and the economy are interrelated, the economic trend towards increased globalization, the new emphases on the quality of output, accountability, and the role in production of competitiveness, modern leadership and management styles, can be expected to influence school systems. These may well become affected by the perception that education is a domain where gains in the competitive advantage of European nations can be realized. This explains why education is a burning question on the agenda of European governments and in high demand in modern economies in Europe at present. Governments seek to realize quality improvements in education which are expected to lead to increased effectiveness and competitiveness on the international markets. This search for quality and excellence is international in scope. Comparativists have begun to examine innovations which have proven to be successful in other countries. Hence, the globalization of the economy may well, in due course, bring about a harmonization not only in the core elements of the school curriculum but also in the approach taken to govern educational systems and to evaluate the student achievement and school performance. The writing on the wall is already visible in some European countries today.

Europe is undergoing a radical process of change in terms of trade and labour market integration and increased communication. Several important developments can be identified that change the context of educational policy. Prominent among these are: progress towards European integration, the stiffening of international competition, concern with the global environment, increasing economic interdependence and recent "high tech" advances not least in telecommunications. It is also important that Europe is becoming more and more aware of its cultural identity, of its being different from the United States and Japan, although these share the scientific and technical rationality specific to "Western" industrial civilization. Europe is not willing to follow any "Big Brothers" unconditionally, or so it would seem.

Promising changes have also taken place in education, even though these are more difficult to document. Of importance is the fact that citizens in European countries have been required by circumstances to ponder what a common Europe would be like, whether they want to or not. This has created a new awareness for change as well as a sense of history and future, and has served to broaden perspectives.

Paradoxically, however, the trend towards increased internationalization is accompanied by an increase in provincialism, cultural parochialism and ethnic conflicts, showing that education is provincial and nationalistic in orientation. These partly contradictory tendencies are bound to have repercussions on schools and the education they can offer. It will become much less fragmented in national school systems that today are characterized by widely different curricula, examination models and qualifications. A process of harmonization is already going on. This brings new challenges for both the nation-states with their different educational traditions and for the various European intergovernmental agencies that have been established during the last few decades.

Education as an area of government competence is, as pointed out above, nationalistic in orientation. This is increasingly becoming a challenge for educational systems. Yet there is hardly any awareness today that an international framework for school education is needed. The development of education is not necessarily going in the same direction. The traditions in the school system are so strong that even certain convergence in economic and political matters would not necessarily have to result in an immediate convergence in educational policy and in strategies of educational evaluation. Perhaps the resistance to change caused by traditions is too strong to let educational systems evolve in the same direction. Change involving harmonization and convergence in European school education, given the opposition to change driven by strong traditions and vested interests, therefore has to be planned, tried out and finally implemented over a long time span. A European school system cannot be established overnight.

The "European dimension" cannot be dealt with solely in terms of school subjects, as was the case when "international education" was introduced in the curriculum in some countries. The question of what Europe is at present, and what it may well become, is central not only to decisions about trade, industry, and monetary and labour market policy. It has, of course, practical implications for the challenges and problems that will beset our schools, not least in terms of what they should teach. As the Europe of the future is taking shape at present, it is imperative that the question of the "European dimension" of school policy be put on the agenda now.

Formal education today, as carried out in primary and secondary school, is not taking place in a social vacuum. It has to be considered with all its educative interconnections. The school is but *one* agent or institution among several others which together constitute the "ecology of education".[1] The role of the family and what the parents actually are doing with the child often tends to be neglected in the reform debate, whereas attention has mainly been focused on the role of the school. The media and the peer group should also be mentioned.

It is tempting for professional educators employed as teachers to consider educational problems as those which make their appearance within the classroom walls. Thus problems are often conceived of in didactic terms, and solutions are framed in terms of finding the "right method to teach". Problems occur even with respect to the latter. Teachers often tend to underestimate the difficulty of finding and applying a pedagogically adequate teaching method, and overestimate the subject matter they are supposed to deal with. This holds equally well for primary and university teachers.

But in considering the overriding problems and issues of school education in modern society, such as the ones we are confronted with in the emerging new Europe, we must take a holistic and broad perspective approach, i.e. we have to consider and analyze the problems of schooling in their total social context, with their interdependence and their interconnections. Such an approach is particularly appropriate in

---

[1] In *Public Education* (1976), L. Cremin provided a definition of education in which "deliberateness" and "intentionality" were central notions. Even though it is acknowledged that learning could take place in many situations where intentionality is not present, education is defined "as a process more limited than what the sociologist would call socialization or the antropologist enculturation" (Cremin, 1979, p. 18). In delineating the boundary between education and socialization, Cremin coined the phrase "ecology of education". This concept "projects us to beyond the schools and colleges to the multiplicity of individuals and institutions that educate – parents, peers, siblings and friends, as well as families, churches, synagogues, libraries, museums, summer camps, benevolent societies, agricultural fairs, settlement houses, factories, publishers, radio stations, and television networks" (Cremin, 1979, p. 19).

studying the relationships between home and school and between school and work.

It would, indeed, be presumptuous to try to draw up a priority list of the problem areas which pertain to, for example, the relationship between home and school, not least because problems vary in importance and urgency from country to country. But the study group considers that one can identify a common core of problems, which, although they vary in importance between countries, are found in most of them. Therefore, the Task Force began its work by identifying common problems, broadly conceived, and spelling out their ramifications. Without trying to put these problems in any order of priority, we would like to point to the issues mentioned below.

3. Present and Future Challenges in School Education

School education in Europe today, and to an even greater extent in the future, faces challenges of three kinds:

First, challenges connected with the *framework* created by societal changes, such as the growth of the knowledge industry, urbanization, migration, shared ecology, integrated trade and labour markets, and increased communications.

Second, challenges dealing with the *goals* of basic education, such as quality versus quantity, élite versus comprehensive education, equality of opportunity, the establishment of a bigger common core of general education.

Lastly, challenges connected with the *means* of achieving the goals emerging from the new framework, such as teacher training, accountability, and national assessment and increased "grassroot" participation in the governance of the local schools.

Framework

As a result of urbanization and migration, social pathologies, such as racism, bigotry and discrimination, and homelessness and cultural alienation, have been moving into the school to an extent that was unthinkable in the small rural or provincial setting, characterized by tight social control and the strong impact of family life. Many people uprooted and implanted into towns and big cities meet a new climate of values and the reduced possibility of maintaining family cohesiveness. They are often faced with serious social handicaps with respect to education. Spain in the last few decades provides an example of a country where certain poor rural areas have become almost deserted, and where rural workers and small farmers have poured into the big cities, such as Madrid and Barcelona.

Many European countries have experienced a massive influx of immigrants. Problems arising as a result of migration have always existed in Europe, but have tended to become more acute during the last few decades. The mobility across borders in Europe has increased immensely in two respects during the last few decades. Irrespective of the facilitating steps taken by the European Community there has been an enormous increase in the movement of labour, which often has lead to a permanent resettlement in a new country. Occasional visits to other countries, be they called tourism or not, have grown exponentially. Europe is now experiencing what the United States did in the late 19th and well into the 20th century.

There has been a stream of employment-seeking migrants to Western and Northern Europe, in the 1960s, mainly from the Mediterranean area. There is no doubt that the recent removal of the Iron Curtain is an important step in the remaking of the European map. Many expect that a flow of migrant labour from Eastern to Western Europe will be the result. Some of the social pathologies mentioned previously are related to the international migration of workers. Their assimilation into the culture of the host country poses problems, not least for the schools. These are confronted with the task of providing the children of migrants with a new language in addition to the one spoken at home. The extent to which the host country should also seek to promote the home language is subject to different opinions. In the United States the policy both at school and at home for a long time was to have the migrant children acquire the means of communication that would be beneficial for their career in the new country, i.e. to learn English, and to forget the mother tongue as soon as possible. A few decades ago a wave of ethnicity gave rise to demands for bilingual teaching. The preservation of cultural origin was regarded as a value in itself.

Some countries have legislated for bi- or multilingual teaching, which facilitates the assimilation of immigrant families into the culture of the host country. Others regard the immigrant workers more as "guest workers" and have taken few, if any, steps to facilitate adaptation to the new environment. How does the educational system handle the new influx of migrants and, especially, the schooling of children of immigrant parents? This is one of the overriding questions that must be dealt with in the new Europe.

Production systems in a future Europe will be faced with a predominantly international trade market. Their success will depend on quality and adaptability and, hence, on the application of new management styles, the measures taken to ensure quality control, the use of new information technologies and, above all, the strengthening of technology. The competitive edge of European nations is increasingly a function of the competence of their labour force. This competence is, in its turn, dependent on the quality of school education. Technological competence is by and large determined by the standards achieved in science and mathematics teaching. The level of achievement in these subjects has been low in several countries, not least because of the difficulty of recruiting good science and mathematics teachers in competition with industry.

Early childhood education, before the child enters formal schooling, has taken on a particular importance due to the changes in the structure and role of the family during the last few decades. It suffices to mention a few facts. The number of mothers working full time outside the home has increased many times over the last fifty years. The number of divorces has also increased. A growing number of children spend the day in pre-school institutions. The entry age into formal schooling is an important issue in most countries. France and Hungary, Belgium and the Netherlands, have for a long time had nursery schools catering for most infants from two to five or six years of age. In the United Kingdom children generally enter compulsory schooling at the age of five, whereas, in the case of the Netherlands, almost all children in an age group are already in primary schools at four. However, children who in other educational systems enter at the age of seven years do not appear, say at the age of 10, to perform worse in elementary reading, writing and numeracy skills than children entering at five years of age in the United Kingdom or the Netherlands. [2] Naturally, this raises the question as to what extent pre-schools matter for a child's cognitive development. It also brings into focus the necessity of questioning the appropriateness and usefulness of the fixed-age admissions scheme employed in many European countries today, as such schemes may neglect the great degree of variability in intellectual, physical and emotional development in school children.

The decentralization and extension of the autonomy of the local school presents a major challenge for local communities and the school leadership, as new management styles in education are needed. Young workers in the labour force today have, on average, received a better formal education than in earlier days. There is a discrepancy between the old 'top-down' autoritarian management style and the people who are to be managed. They demand participation in decision-making. The catchword is democratization in working life, which is to be achieved by removing power hierarchies at the workplace. There is a relationship

---

[2] R.L. Thorndike, *Reading Comprehension Education in Fifteen Countries,* Stockholm, 1973.

between educational attainment and public demand for new management styles.

Other challenges confronting school education today, against the background of European integration and other recent developments in the economics and politics of the region, concern the diversification of learning opportunities, the tripartite and parallel systems of vocational and general secondary education, the demise of overspecialized vocational education and, especially, the need for lifelong education. Here we are faced with a serious dilemma. On the one hand, future society will demand a well-educated population. The idea is that the level of educational attainment of the population should be increased. On the other hand, large groups of young people are not interested in staying on in school once attendance is no longer compulsory – usually after 16 years of age. How can we motivate and prepare children for lifelong learning if they reject school at an early age? If the period of compulsory schooling cannot be lengthened, then the school must be made into a place where children want to go voluntarily. Schools should not present a conflict with the world, as is unfortunately all too often the case.

Goals

We are confronted with challenges in conceiving the objectives of school education. One of the key issues here is the rising tide of meritocracy. Another is the return of neo-conservative attitudes to education. Challenges caused by the emergence or the reinforced importance of values ought to enter the discussion. Examples are ideas about pluralist democracy, human rights, mutual respect, the new liberalism and autonomy in the management of schools. Challenges caused by the competition schools encounter from other educative agents in society also need to be given attention.

The principal goal of universal school education was for a long time the building and consolidation of the nation-state. The school was there to inculcate values and attitudes which constituted the backbone of citizenship and to establish a uniform national language to replace minority languages and dialects. School education was charged with the task of establishing national consciousness and identity. The teaching of national history traditionally not only boosted such an identity but also emphasized differences and tensions between countries. The encouragement of a feeling of a new identity, that of being Europeans rather than Germans or Slovenians, Poles or Danes, is evidently a crucial challenge posed by the unprecedented situation at present, that of a Europe growing closer together. The school may provide a context in which the building of such a European identity can take place.

Accountability as regards what the school is doing and achieving has become a growing concern in many countries. Typically, national programs of assessment of the progress students are making are either under preparation or have already been launched, for instance in Britain and France, Norway and Sweden, and the Netherlands. National assessment was introduced in the United States of America in the 1960s as a means of arriving at what was referred to as an "educational GNP". Behind this was a concern with the loss, real or perceived, of competitive power in international trade relations between countries – as is the case in Europe at present.

For a long time after World War II increased educational opportunities and equality of access to further education following compulsory schooling were the main policy objectives, as reflected in, for instance, the educational program of the OECD. An increasing number of young people should be given the opportunity on equal terms to enjoy an increasing number of years of schooling. But in recent years quality more than equality has tended to come to the forefront in the policy debate. The improvement of the image of schools has become an important issue. Compensatory education is again being advocated. Other ideas are also being advanced, for example the idea of accelerated schooling. Characteristic is the attempt to involve parents in the teaching and learning process in school. The drawback is that a big commitment is required on the part of both the school and parents.

The goals of formal schooling in modern Europe need to be better clarified. Originally, universal elementary schooling was legislated because schools were more competent to instill certain cognitive abilities and skills, such as reading, writing and arithmetic, than the family. But given the diminished role of the family due to reduced contacts between parents and children, gradually schools have been expected to equip young people with certain non-cognitive competencies as well, such as those having to do with character formation (cf. Chapter 3). Employers, both public and private, often complain about lack of ability of the latter kind, such as taking responsibility, working independently, or working together with others. The provision of cognitive abilities and skills remains, however, the centerpiece of school education and has tended to become even more so in a society where success in formal schooling has increasingly become a decisive factor in social mobility and achieved status. In brief, the role of the school in a meritocratic and competitive society in relation to the family needs to be considered anew. The crucial question is whether the family as an educator can largely be replaced by the school. To this a negative answer is given in Chapter 4.

Means

What is important for the school of tomorrow? It would seem highly important to acknowledge that schools cannot teach everything and that many subjects, even highly crucial ones will, by neccessity, remain outside the school curriculum. In fact, the complexity of modern society makes it impossible, even for a team of sophisticated curriculum specialists, to acquire an adequate overview. The inertia in the system and the increased complexity of the tasks schools are supposed to address tend to serve as a barrier against change. There are tendencies for what could be labelled "crystallization", i.e. to keep the system as it was. Hence it can be concluded that the school of tomorrow, in the face of ever increasing demands being made on the time budget, are advised to place a high priority, first, on learning to like learning and, second, on teaching how to learn. This will require a return to personal commitment and individuality.

In a pluralist society it is easy to draw up a long list of tasks that schools ought to address. But time, in spite of prolonged formal schooling, is limited. Adding burdens to the school's mandate is most likely not going to make the school a better place for children. It is more likely that the teachers will be blamed because the school will not succeed in responding to a widened mandate. Even in a pluralist democratic system the school can hardly be expected to cater effectively for all wishes. How much can the school cope with? The problem implicit in the above is not only that effective schools can carry only a limited burden of tasks but also that a good school is a school that follows its time. Hence the dilemma is that, on the one hand, a good school is required, and rightfully so, to take a stance on current issues and to respond adequately to changes in society that form demands on schooling, while it is also recognized that it is problematic to continue widening the scope of education by piling new burdens on top of the others.

Does the historically unique situation in Europe at present demand deliberate and concerted action to create and cultivate talent? The creation of an élite does not seem to be a topic in the same order of magnitude as before. The risk that talent is not spotted and will be lost to the nation is now less than before, because primary and secondary schooling in Europe has become nearly universal. Instead the problem concerns the perhaps 20 per cent children at the top and bottom end of the achievement scale. For them, the standard of teaching is often dictated by the middle-ground approach that characterizes frontal teaching which, although outdated, is still the main strategy of pedagogics in our schools. Does this approach favour the low achievers to the detriment of their more gifted classmates, as is often claimed in the debate on which teaching methods and pedagogical strategies we should opt for, and whether we should go for early school differentiation? Many Swedish researchers tend to interpret the findings emerging from, for example, the IEA studies in mathematics and science, as indicating that the middle-ground approach tends to raise the achievement level of slow learners and low performers much more than the top achievers will lose, so that there is a net increase in yield to be gained. [3] Hence, great care has to be taken in putting too much emphasis on the cultivation of talent to the detriment of the rest.

Another challenge, one that ties in with the presence of poverty among European children, has to do with the persistent problem of wastage in education, for example the wastage of talent and resources that may result from school repetition and drop-out. Grade repetition still exists in a number of countries in both Western and Eastern Europe. In Belgium about 20 per cent of the children repeat the 5th grade of primary schooling; in France the figure is 10 per cent for the 6th grade. [4] In most countries the wastage rate is higher at the post-primary level, and higher in vocational schools than in general secondary schools. The drop-out rate tends to be particularly high in the first grade of the upper secondary school. What are the gains of systems with and without repetition? The system with repetition does seem to increase yield, but to what cost? There is little evidence that grade repetition is pedagogically efficient, and some evidence to the contrary. There is moreover the phenomenon of "sudden death" in systems with automatic promotion. In asking about the purposes of extended schooling, the issue of terminal points in the system must be raised.

"Grassroots" participation, both at an experimental stage and during the final implementation of a major educational innovation, is crucial to its eventual success. Reforms launched over the heads of teachers and school administrators have consistently failed. Hence the important lesson is that the objective of bringing about change in European school education, for example by encouraging the harmonization of school curricula and the standardization of examination practices across countries, is not likely to be achieved unless understanding is solicited of teacher unions, professional associations and administrative and political bodies and their support is ensured.

Recruitment and training of teachers has become a major problem in several countries in Europe. Until the mid-20th century school teaching was an important avenue of social mobility for young people with a

---

[3] Cf. S. Marklund, *The IEA Project: An Unfinished Audit*, Report No. 64, Stockholm University, 1983.
[4] Source: UNESCO, *The Evolution of Wastage in Primary Education in the World*, Paris, 1988a.

lower class background or coming from rural families. Training institutions for primary teachers, in particular, could recruit able young people who did not have the opportunity to enter professions where upper secondary school followed by university attendance was required. The situation has changed markedly during the last few decades. With the greatly increased provision of post-school education and steps taken to promote young people of lower class background to enter upper secondary schools and higher education institutions, teaching has to compete with other professions for able recruits. Secondly, many more teachers than previously are now needed, and this has meant less competition in entering teacher training institutions. Thirdly, the very fact that teaching has been confronted with social problems, which previously did not affect the schools to the same extent, has made teaching less attractive than before.

During the 1970s and 1980s there was a gradual deprofessionalization of the teaching force. There is an indication that professionalization will become dominant in the 1990s. The role of the headteacher is also changing. They are being transformed into administrative specialists. How does professionalization affect the teacher's job? What are the expected outcomes of acquiring an additional qualification? At present, the reward for being a good pedagogue is to be made into a paper shuffler. How does one use incentives to create a diversified hierarchy of teachers, so that the good ones will remain in the classroom?

There is no doubt that in several European countries the professional status of teachers has been going downhill during the period of enrolment explosion, particularly at the lower secondary level. Primary school teachers have suffered much less "de-professionalization" than secondary school teachers, who previously were teaching a small select social and intellectual élite. Irrespective of the competition with other professions, the sheer size of the demand has made it difficult to maintain the high overall standard of the teaching profession. An issue one could mention in this connection is the tendency to employ teachers on a temporary basis in order to fill in short term vacancies. In extreme cases, one out of three teachers in a school can be a stand-in for an absentee. This development will have to be watched if teaching standards in Europe are to remain high.

High quality science teaching is crucial for a Europe which on the international market is competing with Japan, where since the 1950s a deliberate emphasis in school policy and planning has been put on good grounding in science and mathematics. Cross-national comparisons have consistently shown Japanese pupils superior to the European ones.

Mathematics and the sciences form the basis of all technology. Thus they ought to be adequately represented in school curricula. For reasons spelled out in an earlier chapter of this report, these subjects are at present failing to attract enough young, talented students – particularly girls in some EC countries. Hence one challenge posed to European school authorities and schools themselves is the design of a strategy for revitalizing the teaching and curricula of mathematics and science.

A distinctive feature in European culture is multilingualism. According to the recommendations of European employers, [5] school curricula ought "to contain a minimum of three foreign languages in every European country". Few students in our secondary schools measure up to this demand at present, even in the case of the French *lycée* or the prestigious German *Gymnasium*.

4. Education in a North-South Perspective

One overriding challenge deserves to be given special consideration: the imperative of improving the situation of school education in the developing countries. In spite of the generally improved intake capacity of primary and lower secondary education achieved in the Third World over the past 20 years, enrolment rates are now actually declining in some countries, mainly as a consequence of a high birth rate. Access problems have become acute in many developing countries. As shown in Table 11.1, millions of children in the ages 6 to 11 years are not in school, despite the fact that primary education is a crucial factor in economic and social development. A second matter of concern is that the quality of schooling is often low, which results in the paradoxical situation that many of those who have gone through a full cycle of four or even six years of schooling lack useful literacy, numeracy, and problem-solving skills. [6] Schooling in the South is taking place in a resource-poor environment; a European effort may be most needed in this context of development needs. How can Europe express solidarity in a non-ideal and competitive world, and what should be the role of the school system?

The continued deterioration in the economic, financial and social situation of countries in the Third World, particularly of the low-income countries in sub-Saharan Africa and Asia, is undoubtedly the main "challenge" Europe will have to face, now that the dissolution of the

---

[5] Quoted from a study on education and European competence conducted for the European Round Table of Industrialists, by K. Kairamo (Ed.), 1989, p. 10.
[6] Cf. The World Bank, *Improving the Quality of Primary Education in Developing Countries,* Washington, D.C., 1989.
[7] Cf. The World Bank, *Sub-Saharan Africa: From Crisis to Sustainable Growth,* 1989.
[8] Source of data: UNESCO, *Trends and Projections of Enrolment by Level of Education and Age, 1960-2025,* Paris, 1989.

Warsaw Pact has caused the threat of an East-West military confrontation to recede. North-South inequalities are reinforced by the fact that the gap in living standards is increasing. This trend is also noticeable between countries in the South. As a background to the problem, it should be mentioned that, despite a general increase in domestic agricultural production, the balance of trade of many low-income countries, which deteriorated from 1975 to 1985, has not significantly improved since the world economic crisis came to an end in the West. [7] The prospect of achieving sustainable economic growth is, in poor developing countries, being seriously impaired as a consequence of poor education and underinvestment in human resources. This makes it even more imperative that Europe lives up to its responsibility. The EC countries have shown that they are ready to adopt innovatory aid policies, for example in the context of the so-called "Lomé Agreements", which involve an increasing number of the countries of the South. But even if the countries of Europe embark upon the path of a massive and more effective aid programme than hitherto, its positive effects will only be long term, as is the case with all investment in human capital.

Table 11.1: *Out-of-School Youth (millions)* [8]

Region	Age 6 to 11 years		Age 12 to 17 years	
	1970	1985	1970	1985
Developed countries	9,2	9,7	27,3	16,1
Developing countries	173,8	133,7	223,4	283,4
of which Africa	34,2	37,4	35,0	40,8
Latin America and the Caribbean	13,8	8,8	19,5	18,4
Asia and the Pacific	125,1	86,6	168,2	223,9

Moreover, it is acknowledged that beyond immediate problems and emergency solutions, it is only the reduction of the gap between a rich, dynamic Europe and the countries of the South that will allow lasting solutions to be envisaged. The immense effort of solidarity and innovation that the elaboration and the realization of such a policy of assistance and cooperation will require cannot be the task of the political authorities alone. It will have to be understood and sustained by public opinion, which will be all the more difficult because the crises and the tensions in which the urgency of the measures to be taken will become manifest will do harm to the image of the Third World put out by the media. In the most favourable case, the media can excite a wave of solidarity for help in situations of extreme distress. But the world dimension of questions of economic and social development can only be understood, in all its complexity, by an internationally oriented education; only education can respond to this challenge. For some years several measures have been taken to introduce into school syllabuses elements of educating for dealing with international questions. Examples are "international education", "multicultural education", "education for development", and education for "peace and international understanding". These experiments are still limited and very inadequate, but can be developed if "stimulatory programmes" are undertaken to evaluate the experiments, to make known their results and to seek how to insert this international dimension in all the existing initiatives rather than specifically to add on a new programme.

Furthermore, we may wonder whether better knowledge of the problems of development would be sufficient to exert an influence over attitudes and behaviour. Apart from cognitive objectives, it would be appropriate to envisage forms of extra-curricular activities capable of involving young people in actions in which could be manifested in concrete terms a solidarity with a human dimension. The Unesco Clubs and other voluntary associations have experience that might be even more useful because several United Nations organizations support actions of this kind, and because their multilateral character is particularly well adapted to guide such initiatives within a coherent framework.

The constitution of a united Europe cannot be realized without active participation in policies of world development, and in the training of minds within this perspective.

5. Some Concluding Observations

The challenges dealt with in the previous paragraphs are calling for changes which will be met with strong resistance, simply because school practices are deeply rooted in national traditions. Changes in education take time and cannot be expected to occur over night. But some of them are bound to occur with a reasonable delay if the overriding idea of a European identity is to stand a chance of being realized. An interaction of Eastern and Western traditions in education is probably a prerequisite for this. The opening up of the political situation in Europe may well have given us a unique opportunity of achieving a combination of the best features of different school systems.

In entering a high-technology, post-industrial, information society, where education increasingly tends to become a vehicle for both individual career and national progress, it also becomes increasingly important to exchange experiences gained in adapting the school both to individual needs and to the ones of society. It is close at hand for a country to regard its own system of education as a model for other countries. The opposite tendency is much less frequent. Even though the historical development of each nation is unique and profoundly affects the character of its school system, the common core of culture created not least by modern science, technology and literature is growing. This does not necessarily lead to the establishment of a common, cross-national core curriculum, but to greater similarities between school structures and curricula. There is no doubt that structural reforms in several European countries have moved in the same direction, for instance by postponing the organizational differentiation of pupils into different tracks of study. The same applies to initiatives by central governments to evaluate or assess on a national scale to what extent the goals for basic school education have been achieved. A greater *rapprochement* not least under the impact of exchange of university students can be envisaged.

However, a European road towards cooperation in economics and education cannot be conceived in isolation from the rest of the world. Beyond its relationships with the United States, Canada and the Far East, the problems posed by the Third World and its development will be a major challenge for Europe. Europe has to contribute meaningfully to this worldwide issue; this will require not only international policy measures very different from those taken until now by the international community, [9] but also a profound change in mentality and attitudes, to which education will have to bring its own contribution.

Typically, comparative studies in education did not have a real breakthrough until the early 1960s, when education began to be conceived as an instrument of national and economic development, and when major reforms aiming at establishing more equality of opportunity were launched in several countries. The interest in scholarly comparative studies in education by policymakers has not so much been determined by intentions to "borrow" from other countries but to profit from experiences gained from "best practices" employed, in particular those related to educational issues in their own country.

Europe has historically been faced with the persistent challenge of managing pluralism. This takes us back to the subject of values and goals in school education. It is not sufficient to advance the sweeping claims that children should learn three foreign languages or that schools ought to provide a democratic education. After all, what does the term really

---

[9] One could think here of international proposals such as those presented by the Club of Rome and the Brandt Report.

mean? Tolerance is central to the understanding and practice of a pluralistic democracy. Until recently, the democratic idea was not global in orientation but given meaning in the context of the nation-state. Democracy was something applied within the boundaries of a particular country. As this situation has changed drastically, educational systems may have to take account of practical experiences and outcomes of pluralist democracy in neighbouring countries, which may adhere to other value systems.

European school education still presents a picture of diverging models and ideas. This is reflected in the diversity of provisions of secondary education, even though certain similarities can be noted, for instance that education at this level has become more or less universal. Primary schooling was in the 19th century legislated all over Western and Northern Europe with the dual purpose of establishing a common frame of reference for the citizens and in building nation-states with national languages substituted for local languages and dialects. We can expect that in the new, eventually integrated, Europe the goals and content of schooling will be gradually replaced by a European concept. Yet, we should be cautious in dealing with questions of harmonization, given the diversity of cultures and languages, views and interests. It is difficult to imagine what a "common core" curriculum of a United Europe would contain. On the other hand, European integration offers a challenge to establish certain basic elements in school curricula, constituting a shared frame of reference and a new European identity.

Statistical Supplement

Table S.1: Demographic and Economic Indicators of European Countries (a)

Country	Population in age group 0-14 years %	Population (100,000)	Population density (sq. km)	Dependancy ratio [b] %	Population of foreign citizenship %	Total GNP [d] (billion $)	Per capita [e] income 1980 US $	Growth of real GDP/GNP Average [f] 1979-1987	Growth of real GDP/GNP % change 1989-1990	Per capita [g] volume index for GDP 1988
Austria	17.5	76	91	85.9	3.9	117,6	15,560	1.6	3.4	65
Belgium	18.3	99	323	78.7	8.7	143,6	14,550	1.6	3.3	66
Denmark	17.3	51	119	89.5	2.8	94,8	18,470	2.0	1.1	70
France	20.2	560	103	68.2	6.6	898,7	16,080	1.7	3.1	70
Germany	14.7[c]	774	246[c]	102.8	7.6[c]	1252,1	18,530[c]	1.4	3.9	74
Greece	20.0	100	76	68.0	2.2	48,0	4,790	1.2	1.4	35
Ireland	28.2	35	52	39.2	—	26,8	7,480	1.3	3.8	40
Italy	17.8	574	191	77.2	8.1	765,3	13,320	2.2	3.1	67
Netherlands	18.5	149	434	67.7	4.0	214,5	14,530	1.2	3.3	66
Norway	18.9	42	14	85.6	3.2	84,2	20,020	3.6	2.7	84
Portugal	22.1	103	112	57.1	0.1	37,3	3,670	2.3	4.0	35
Spain	21.1	392	78	61.0	0.1	301,8	7,740	2.1	4.2	48
Sweden	17.8	85	20	100.0	5.0	160,0	19,150	2.0	0.7	75
Switzerland	17.0	66	160	85.2	15.6	178,4	27,260	2.0	2.6	86
Turkey	36.6	540	70	10.9	9.4	68,6	1,280	4.6	4.2	23
United Kingdom	18.8	570	233	82.0	3.1	730,0	12,800	2.0	0.9	69

Notes: (a) Population estimates as on 1st January 1988. (b) Number of persons of 65 years and over per 100 persons below 15 years of age. (c) Refers to the situation in former West-Germany. (d) GNP = gross national product. GDP = gross domestic product. (e) 1988 estimates, according to World Bank Annual Report 1990. (f) Averages are computed on the basis of 1987 values expressed in 1987 U.S. dollars. (g) The index shows the size of each country's GDP per capita relative to that of the U.S. when converted to U.S. dollars using purchasing power parities (comparative price levels).

Source: Council of Europe (1990) Recent Demographic Developments in the Member States of the Council of Europe, 1989 Edition; World Bank (1990) Annual Report; OECD (1990) Economic Outlook (No. 47) and OECD (1989) Main Economic Indicators.

Table S.2: *Eastern Europe, Basic Economic Indicators (1988-1989 data)*

	Population (100,000)	Total GDP (billion $)	Per capita income 1988 US $	Hospital beds per 10,000 inhabitants	Cars per 1000 inhabitants	TV-sets per 1000 inhabitants
Bulgaria	90	51	5,630	94	115	190
Czechoslovakia	155	118	7,600	102	169	282
Germany, former East	167	156	9,360	102	188	365
Hungary	106	69	6,500	90	132	274
Poland	373	207	5,540	70	98	253
Romania	227	94	4,120	90	—	171
For comparison:						
Soviet Union	2,788	1,547	5,550	—	—	—
Spain	390	340	8,716	—	250 [a]	290 [a]
Portugal	103	41	3,987	—	165 [a]	170 [a]
Germany, former West	614	1,202	19,575	—	412 [a]	443 [a]
EC 12	3,248	4,745	14,609	—	310 [a]	309 [a]

Source: Eurofile, *Annex Bank Review*, 1990; EUROSTAT, *The EC Twelve: Key Figures*, March 1989.
Note: (a) 1986 figures.

Table S.3: *Government Spending on Education, European Countries (in per cent of gross national product, GNP)*

	1980	1986	Trend
Austria	5.6	6.0	+
Belgium	6.0	5.5	-
Bulgaria	5.6	7.1	+
Czechoslovakia	4.8	5.2	+
Denmark	6.9	7.7	+
Finland	5.5	5.9	+
France	5.1	6.1	+
Germany	4.7	4.6	-
Greece	2.2	2.6	+
Hungary	4.7	5.7	+
Iceland	4.0	4.0	x
Ireland	6.6	6.7	+
Italy	4.4	4.0	-
Luxemburg	6.0	5.2	-
Netherlands	7.9	6.9	-
Norway	7.2	6.9	-
Portugal	4.4	4.4	x
Romania	3.3	2.1	+
Spain	2.6	3.2	-
Sweden	9.1	7.6	-
Switzerland	5.0	4.8	-
Turkey	3.4	2.1	-
United Kingdom	5.6	5.2	-
Soviet Union	7.3	7.0	-
Yugoslavia	4.7	3.8	-

Source: UNESCO, *Statistical Yearbook,* 1990.

Table S.4: *Length of Compulsory and Secondary Education in Europe (1987-1989 data)*

Country	Compulsory Education		Entrance age and duration of primary education	Entrance age and duration of secondary education
	Age limits	Duration		
Albania	6-13	8	6-13	14-17
Andorra:				
French Schools	6-16	10	6-10	11-14 / 17*
Spanish Schools	6-15	10	6-10	11-13 / 17
Austria	6-15	9	6-9	10-13 / 17
Belgium	6-18	12	6-11	11-14 / 18
Bulgaria	7-14	8	7-14	15-17
Czechoslovakia	6-15	8	6-13	14-17
Denmark	7-15	9	7-12	13-15 / 18
Finland	7-16	9	7-12	13-15 / 18
France	6-16	10	6-10	11-14 / 17
Germany	6-18	12	6-9	10-15 / 18
Gibraltar	4-15	12	4-11	12-15 / 17
Greece	6-15	9	6-11	12-15 / 17
Hungary	6-16	10	6-13	14-17
Iceland	7-15	8	7-12	13-15 / 19
Ireland	6-15	9	6-11	12-14 / 16
Italy	6-13	8	6-10	11-13 / 18
Liechtenstein	6-14	8	6-10	11-13 / 16
Luxemburg	6-15	9	6-11	12-14 / 18
Malta	5-16	10	5-10	11-15 / 17
Monaco	6-16	10	6-10	11-14 / 17
Netherlands	5-16	11	5-11	12-14 / 17
Norway	7-15	9	7-12	13-15 / 18
Poland	7-14	8	7-14	15-18
Portugal	6-14	6	6-11	12-14 / 17
Romania	6-16	10	6-13	14-17
San Marino	6-13	8	6-10	11-13 / 18
Spain	6-15	10	6-10	11-13 / 17
Sweden	7-16	9	7-12	13-15 / 18
Switzerland	7-16	8,9	7-12	13-15 / 19
United Kingdom	5-16	11	5-10	11-13 / 17
Soviet Union	7-17	10	7-11	12-14 / 16
Yugoslavia	7-15	8	7-10	11-14 / 18

*) The higher age refers to leaving-age after upper secondary school.
Source: UNESCO *Statistical Yearbook*, 1990.

Table S.5: Selected Education Indicators by European Country (around 1987-1989)

Country	Compulsory school Starts at age	Compulsory school duration (years)	Total enrolment in education for 4-24 year age group (%)	Pre-primary level (%)	Gross enrolment rates Second level (%)	Gross enrolment rates Third level (%)	Vocational share in second level education (%)	Government spending on education (% GNP)	Teacher: pupil ratio first level education
Albania	6	8	64.0	55	76	8	70	—	20
Belgium	6	10	75.7	104	99	32	46	5.1	17
Bulgaria	7	8	74.1	95	76	23	60	6.9	18
Czechoslovakia	6	8	64.0	87	38	16	59	5.3	21
Denmark	6/7	9	70.5	97	107	29	32	7.9	12
Finland	7	10	73.5	77	106	35	28	5.9	15
France	6	10	78.9	116	92	30	40	5.7	19
Germany	6	10	73.0	95	94	30	48	4.4	17
Greece	5.5	9	67.6	56	91	24	13	2.7	23
Hungary	6	10	70.0	94	70	15	74	5.6	14
Iceland	7	8	67.8	51	92	23	37	3.7	21
Ireland	6	9	69.8	104	98	22	6	7.1	27
Italy	6	8	65.1	86	74	25	34	5.1	14
Netherlands	5	11	72.2	106	104	32	57	6.8	17
Norway	7	9	71.3	74	95	28	28	6.8	16
Poland	7	8	70.4	51	80	17	76	4.4	16
Portugal	6	8	58.5	31	56	13	1	4.6	17
Romania	6	10	67.1	75	79	11	93	2.1	21
Spain	6	8	77.9	68	102	32	4	3.2	26
Sweden	7	9	74.6	101	91	37	50	7.4	16
Switzerland	6/7	8/9	57.9	66	—	23	7	4.8	24
Turkey	7	8	45.0	4	46	10	22	2.1	31
Soviet Union	7	10	67.0	62	98	22	16	7.5	17
United Kingdom	5	11	63.6	49	83	22	26	5.0	20
Yugoslavia	7	8	58.0	26	80	19	29	3.8	23

Source: UNESCO, *Statistical Yearbook* 1989 and 1990; and OECD; *Education in OECD Countries, 1990 Special Edition.*

Table S.6: *Number of Children Less than 12 Years of Age in European Countries (growth rates 1975-1987)*

	Rate 1975-1982	Rate 1983-1987	Rate 1975-1987
Austria	- 19.5	- 2.7	- 23.1
Belgium	- 11.4	- 3.4	- 15.5
Denmark	- 14.8	- 8.9	- 24.4
Finland	- 6.0	1.1	- 3.7
France	- 5.8	- 2.4	- 9.4
Germany	- 25.7	- 1.8	- 29.3
Italy	—	- 12.8	-28.5
Netherlands	- 16.1	- 2.6	- 20.6
Norway	- 12.4	- 5.2	- 18.8
Portugal	- 9.0	- 9.1	- 18.3
Spain	- 2.4	- 9.9	-16.8
Sweden	- 10.7	- 2.6	- 14.6
Switzerland	-17.7	- 2.2	- 19.5
Turkey	5.2	7.7	14.6
United Kingdom	- 15.4	1.4	- 15.6

Source: *OECD Employment Outlook*, 1990, p. 148.

Table S.7: *Development of the Cohort Aged 16 to 18 Years, 1968-2000 (1970 = index 100)*

	1968	1978	1988	1994	2000
France	103	100	102	86	90
Germany	103	120	100	69	72
Netherlands	100	110	107	79	76
Sweden	104	95	101	91	86
United Kingdom	102	116	114	87	94

Source: United Kingdom, Department of Education and Science, 1990

Table S.8: *Total Unemployment Rate, Youth Unemployment Rate, and Female Labour Force Participation Rate in European Countries*

Country	Total unemployment rate [a] 1983	1990	Youth unemployment rate [b] 1983	1989	Female labour force [c] participation rate 1983	1988
Austria	3.7	3.2	—	—	49.7	53.7
Belgium	12.9	8.7	27.6	17.4	48.8	51.4
Denmark	10.4	9.3	17.9	10.2	74.2	78.3
Finland	5.4	3.8	—	—	72.7	73.0
France	8.4	9.3	21.5	21.2	54.4	55.7
Germany	8.2	6.1	11.6	5.2	49.7	54.4
Greece	7.9	7.9	28.5	24.8	40.4	43.4
Ireland	14.0	14.9	22.3	22.2	37.8	37.6
Italy	9.2	12.2	30.9	32.3	40.3	43.9
Netherlands	15.0	6.8[d]	22.7	16.0[d]	40.3	51.6[d]
Norway	3.4	5.0	—	—	65.5	72.8
Portugal	7.9	5.3	18.6	11.3	59.1	59.1
Spain	18.2	16.1	42.6	35.7	33.0	39.4
Sweden	2.9	1.6	46.6[e]	36.1[e]	76.6	80.1
Switzerland	0.8	0.6	—	—	53.3	57.9
Turkey	16.1	10.9	—	—	—	—
United Kingdom	11.2	6.1	20.5	8.8	57.2	63.5

Notes: (a) Per cent of labour force age 15-64. (b) Share of youth under 25 years in total unemployment. (c) Per cent of female labour force age 15-64. (d) Data especially supplied to OECD by the Netherlands authorities. (e) 1980 and 1986 estimates.
Source: EUROSTAT and OECD.

Table S.9: *Total (TE) and Direct (DE) Effects of Four Predictor Variables on Life Career Outcomes from Age 25 to 56 (N = 671 men; standard errors, s.e.)*

	Home Background				Cognitive Ability				Initial (Youth) Education				Adult Education and Training			
	TE	s.e.	DE	s.e.	TE	s.e.	DE	s.e.	TE	s.e.	DE	s.e.	TE	s.e.	DE	s.e.
Occupational Status:																
Age 25	0.49	(.05)	0.12	(.05)	0.46	(.06)	0.22	(.06)	0.40	(.05)	0.40	(.05)	(n.a.)		(n.a.)	
Age 30	0.55	(.05)	0.13	(.05)	0.52	(.05)	0.22	(.05)	0.46	(.04)	0.46	(.05)	(n.a.)		(n.a.)	
Age 35	0.55	(.05)	0.13	(.05)	0.55	(.05)	0.21	(.05)	0.45	(.04)	0.45	(.05)	0.13	(.04)	0.13	(.04)
Age 40	0.56	(.04)	0.18	(.05)	0.50	(.05)	0.20	(.05)	0.39	(.04)	0.39	(.05)	0.21	(.04)	0.21	(.04)
Age 43	0.57	(.05)	0.18	(.05)	0.50	(.05)	0.15	(.05)	0.34	(.04)	0.34	(.05)	0.32	(.05)	0.32	(.05)
Age 52	0.53	(.05)	0.13	(.04)	0.50	(.05)	0.11	(.06)	0.34	(.04)	0.34	(.05)	0.43	(.05)	0.43	(.05)
Age 56	0.52	(.05)	0.13	(.04)	0.49	(.05)	0.14	(.05)	0.32	(.04)	0.32	(.05)	0.44	(.05)	0.44	(.05)
Earned Income:																
Age 25	0.24	(.03)	*		0.23	(.03)	0.11	(.03)	*		*	(.03)	(n.a.)		(n.a.)	
Age 30	0.41	(.04)	*		0.43	(.03)	0.13	(.03)	0.11	(.04)	0.11	(.04)	(n.a.)		(n.a.)	
Age 35	0.53	(.05)	0.14	(.05)	0.41	(.06)	0.22	(.06)	0.12	(.04)	0.12	(.05)	0.06	(.02)	*	
Age 40	0.47	(.04)	*		0.53	(.03)	0.10	(.03)	0.28	(.03)	0.28	(.04)	0.11	(.02)	*	
Age 43	0.49	(.04)	*		0.52	(.03)	0.12	(.03)	0.23	(.04)	0.23	(.03)	0.19	(.03)	*	
Age 52	0.39	(.04)	*		0.41	(.04)	0.14	(.04)	0.10	(.04)	0.10	(.04)	0.26	(.03)	*	
Age 56	0.36	(.04)	*		0.37	(.03)	0.16	(.03)	0.08	(.04)	0.08	(.04)	0.26	(.03)	*	

Notes: (a) Trivial direct effects are denoted with asterisks. An effect is considered trivial if the standardized path coefficient fails to reach statistical significance at least at the five per cent threshold level. (b) Because the total effect coefficient is equal to the sum of the direct and indirect effects, indirect effect coefficients can be inferred from the data. (c) Estimates are calculated using data derived from the longitudinal Malmö Study, which has been in operation in Sweden since 1938.

Source: A.C. Tuijnman, *Recurrent Education, Earnings, and Well-being*, 1989, p. 188.

Table S.10: *Mean Scores of Student Populations in European Countries and Japan on the IEA Scale "Like School" (1983/1984 data, see notes)*

Educational System	Like School		
	10 year-old level	14 year-old level	Final grade level
England	2.19	2.00	2.13
Finland	2.00	1.63	1.75
Hungary	2.31	1.88	1.81
Italy	2.31	2.13	2.00
Poland	2.38	2.19	1.88
Sweden	1.56	1.81	2.06
Japan	2.00	2.06	2.00

Notes: (a) The IEA scale "like school" consists of seven measures tapping the students attitude to school. (b) The data are derived from the IEA Second International Science Study.
Source: T.N. Postlethwaite and D.E. Wiley (1992), pp. 212-219.

Table S.11: *Attitudes Towards Cooperation With The European Community*

In the field of:	Czechoslovakia			Poland		
	Welcome %	Not welcome %	Don't know %	Welcome %	Not welcome %	Don't know %
Culture	78	5	17	84	4	12
Agriculture	82	3	15	89	3	8
Industry	88	2	10	91	1	8
Higher Education	87	2	11	79	6	15
Defence	41	23	36	63	13	24
Foreign Policy	63	11	27	70	9	21
Environment	91	2	8	90	2	8
Human Rights	79	4	17	84	4	12
Professional Training	86	2	11	81	4	11
Youth Exchange	85	3	12	85	4	11

Source: EUROSTAT

Table S.12: *Country Scores, International Physics Olympiad, 1989*

Germany, West	196.8	Czechoslovakia	152.7
China P.R.	186.5	Austria	140.2
United Kingdom	179.3	Sweden	133.3
Poland	176.0	Finland	126.2
Romania	175.8	Norway	113.8
Germany, East	172.5	(Lituania	108.7)
Hungary	168.7	Belgium	104.0
Netherlands	166.5	Cyprus	88.7
Soviet Union	163.5	Italy	88.2
United States	155.5	Turkey	74.3
Bulgaria	155.4	Iceland	59.7
Yugoslavia	153.0		

Source: Secretariat for the International Physics Olympiad, International Commission on Physics Education.

Table S.13: *Percentage of Mathematics and Physics Lessons Compared to all School Subjects in Some European Countries (1988, per cent)*

Country	Mathematics %	Physics %
Austria	14.2	2.7
Czechoslovakia	18.9	2.6
Germany, former East	19.6	3.0
Germany, former West	16.0	3.4
Hungary	17.6	2.2
Poland	17.3	3.4
Romania	19.9	3.0
Soviet Union	22.1	3.2
Yugoslavia	16.7	1.8

Source: Szabo, *Physics Teaching in Socialist Countries*, 1990.

Table S.14: *L'Enseignement des langues vivantes*

Type d'enseignement Langues proposées	Statut première langue	Horaire	Statut hebdomadaire	Horaire deuxième langue	Statut	Horaire hebdomadaire	troisième langue	hebdomadaire
Belgique								
Secondaire inférieur 4 années	Obligatoire	2 - 6	Obligatoire	2 - 3	Facultative	2	L1: Allemand, anglais, français, néerlandais L2: Allemand, anglais, arabe, espagnol, français, italien, néerlandais L3: Mêmes possibilités de choix que pour L2	
Secondaire supérieur général 3 années	Obligatoire	3 - 4	Obligatoire	3 - 4	Facultative	3 - 4	Mêmes possibilités de choix que pour L1, L2, L3	
Enseignements professionnels	Obligatoire	2 - 6	Facultative	Très variable			Mêmes possibilités de choix que pour L1, L2, L3	
Danemark								
Premier cycle 3 années	Obligatoire	2h 1/2	Facultative	2h 1/2 - 3			L1: Anglais L2: Allemand ou français	
Second cycle général 3 années	Obligatoire	3 - 5	Obligatoire	3 - 5	Obligatoire pour certaines sections		L1: Anglais L2: Allemand ou français L3: Espagnol, français, Italien, russe	
Second cycle professionnel 3 années	Obligatoire (1re année)	1,5 - 6	Facultative (2e et 3e années)				L1: Anglais, allemand L2: Anglais, allemand, espagnol, français, italien, russe	

1 (5)

Supplementary Table 14 (suite)

Type d'enseignement Langues proposées	Statut première langue	Horaire hebdomadaire	Statut deuxième langue	Horaire hebdomadaire	Statut troisième langue	Horaire hebdomadaire
RFA						
Hauptschule 5 - années	Obligatoire	3 - 5				L1: Anglais, français
Realschule 6 années	Obligatoire 6 années	3 - 5	Obligatoire 4 années			L1: Anglais, français L2: Anglais, français ou autre langue
Gymnasium 9 années	Obligatoire (9 années)	3 - 6	Obligatoire (3e à 9e année)	3 - 6	Obligatoire (6e à 9e année)	3 - 6	L1: Anglais, français, latin ou autre langue L2: *Idem* L3: *Idem* plus grec?
Enseignements professionnels Anglais, espagnol,		Obligatoire 1 - 6	Obligatoire pour les filières commerciales			Obligatoire pour les filières commerciales	L1:
Grèce						
Premier cycle 3 années	Obligatoire	3				L1: Allemand, anglais, français
Second cycle 3 années	Obligatoire	3				Mêmes possibilités de choix
Enseignements professionnels	Obligatoire 1 - 2					Mêmes possibilités de choix

Supplementary Table 14 (suite)

	Type d'enseignement	Statut première langue	Horaire hebdomadaire	Statut deuxième langue	Horaire hebdomadaire	Statut troisième langue	Horaire hebdomadaire	Langues proposées
Espagne	Premier cycle 3 années	Obligatoire	3					L1: Anglais, français
	Deuxième cycle général 4 années	Obligatoire	3					L1: Allemand, anglais, français, italien, portugais
	Enseignements professionnels 2 - 4 années	Obligatoire	2					L1: Anglais, français
France	Premier cycle 4 années	Obligatoire	3	Obligatoire (3e et 4e années)	3			Anglais, allemand, espagnol, italien, russe, portugais, arabe, chinois, hébreu, polonais, néerlandais, japonais, danois, grec moderne
	Second cycle général 3 années	Obligatoire	2 - 3	Obligatoire	2 - 3	Facultative	3	
Irlande	Premier cycle 3 années	Facultative	3	Facultative	3			
	Deuxième cycle 2 années	Facultative	3	Facultative	3	Facultative	3	Allemand, espagnol, français, italien
	Enseignements professionnels	Facultative	3	Facultative	3			

3 (5)

350 Schooling in Modern European Society

Supplementary Table 14 (suite)

	Type d'enseignement Langues proposées	Statut première langue	Horaire hebdomadaire	Statut deuxième langue	Horaire hebdomadaire	Statut troisième langue	Horaire hebdomadaire
Italie	Premier cycle 3 années	Obligatoire	3				
	Second cycle classique 5 années	Obligatoire (2 premières années)	3			Allemand, anglais, espagnol, français	
	Second cycle scientifique 5 années	Obligatoire (3 dernières années)	3				
	Enseignements professionnels 3 - 5 années	Facultative		Facultative sections commerciales	2 - 4	Facultative sections commerciales	3 - 6
Luxembourg	Premier cycle 3 années	Obligatoire	3 - 5	Obligatoire	3 - 5	Obligatoire	3 - 5
	Deuxième cycle général 4 années	Obligatoire	3 - 5	Obligatoire	3 - 5	Obligatoire	3 - 5 Allemand, anglais, français
	Enseignements professionnels	Obligatoire	2 - 6	Obligatoire	2 - 6	Obligatoire	2 - 6
	Premier cycle général 4 - 5 années	Obligatoire	3 - 4	Obligatoire	3 - 4	Obligatoire	3 - 4	L1: Anglais, français L2: Anglais, français L3: Allemand, espagnol, russe
Pays-Bas	Deuxième cycle général 2 années	Facultative	3 - 6	Facultative	3 - 6	Facultative	3 - 6

Supplementary Table 14 (suite)

	Type d'enseignement	Statut première langue	Horaire hebdomadaire	Statut deuxième langue	Horaire hebdomadaire	Statut troisième langue	Horaire hebdomadaire	Langues proposées
	Enseignements	Facilitative	3-17	Facultative	3-17	Facultative	3-17	L1: Anglais L2: Français L3: Allemand
	Premier cycle 3 années	Obligatoire	2 - 3	Obligatoire	3			
Portugal	Deuxième cycle général 3 années		Obligatoire	2 - 4	Obligatoire	4	Allemand, anglais, français	Allemand, anglais, français
	Enseignement professionnels	Facultative		Facultative				
	Premier cycle 5 années	Facultative	2h 1/2 2 - 3	Facultative	2h 1/2 - 3			L1: Allemand, espagnol, français L2: Allemand, espagnol, italien ou autres
Royaume-Uni	Deuxième cycle 2 années	Facultative	Facultative	2 h 1/2 - 4	Facultative	2 h 1/2 - 4	Facultative	2 h 1/2 - 4	L1: Allemand, espagnol, français L2: Allemand, espagnol, italien L3: Allemand, espagnol, italien, russe ou autres

Source: Council of Europe (1988).

Table S.15: *Science Achievement of 10 and 14 Year-olds (mean scores and standard scores sex difference)*

Country	10 year-olds mean	10 year-olds sex difference [a]	14 year-olds mean	14 year-olds sex difference [a]
England	48.8	0.23	55.8	0.37
Finland	63.8	0.31	61.7	0.29
Hungary	60.2	0.15	72.2	0.16
Italy (grade 8)	55.8	0.17	59.6	0.39
Netherlands	—	—	65.8	0.52
Norway	52.9	0.41	59.8	0.39
Poland	49.7	0.26	60.4	0.36
Sweden [b]	53.4	0.30	61.4	0.41
Japan	64.3	0.12	67.3	0.31
United States [c]	54.8	0.20	54.8	0.25

Notes: (a) The standard score 'sex difference' is obtained by subtracting the girls' score from the boys' score and dividing the result by the average standard deviation. The greater the value the greater is the difference between the sexes. (b) Sweden: grades 4 and 8. (c) The United States scores are based on the IEA core test scores only.
Source: IEA, and T.N. Postlethwaite and D.E. Wiley (1991).

Table S.16: *Linguistic Relations with Other European Countries*

Language	Official language of a neighbouring state	Official language of a state which is not neighbouring	Minority language in another neighbouring state	Minority language in another state which is not neighbouring	Minority language in a strict sense
Albanian (Italy)	*	yes	*	*	*
Basque (France)	*	*	yes	*	*
Breton (France)	*	*	*	*	yes
Catalan (France)	*	*	yes	*	*
Catalan (Italy)	*	*	*	yes	*
Corsican (France)	*	*	*	*	yes
Croatian (Italy)	*	yes	*	*	*
Danish (Germany)	yes	*	*	*	*
Flemish (France)	yes	*	*	*	*
Franco-Provençal (Italy)	*	*	*	*	yes
French (Italy)	yes	*	*	*	*
Frisian (Germany)	*	*	yes	*	*
Frisian (Holland)	*	*	yes	*	yes
Friulan (Italy)	*	*	*	*	*
German (Belgium)	yes	*	*	*	*
German (Denmark)	yes	*	*	*	*
German (France)	yes	*	*	*	*
German (Italy)	*	yes	*	*	*
German (Italy) (Alto Adige/Südtirol) [a]	yes	*	*	*	*
	*	*	*	*	*
Greek (Italy)	*	yes	*	*	*
Irish (U.K.)	yes	*	*	*	*
Irish (Eire) [b]	*	*	*	*	*
Ladin (Italy)	*	yes	*	*	*
Occitan (France)	*	*	yes	*	*
Occitan (Italy)	*	*	yes	*	*
Polish (Germany)	*	yes	*	*	*
Sardinian (Italy)	*	*	*	*	yes
Scottish Gaelic (U.K.)	*	*	*	*	yes
Slovene (Italy)	yes	*	*	*	*
Welsh (U.K.)	*	*	*	*	yes

Notes: (a) Official language of the Swiss Confederation. (b) This is a national language of the Republic since 1922 and an official language since 1937. (c) *: not applicable.
Source: Commission of the EC, *Linguistic Minorities in Countries Belonging to the EC*, 1986.

Table S.17: Foreign Students in Eleven EC Countries by Country of Origin, c. 1986

Host Country	Total	Australia	Belgium	Canada	Denmark	France	Greece	Ireland	Italy
France	123,978	72	890	1,067	219	0	2,663	197	1,492
West Germany	81,724	136	575	436	266	2,603	6,454	188	2,458
United Kingdom	53,694	455	304	1,014	116	1,035	2,285	809	506
Netherlands	7,873	39	569	70	46	138	74	31	266
Belgium	22,555	6	0	83	26	1,620	808	22	3,144
Ireland	2,684	19	18	59	7	51	6	0	8
Denmark	3,400	13	7	25	0	67	15	16	30
Italy	28,068	32	95	93	17	336	12,222	7	0
Spain	—	—	—	—	—	—	—	—	—
Portugal	2,407	3	4	20	—	16	3	—	6
Greece	6,683	5	2	48	4	8	0	—	19
Total EC	333,066	780	2464	2,915	701	5,874	24,530	1,270	7,929
Non-EC	615,605	1,982	1226	16,347	1,178	7,431	9,519	1,565	10,549
TOTAL	948,671	2,762	3690	19262	1879	13,305	34,049	2,835	18,478

Continued

Table S.17 Continued

Host Country	Japan	Netherlands	Portugal	Spain	Turkey	United Kingdom	United States	West Germany	Other
France	851	560	2,634	2,768	763	2,086	3,473	3,660	100,583
West Germany	1,150	1,974	542	1,815	10,215	2,042	4,253	0	46,617
United Kingdom	423	429	137	268	186	0	4,166	1,777	39,784
Netherlands	46	0	62	219	631	435	288	1,449	3,510
Belgium	24	1,689	266	1,218	379	242	220	565	12,243
Ireland	3	10	2	16	3	982	281	--	1,213
Denmark	12	57	4	19	80	211	166	360	2,368
Italy	—	55	23	65	135	174	976	1,740	12,098
Spain	—	—	—	—	—	—	—	—	—
Portugal	--	3	0	46	--	9	40	23	2,234
Greece	4	4	—	—	427	58	252	42	5,810
Total EC	2,513	4,781	3,670	6,434	12,819	6,239	14,121	9,616	226,460
Non-EC	20,285	2,199	1,041	3,876	4,342	9,115	6,805	15,251	502,844
TOTAL	22,798	6,980	4,711	10,310	17,161	15,354	20,926	24,867	729,304

Source: UNESCO, *Statistical Yearbook*, 1990.

Table S.18: *Trends and Projections of School Enrolment in Europe (enrolment ratios by age group, in percent, regions according to Unesco definitions)*

Region	Year	Age 12-17 %	Age 18-23 %
Eastern Europe [a]	1960	69.7	15.1
	1980	83.5	22.5
	2000	82.2	27.9
Northern Europe [b]	1960	66.0	13.3
	1980	81.0	26.0
	2000	88.9	38.6
Southern Europe [c]	1960	40.6	10.9
	1980	74.6	30.7
	2000	91.7	42.3
Western Europe [d]	1960	62.3	10.6
	1980	77.3	28.6
	2000	88.0	51.4

Notes: (a) Eastern Europe = Bulgaria, Czechoslovakia, Hungary, Poland, Romania. Includes also the former German Democratic Republic. (b) Northern Europe = Denmark, Finland, Iceland, Ireland, Norway, Sweden, United Kingdom. (c) Southern Europe = Albania, Greece, Italy, Malta, Portugal, Spain, Yugoslavia. (d) Western Europe = Austria, Belgium, France, West Germany, Luxemburg, Netherlands and Switzerland.
Source: UNESCO, *Trends and Projections of Enrolment by Level of Education and Age*. Paris: Office of Statistics, 1989.

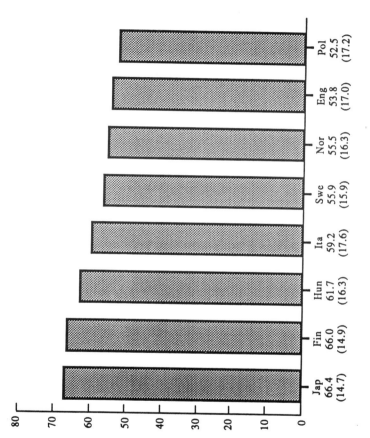

Supplementary Fig. 1 Science Achievement of 10 Year-olds on 40 Test Items in Eight Countries (per cent, mean scores, standard deviations).
Source: IEA; T.N. Postlethwaite and D.E. Wiley (1992).

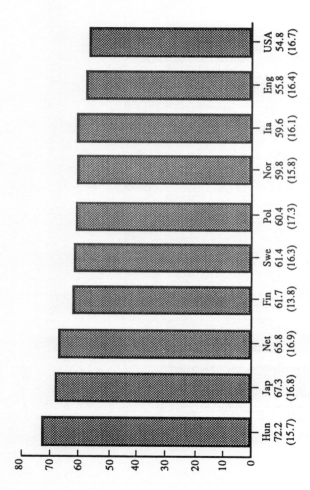

Supplementary Fig. 2 Science Achievement of 14 Year-olds on the Core Test in Ten Countries (per cent, mean scores, standard deviations).
Source: IEA; T.N. Postlethwaite and D. E. Wiley (1992)

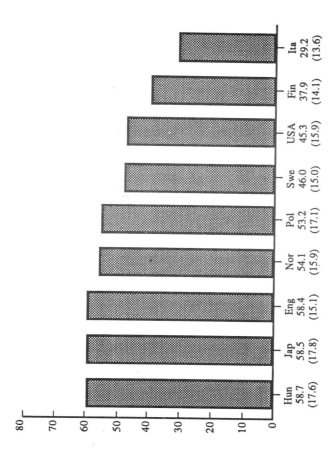

Supplementary Fig. 3 Science Achievement of 18 Year-olds on Reduced Number of Test Items (as taken by United States' Students) in Nine Countries (per cent, mean scores, standard deviations).
Source: IEA; T.N. Postlethwaite and D.E. Wiley (1992)

References

Abdallah-Pretceille, M. (1986). *Multicultural Policies and the Consequences for Teacher Education.* (OECD working document). Paris: OECD.

Absalom, R. (1990). Practical rather than Declamatory Co-operation: ERASMUS in 1990, An Appraisal. *European Journal of Education,* Vol. 25 (1), 39-54.

Ahamad, B. and Blaug, M. (1973). *The Practice of Manpower Forecasting: A Collection of Case Studies.* San Francisco, California: Jossey-Bass.

Anderson, L.W. and Postlethwaite, T.N. (1988). What IEA Studies Say about Teachers and Teaching. In: Alan C. Purves (Ed.), *International Comparisons and Educational Reform* (pp. 75-76). Alexandria, Virginia: Association for Supervision and Curriculum Development.

Anderson, C.A. and Bowman, M.J. (1967). Theoretical Considerations in Educational Planning. In: Bereday, G.Z.F. and Lauwerys, J.A. (Eds), *The World Yearbook of Education 1967: Educational Planning.* London: Evans.

Anderson, L.W., Ryan, D.W., and Shapiro, B.J. (1988). *The Classroom Environment Study: Teaching for Learning.* Oxford: Pergamon Press.

Anweiler, O. (1985). Die sowjetische Schul- und Bildungsreform von 1984. *Osteuropa,* 343, 839-860.

Apple, M.W. (1988). Redefining Equality: Authoritarian Populism and the Conservative Restoration. *Teachers College Record,* Vol. 90, 167-184.

Arnberg, L. (1987). *Raising Children Bilingually: The Pre-school Years.* Clevedon, Avon: Multilingual Matters Ltd.

Arvidson, L. and Rubenson, K. (1990). Education and Training of the Labour Force in EFTA Countries: Summary and Comments. Paper presented at Conference "New Challenges in Education and Training of the European Labour Force", Stockholm, June 13-14, 1990.

Axelrod, R. (1984). *The Evolution of Cooperation.* New York: Basic Books.

Baumgratz-Gangl, G. and Deyson, N. (1990). *Mobility of Students in Europe: Linguistic and Socio-cultural Conditions.* Luxemburg: Office for Official Publications of the European Communities.

Becker, G.S. (1964). *Human Capital.* New York: National Bureau of Economic Research.

Bednarzik, R.W. and Shields, C.R. (1989). Labor Market Changes and Adjustments: How do the United States and Japan Compare? *Monthly Labor Review,* February, 31-42.

Behrman, J.R., Hrubec, Z., Taubman, P. and Wales, T. (1980). *Socio-economic Success: A Study of the Effects of Genetic Endowments, Family Environment, and Schooling.* Amsterdam: North-Holland Publishing Company.

Bell, D. (1973). *The Coming of Post-industrial Society: A Venture in Social Forecasting.* New York: Basic Books.

Bereday, G.Z.F. (1977). Social Stratification and Education in Industrialized Countries. *Comparative Education Review,* Vol. 21 (2), 195-210.

Bernstein, B. (1973). *Class, Codes and Control, Vol. 1.* London: Paladin.

Blau, P.M. and Duncan, O.D. (1967). *The American Occupational Structure.* New York: Wiley.

Blaug, M. (1977). The Empirical Status of Human Capital Theory: A Slightly Jaundiced Survey. *Journal of Economic Literature,* Vol. 14, 827-855.

Blaug, M. (1985). Where are we Now in the Economics of Education? *Economics of Education Review,* Vol. 4, (1), 17-28.

Bloom, B.S. (1964). *Stability and Change in Human Characteristics.* New York: Wiley.

Bloom, B.S. (1981). *All our Children Learning: A Primer For Parents, Teachers, and Other Educators.* New York: McGraw-Hill.

Bloom, B.S. (1984). The 2 Sigma Problem: The Search for Methods of Instruction as Effective as One-to-one Tutoring. *Educational Researcher,* Vol. 13 (6), 4-16.

Bloom, B.S. (Ed.) (1956). *Taxonomy of Educational Objectives: The Classification of Educational Goals. Handbook 1: Cognitive Domain.* New York: Longmans and Green.

Bock, R.D. and Mislevy, R.J. (1988). Comprehensive Educational Assessment for the States: The Duplex Design. *Educational Evaluation and Policy Analysis,* Vol. 10 (2), 89-105.

Boissiere, M., Knight, J.B. and Sabot, R.H. (1985). Earnings, Schooling, Ability, and Cognitive Skills. *American Economic Review,* Vol. 75 (4), 1016-1030.

Botkin, J.W., Elmandjra, M., and Malitza, M. (1979). *No Limits to Learning: Bridging the Human Gap.* A Report to the Club of Rome. Oxford: Pergamon Press.

Bottani, N. (1990). Les nouvelles mesures de l'éducation: illusion de la recherche ou recherche d'une illusion? Le cas des indicateurs internationaux de l'enseignement. Exposé préparé pour un séminaire organisé par la SICESE à Frascati, Villa Falconieri, 3-6 avril 1990.

Boucher, L. (1982). *Tradition and Change in Swedish Education.* Oxford: Pergamon Press.

Boyd, W. and King, E.J. (1977). *The History of Western Education. Eleventh Edition.* London: Adam and Charles Black.

Boyle, E., Crosland, A., and Kogan, M. (1971). *The Politics of Education.* Harmondsworth, Middlesex: Penguin.

Braddock II, J.H. (1981). The Issue is Still Equality of Educational Opportunity. *Harvard Educational Review,* Vol. 51, 490-496.

Brock, W.H. (1975). From Liebig to Nuffield: A Bibliography of the History of Science Education 1839-1974. *Studies in Science Education,* Vol. 2, pp. 67-99.

References

Brophy, J.E. and Good, T.L. (1986). Teacher Behavior and Student Achievement. In: M.C. Wittrock (Ed.), *Handbook for Research on Teaching, Third Edition*. New York: Macmillan.

Bruner, J.S. (1960). *The Process of Education*. New York: Vintage Books.

Bruner, J.S., Goodnow, J.J., and G.A. Austin (1967). *A Study of Thinking*. New York: Science Editions.

Bullivant, B. (1981). *The Pluralist Dilemma in Education: Six Case Studies*. Sydney: Allen and Unwin.

Bulmer, S. and Wessels, W. (1987). *The European Council: Decision-making in European Politics*. London: Macmillan.

Burstein, L. (1992). *Second International Mathematics Study: Student Growth and Classroom Process*. Oxford: Pergamon Press.

Bykov, A. (1989). General Frames of EC-CMEA Economic Relations from the Point of View of Perestroika in the U.S.S.R. In: M. Maresceau (Ed.), *The Political and Legal Framework of Trade Relations Between the European Community and Eastern Europe*. Dordrecht: Martinus Nijhoff.

Byram, M. (1986). *Minority Education and Ethnic Survival: Case Study of a German School in Denmark*. Clevedon, Avon: Multilingual Matters Ltd.

Caillods, F. (Ed.) (1989). *The Prospects for Educational Planning*. Paris: UNESCO, International Institute for Educational Planning.

Capelle, J. (1964). *L'École de demain reste à faire*. Translated with an Introduction by W.D. Halls (1965). *Tomorrow's Education: The French Experience*. Oxford: Pergamon Press.

Carnegie Forum on Education and the Economy. (1986). *A Nation Prepared: Teachers for the 21st Century*. Washington, D.C.: The Carnegie Foundation.

Carnegie Foundation, International Institute Examinations Inquiry (1936). *Essays on Examinations*. London: Macmillan.

Cecchini, P., with M. Catinat and A. Jacquemin (1988). *The European Challenge: 1992. The Benefits of a Single Market*. Brussels: The Commission of the European Communities; and Aldershot: Wildwood House.

Cerych, L. (1989). University-Industry Collaboration: A Research Agenda and Some General Impacts on the Development of Higher Education. *European Journal of Education*, Vol. 24 (3), 309-314.

Cerych, L. and Jallade, J.-P. (1986). *The Coming Technological Revolution in Education*. Paris: Institut Européen d'Éducation et Politique Sociale.

Cha, Y.-K. (1991). Effect of the Global System on Language Instruction, 1850-1986. *Sociology of Education*, Vol. 64 (1), 19-32.

Chamouni, R. (1989). Evaluation de l'enseignement de la langue nationale. In: *Les Algériens et l'enseignement de l'arabe en France*. Paris: Centre culturel algérien.

Chevènement, J.-P. (1985). *Apprendre pour entreprendre*. Paris: Le Livre de Poche.

Chomsky, N. (1957). *Syntactic Structures*. The Hague: Mouton.

Coakley, J. (1990). National Minorities and the Government of Divided Societies: A Comparative Analysis of some European Evidence. *European Journal of Political Research,* Vol. 18 (4), 437-456.

Coens, D. (1989). *Met een nieuwe lei: over de toekomst van het onderwijs in Vlaanderen.* Brugge: Desclée DeBrouwer.

Cohn, E. and Geske, T.G. (1990). *The Economics of Education, Third Edition.* Oxford: Pergamon Press.

Coleman, J.S. (1961). *The Adolescent Society: The Social Life of the Teenager and Its Impact on Education.* London: The Free Press.

Coleman, J.S. (1982). *The Asymmetric Society.* New York: Syracuse University Press.

Coleman, J.S. (1987). Families and Schools. *Educational Researcher,* Vol. 16 (6), 32-38.

Coleman, J.S. (1990). *Foundations of Social Theory.* Cambridge, Massachusetts: Harvard University Press.

Coleman, J.S., Campbell, E., Hobson, C., McPartland, J., Mood, A., Weinfeld, F., and York, R. (1966). *Equality of Educational Opportunity.* Washington, D.C.: U.S. Government Printing Office.

Coleman, J.S., Hoffer, T. and Kilgore, S. (1981a). *Public and Private Schools.* Chicago, IL: National Opinion Research Center and University of Chicago.

Coleman, J.S., Hoffer, T. and Kilgore, S. (1981b). Questions and Answers: Our response. *Harvard Educational Review,* Vol. 51, 526-545.

Coleman, J.S., Hoffer, T. and Kilgore, S. (1982). *High School Achievement: Public, Catholic, and Private Schools Compared.* New York: Basic Books.

Comber, L.C. and Keeves, J.P. (1973). *Science Education in Nineteen Countries: An Empirical Study.* New York: Wiley.

Commission of the European Communities (1982). *A European Policy for Europe.* Luxemburg: Office for Official Publications of the EC.

Commission of the European Communities (1984). Draft Treaty Establishing the European Union. *Bulletin of the European Communities,* Vol. 17 (2), 7- 28.

Commission of the European Communities (1985). *Completing the Internal Market: White Paper from the Commission to the European Council* (The Cockfield White Paper of June 14, 1985, COM(85)310). Luxemburg: Office for Official Publications of the EC.

Commission of the European Communities (1986). *Linguistic Minorities in Countries Belonging to the European Community,* Luxemburg: Office for Official Publications of the EC.

Commission of the European Communities (1987). *Stability, Equity and Efficiency* (The T. Padoa-Schioppa Report). Brussels and Luxemburg: Office for Official Publications of the EC.

Commission of the European Communities (1988a). *Education in the European Community: Medium-term Perspectives 1989-1992.* (COM(88)280). Luxemburg: Office for Official Publications of the EC.

Commission of the European Communities (1988b). *Entschliessung des EG-Rates über Umweltbildung vom 24.5.1988.* Brussels: EG-Rat der Vereinigten Minister für das Bildungswesen.

Commission of the European Communities (1989a). *Racism, Xenophobia and Intolerance: An EC Survey.* Luxemburg: Office for Official Publications of the EC.

Commission of the European Communities (1989b). *Who Cares for Europe's Children? The Short Report of the European Childcare Network.* Luxemburg: Office for Official Publications of the EC.

Commission of the European Communities (1990a). *Eurobarometer: Public Opinion in the European Community, 2 Vols.* Brussels: Directorate-General of Information, Communication and Culture.

Commission of the European Communities (1990b). Council Decision for Establishing an Action Programme for the Development of Continuing Vocational Training in the European Community. *Education Document SOC 6276/90.* Brussels: Commission of the EC.

Conant, J.B. (1959). *The American High School Today.* New York: McGraw-Hill.

Coombs, P.H. (1968). *The World Educational Crisis: A Systems Analysis.* New York: Oxford University Press.

Coombs, P.H. (1985). *The World Crisis in Education: The View from the 1980s.* New York: Oxford University Press.

Cooney, J.A. (1978). *The Future of Institutionalized Schooling.* Rapporteurs' report of an international conference held at the Aspen Institute for Humanistic Studies, Berlin, November 24-27, 1978.

Corner, T. (1988). The Maritime and Border Regions of Western Europe. *Comparative Education* (Oxford), Vol. 24, (2), pp. 229-245.

Council for Cultural Cooperation (1986). *The Education and Cultural Development of Migrants.* Final report of CDCC Project No. 7. Strasburg: Council for Cultural Cooperation.

Council of Europe (1970). *Permanent Education.* Strasburg: Council of Europe.

Council of Europe (1982). *Sex Stereotyping in Schools.* (A Report of the Educational Research Workshop held in Honefoss, May 5-8, 1981). Lisse, the Netherlands: Swets and Zeitlinger.

Council of Europe (1985). *Divorce, Judicial Separation and Re-marriage: Recent Trends in Member States.* Population Studies, No. 17. Strasburg: Council of Europe.

Council of Europe (1987a). *Interculturalism and Education.* Document DECS/EGT (87) 26. Strasburg: Council of Europe.

Council of Europe (1987b). *New Challenges for Teachers and Their Education.* Report of the Fifteenth Session of the Standing Conference of European Ministers of Education, Helsinki, May 5-7, 1987 (Document M-ED-15-9). Strasburg: Council of Europe.

Council of Europe (1987c). *National Reports on Teacher Education.* Working document prepared for the Fifteenth Session of the Standing Conference of European Ministers of Education, Helsinki, 5-7 May 1987 (Document M-ED-15-4). Strasburg: Council of Europe.

Council of Europe (1988a). *Report on the Regional or Minority Languages in Europe.* Explanatory memorandum ACPL8PII.23. Strasburg: Council of Europe.

Council of Europe (1988b). *Resolution 192 on Regional or Minority Languages in Europe.* Draft resolution presented by the Committee on Cultural and Social Affairs. Strasburg: Council of Europe.

Council of Europe (1989). *Report on Family Policies in Europe.* (Steering Committee on Social Policy, Fourth Meeting, October 1989). Strasburg: Council of Europe.

Council of Europe (1990a). *Recent Demographic Developments in the Member States of the Council of Europe, 1989 Edition.* Strasburg: Council of Europe.

Council of Europe (1990b). *Cohort Fertility in Member States of the Council of Europe.* Population Studies, No. 21. Strasburg: Council of Europe.

Council of Europe (1990c). *Household Structures in Europe.* Population Studies, No. 22. Strasburg: Council of Europe.

Cox, C.B. and Dyson, A.E. (Eds) (1970). *Black Paper Two: The Crisis in Education.* London: The Critical Quarterly Society.

Creemers, B.P.M. and Scheerens, J. (Eds) (1988). Developments in Middle-school Education in Western Europe. *International Journal of Educational Research,* Vol. 12, 453-574.

Cremin, L.A. (1961). *The Transformation of the School. Progressivism in American Education 1876-1957.* New York: Alfred A. Knopf.

Cremin, L.A. (1976). *Public Education.* New York: Basic Books.

Cremin, L.A. (1979). Changes in the Ecology of Education: The School and Other Educators. In: T. Husén (Ed.), *The Future of Formal Education* (pp. 18-29). Stockholm: Almqvist and Wiksell.

Cyprus, Ministry of Education (1990). *Development of Education 1988-1990.* National Report of Cyprus to the 42nd Session of the International Conference on Education in Geneva. Nicosia: Ministry of Education.

Dahllöf, U. (1963). *Kraven på gymnasiet. Undersökningar vid universitet och högskolor, i förvaltning och näringsliv.* [Demands on the upper secondary school. Studies of higher education, administration and business] SOU: 1963:22. Stockholm: Almänna förlaget.

Debeauvais, M. (1970). *Étude comparative sur les dépenses d'enseignement dans les pays de l'OCDE et leur évolution depuis 1950.* Rapport de Base No. 2 (STP,70,7). Paris: OECD.

Denison, E.F. (1962). *Why Growth Rates Differ: Post-War Experience in Nine Western Countries.* Washington, D.C.: The Brookings Institution.

DiMaggio, P. and Mohr, J. (1985). Cultural Capital, Educational Attainment, and Marital Selection. *American Journal of Sociology,* Vol. 92, 1231-1261.

Dore, R. (1976). *The Diploma Desease: Education, Qualification and Development.* London: Allen and Unwin.

Duncan, W. (1989). *Engendering School Learning.* Studies in Comparative and International Education No. 16. Institute of International Education, University of Stockholm.

Dunkin, M.J. (Ed.). (1987). *The International Encyclopedia of Teaching and Teacher Education.* Oxford: Pergamon Press.

Eckstein, M.A. and Noah, H.J. (1988). Forms and Functions of Secondary School-leaving Examinations. *Comparative Education Review,* Vol. 33 (3), 295-316.

Educational Testing Service (1989). *A World of Differences: An International Assessment of Mathematics and Science.* Princeton, New Jersey: ETS.

Egger, E. (Ed.) (1971). *Secondary School Leaving Examinations.* Strasburg: Council for Cultural Cooperation, Council of Europe.

Eicher, J.-C. (1984). Educational Costing and Financing. *Staff Working Papers No. 550.* Washington, D.C.: The World Bank.

Eicher, J.-C. (1989). La crise financière dans les systèmes d'enseignement. In: F. Caillods (Ed.), *Les perspectives de la planification de l'education* (pp. 76-107). Paris: UNESCO, International Institute for Educational Planning.

Eicher, J.-C. (1990). *Rethinking the Finance of Post-compulsory Education.* (Mimeographed document). Dijon: Université de Bourgogne.

Eijkelhof, H.M.C., Holl, E., Pelupessy, B., Van der Valk, A.E., Verhagen, P.A.J. and Wierstra, R.F.A. (Eds) (1986). *Op weg naar de vernieuwing van het natuurkundeonderwijs* (Towards the renewal of physics education). The Hague: SVO.

Ekström, R. (1959). *Experimental Studies of Homogeneous Grouping: A Review of the Literature.* Princeton, New Jersey: Educational Testing Service.

Elvin, L. (1961). Preface. In: A.H. Halsey (Ed.), *Ability and Educational Opportunity* (pp. 9-12). Paris: OECD.

Ely, D.P. and Plomp, T. (1986). The Promises of Educational Technology: A Reassessment. *International Review of Education,* Vol. 32 (2), 231-249.

Ely, P. and Denney, D. (1987). *Social Work in a Multiracial Society.* London: George Allen and Unwin.

Eulefeld, G. (1989). Umwelterziehung als umfassende Aufgabe – eine Einführung in die Themen der Arbeitsgruppen. In: *Umweltbildung in der EG.* Dokumentation einer internationalen Fachtagung am 23./24 Februar 1988 in Dortmund (pp. 28-38). Bonn: Bundesminister für Bildung und Wissenschaft.

EURIDYCE (1988). *The Teaching of Languages in the European Community.* Brussels: EURIDYCE European Unit.

EUROBAROMETER (1991). Opinions and Attitudes in EC Countries. In: *European Affairs,* Vol. 5 (February/March), pp. 65-68, 1991.

European Parliament. (1981). *A Community Charter of Regional Languages and Cultures and a Charter of Rights of Ethnic Minorities*. Brussels and Strasburg: Parliament of the European Communities, Office for Official Publications of the European Communities.

European Round Table of Industrialists (1989). *Education and European Competence. ERT Study on Education and Training in Europe*. Brussels: European Round Table of Industrialists.

EUROSTAT (1987). *Demographic Statistics. Theme 3, Series 3*. Luxemburg: Statistical Office of the European Communities.

EUROSTAT (1989). *Racism, Xenophobia and Intolerance: An EC Survey*. Luxemburg: Statistical Office of the European Communities.

EUROSTAT (1990). *Inequality and Poverty in Europe 1980-1985*. Population and Social Conditions Report 1990:7. Luxemburg: Statistical Office of the European Communities.

Eyskens, M. (1991). Europa op een tweesprong. Leading article in *NRC Handelsblad*, 15 June 1991.

Fägerlind, I. and Sjöstedt, B. (1990). *Educational Planning and Management in Europe: Review and Prospects*. Working document for UNESCO Congress on Planning and Management of Educational Development, Mexico, March 26-30, 1990. Institute of International Education, University of Stockholm.

Faure, E., Herrera, F., Kaddoura, A.-R., Lopes, H., Petrovsky, A.V., Rahnema, M., and Ward, F.C. (1972). *Learning To Be: The World of Education Today and Tomorrow*. London: Harrap.

Fend, H. (1982). *Gesamtschule im Vergleich: Bilanz der Ergebnisse des Gesamtschulversuchs*. Weinheim: Beltz.

Fend, H. (1984). *Pädagogik des Neokonservativismus*. Frankfurt: Suhrkamp.

Feynman, R. (1965). *The Character of the Physical Law*. London: Penguin Books.

Finland, Ministry of Education (1990). *Developments in Education 1988-1990 in Finland*. Report by the Ministry of Education to the 42nd Session of the International Conference on Education in Geneva. Reference Publications 15. Helsinki: Ministry of Education.

Finocchiaro, M. and Brumfit, C. (1983). *The Functional-Notational Approach: From Theory to Practice*. New York: Oxford University Press.

Fothergill, R. (1988). *Implications of New Technology for the School Curriculum*. London: Kogan Page.

Fourastié, J. (1979). *Les trente glorieuses, ou la révolution invisible de 1946 à 1975*. Paris: Fayard.

Fraser, B.J., Walberg, H.J., Welch, W.W. and Hattie, J.A. (1987). Syntheses of Educational Productivity Research. *International Journal of Educational Research*, Vol. 11, 145-252.

Freeman, R.B. (1976). *The Over-educated American*. New York: Academic Press.

Friedenberg, E.Z. (1963). *Coming of Age in America: Growth and Acquiescence*. New York: Random House.

Gabor, D. (1962). *Inventing The Future*. Cambridge: Cambridge University Press.

Garcia-Garrido, J.L. (Ed.), Primary Education on the Treshold of the Twenty-first Century. *International Yearbook of Education*, Vol. 38. Geneva: UNESCO, International Bureau of Education.

Gardner, J.W. (1961). *Excellence: Can We Be Equal and Excellent Too?* New York: Harper and Brothers.

Germany, Federal Ministry of Education and Science (1986). *Basic and Structural Data 1986-87 Edition*. Bonn: Bundesminister für Bildung und Wissenschaft.

Ginsburg, A.L., Noell, J. and Plisko, V.W. (1988). Lessons from the Wall Chart. *Educational Evaluation and Policy Analysis*, Vol. 10 (1), 1-12.

Glass, G.V., Cohen, L.S., Smith, M.L. and Filby, N. (1982). *School Class Size: Research and Policy*. London: Sage.

Goodman, P. (1962). *Compulsory Miseducation*. New York: Random House.

Gorbachev, M. (1987). *Perestroika. New Thinking For Our Country and the World*. London: Collins.

Gordon, P. (1989). The Rise of Racism in Europe. In: *Fortress Europe? The Meaning of 1992*. London: The Runnymede Trust.

Goulet, H. (1990). Child Care in OECD Countries. In: *Employment Outlook*, July 1990 (pp. 123-151). Paris: OECD.

Grotberg, E.H. (1985). Head Start Program. In: *The International Encyclopedia of Education* (pp. 2138-2141). Oxford: Pergamon Press.

Grubb, W.N. (1985). The Convergence of Educational Systems and the Role of Vocationalism. *Comparative Education Review*, Vol. 29 (4), pp. 15-38.

Guerrieri, P. and Padoan, P.C. (1989). Integration, Cooperation and Adjustment Policies. In: P. Guerrieri and P.C. Padoan (Eds), *The Political Economy of European Integration* (pp. 1-30). London: Harvester Wheatsheaf.

Gunstone, R.F. (1985). Secondary School Programmes in Science Education. *International Encyclopedia of Education*, Vol. 8, pp. 4461-4465.

Guthrie, J.W. (1990). The Industrialized World's Evolving Political Economy and the Implications for Educational Evaluation. (mimiographed paper). University of California, Berkeley.

Haggis, S. and Aday, P.A. (1979). Review of Integrated Science Worldwide. *Studies in Science Education*, Vol. 6, 69-91.

Halls, W.D. (1970). *Foreign Languages and Education in Western Europe*. London: Harrap.

Halls, W.D. (1974). Towards a European Education System? *Comparative Education*, (Oxford), Vol. 10 (3), 211-219.

Halls, W.D. (1980). A Curriculum for Europe. *Prospects*, Vol. 10 (2), 219-229.

Halls, W.D. (1985). The International Baccalaureate. *International Encyclopedia of Education*, Vol. 5, 2647-2648.

Halsey, A.H. (Ed.) (1961). *Ability and Educational Opportunity*. Paris: OECD.

Halsey, A.H. (Ed.) (1972). *Education Priority*, Vol. 1 (pp. 3-12). London: Her Majesty's Stationery Office.

Hanushek, E.A. (1981). Throwing Money at Schools. *Journal of Policy Analysis and Management*, Vol. 1 (1), 19-41.
Hanushek, E.A. (1986). The Economics of Schooling. *Journal of Economic Literature.* Vol. 24, 1141-1177.
Hanushek, E.A. (1989). The Impact of Differential Expenditures on School Performance. *Educational Researcher*, Vol. 18 (4), 45-51.
Hargreaves, A.G. (1990). Language and Identity in Beur Culture. *French Cultural Studies* Vol. 1 (1), 47-58.
Hartog, P.J. and Rhodes, E.C. (1935). *An Examination of Examinations.* London: Macmillan.
Hauser, R.M. and Sewell, W.H. (1986). Family Effects in Simple Models of Education, Occupational Status and Earnings: Findings from the Winconsin and Kalamazoo Studies. *Journal of Labour Economics,* Vol. 4, S83-S115.
Haviland, J. (Ed.) (1988). *Take Care, Mr Baker. A Selection From the Advice on the Government's Education Reform Bill Which The Secretary of State for Education Invited But Decided Not to Publish.* London: Fourth Estate.
Heidenheimer, A.J., Heclo, H. and Adams, C.T. (1990). *Comparative Public Policy: The Politics of Social Choice in America, Europe, and Japan. Third Edition.* New York: St. Martin's Press.
Hoffmann, S. (1991). The Future of Europe. *Bulletin of the American Academy of Arts and Sciences,* Vol. XLIV (5), 15-28.
Hollister, R.G. (1967). *A Technical Evaluation of the First Stage of the Mediterranean Regional Project.* Paris: OECD.
Honout, J. (1989). *Letteren-nota Nederlands Cultuurbeleid en Europese Eenwording.* Rijswijk (The Hague): Ministerie van Welzijn, Volksgezondheid en Cultuur, Directoraat-Generaal Culturele Zaken.
Husén, T. (1959). Interaction between Teacher and Pupils as a Determinant of Motivation and Satisfaction with School Work. *Proceedings of the XIIIth Congress of Applied Psychology* (pp. 471-482). Rome.
Husén, T. (1969). *Talent, Opportunity and Career.* Stockholm: Almqvist and Wiksell.
Husén, T. (1974). *Talent, Equality and Meritocracy.* The Hague: Martinus Nijhoff.
Husén, T. (1979). *The School in Question. A Comparative Study of the School and Its Future in Western Societies.* Oxford: Oxford University Press.
Husén, T. (1987). Policy Impact of IEA Research. *Comparative Education Review,* Vol. 31 (1), 29-46.
Husén, T. (1989). Schools for the 1990s. *Scandinavian Journal of Educational Research,* Vol. 33 (1), 3-15.
Husén, T. (1990). *Education and the Global Concern.* Oxford: Pergamon Press.
Husén, T. and Opper, S. (Eds) (1983). *Multilingual and Multicultural Education in Immigrant Countries.* Oxford: Pergamon Press.
Hymes, D. (1972). On Communicative Competence. In: J.B. Pride and J. Holmes (Eds), *Sociolinguistics* (pp. 269-293). Harmondsworth: Penguin.

Härnqvist, K. (1989). *Background, Education and Work As Predictors of Adult Skills*. Göteburg: Department of Education and Educational Research, University of Göteborg.

Illich, I. (1970). *Deschooling Society*. New York: Harper and Row.

Inoguchi, T. (1991). Japan and Europe: Wary Partners. *European Affairs*, Vol. 5 (1), pp. 54-58.

International Academy of Education (1991). *Home Environments and School Learning*. San Fransisco: Jossey-Bass.

International Association for the Evaluation of Educational Achievement (1988). *Science Achievement in Seventeen Countries: A Preliminary Report*. Oxford: Pergamon Press.

International Labour Organization (1990). *Yearbook of Labour Statistics*, 50th Edition. Geneva: ILO.

Iverson, B.K. and Walberg, H.J. (1982). Home Environment and School Learning: A Quantitative Synthesis. *Journal of Experimental Education*, Vol. 50, 144-151.

Jencks, C.S., Smith, M., Acland, H., Bane, M., Cohen, D., Gintis, H., Hyns, B. and Michelson, S. (1972). *Inequality. A Reassessment of the Effect of Family and Schooling in America*. New York: Basic Books.

Jencks, C.S., Bartlett, S., Corcoran, M., Crouse, J., Eaglesfield, D., Jackson, G., McClelland, K., Mueser, P., Olneck, M., Schwartz, J., Ward, S. and Williams, J. (1979). *Who Gets Ahead? The Determinants of Economic Success in America*. New York: Basic Books.

Jensen, A.R. (1973). *Educability and Group Differences*. London: Methuen.

Judge, H.G. (1988). Cross-national Perceptions of Teachers. *Comparative Education Review*, Vol. 32 (2), 143-158.

Kahle, J.B. (Ed.) (1985), *Women in Science: A Report from the Field*. Basingstoke: The Falmer Press.

Kairamo, K. (Ed.) (1989). *Education for Life: A European Strategy*. London: Butterworths, and Brussels: Round Table of European Industrialists.

Keeves, J.P. (1991). Specialization in Science: 1970 and 1984. In: T. Husén and J.P. Keeves (Eds), *Issues in Science Education* (pp. 65-105). Oxford: Pergamon Press.

Keeves, J.P. (Ed.) (1992). *Changes in Science Education and Achievement: 1970-1984*. Oxford: Pergamon Press.

Keeves, J.P., Morgenstern, Ch., and Saha, L.J. (1991). Educational Expansion and Equality of Opportunity: Evidence from Studies Conducted by IEA in Ten Countries in 1970-71 and 1983-84. *International Journal of Educational Research*, Vol. 15 (1), 61-80.

Kelly, A. (1978). *Girls and Science: An International Study of Sex Differences in School Science Achievement*. Stockholm: Almqvist and Wiksell.

Keppel, F. (1966). *The Necessary Revolution in American Education*. New York: Harper and Row.

Kerr, C. (1990). The Internationalization of Learning and the Nationalization of the Purposes of Higher Education: Two "Laws of Motion" in Conflict? *European Journal of Education,* Vol. 25, 5-22.

Kerr, S.T. (1990). Will Glasnost lead to Perestroika? Directions of Educational Reform in the U.S.S.R. *Educational Researcher,* Vol. 20 (October), 27-31.

Kifer, E. (1988). What IEA Studies Say About Curriculum and School Organization. In: A.C. Purves (Ed.), *International Comparisons and Educational Reform* (pp. 51-72). Alexandria, Virginia: Association for Supervision and Curriculum Development.

Kleinespel, K. (1990). Schule als biographische Erfahrung. Die Laborschule im Urteil ihrer Absolventen. *Studien zur Schulpädagogik und Didaktik. Bd. 3.* Weinheim and Basel: Beltz.

Kogan, M. (1990). Monitoring, Control and Governance of School Systems (mimeographed paper). Faculty of Social Sciences, Brunel University.

Krashen, S.D. (1981). *Second Language Acquisition and Second Language Learning.* Oxford: Pergamon Press.

Kwabena, G.-B. and Gyapong, A.O. (1991). Characteristics of Educational Production Functions: An Application of Canonical Analysis. *Economics of Educational Review,* Vol. 10, (1), 7-17.

Kulik, J.A. and Kulik, C.L.C. (1989). Meta-analysis in Education. *International Journal of Educational Research,* Vol. 13 (3), 221-340.

Lambert, W.E. (1977). The Effect of Bilingualism upon the Individual: Cognitive and Socio-cultural Consequences. In: P.A. Hornby (Ed.), *Bilingualism: Psychological, Social and Educational Implications.* New York: Academic Press.

Landsheere, G. de (1989). *The Information Society and Education: Synthesis of National Reports From the 16th Session of the Standing Conference of European Ministers of Education.* Strasburg: Council of Europe.

Lareau, A. (1987). Social Class Differences in Family-school Relationships: The Importance of Cultural Capital. *Sociology of Education,* Vol. 60, 73-85.

Lawlor, S. (1990). *Teachers Mistaught.* London: Centre for Policy Studies.

Leclercq, J.M. and Rault, Ch. (1989). *Les systèmes éducatifs en Europe: vers un Espace Communautaire?* Notes et Études Documentaires No. 4899. Paris: La Documentation française.

Lesourne, J. (1988). *Éducation et Société. Les Défis de l'an 2000.* Paris: Éditions la Découverte et Le Monde.

Levin, H.M. (1987). Accelerated Schools for Disadvantaged Students. *Educational Leadership,* Vol. 44 (6), 19-21.

Levin, H.M. (1989). Economic Trends Shaping the Future of Teacher Education. *Educational Policy,* Vol. 4 (1), 1-16.

Lewis, J.L. (Ed.) (1986). *Science and Technology Education and Future Human Needs.* Oxford: Pergamon Press.

Lewis, J.L. (1991). Science in Society: Impact on Science Education. In: T. Husén and J.P. Keeves (Eds), *Issues in Science Education: Science Competence in a Social and Ecological Context.* (pp. 153-161). Oxford: Pergamon Press.

Lewis, R. (1988). *Anti-racism. A Mania Exposed.* London: Allen and Unwin.

Linguistic Minorities in England. A report by the Linguistic Minorities Project for the Department of Education and Science. Institute of Education, University of London, July 1983.

Linn, M.C. (1987). Establishing a Research Base for Science Education: Challenges, Trends and Recommendations. *Journal of Research in Science Teaching,* Vol. 24 (3), 191-216.

Little, A. (1975). The Background of Under-achievement in Immigrant Children in London. In: G.K. Verma and C. Bagley (Eds), *Race and Education Across Cultures.* London: Macmillan.

Lockard, J.D. (1975). *Science and Mathematics Curriculum Developments Internationally 1956-1974.* University of Maryland, Center for Science Teaching, College Park, Maryland.

Lodge, J. (Ed.) (1989). *The European Community and the Challenge of the Future.* London: Pinter Publishers.

Lübke, S.-I. and Michael, M. (1989). *Absolventen '85 – Eine empirische Längsschnittstudie über die Bielefelder Laborschule.* Bielefeld: IMPULS 19 (Informationen, Materialien, Projekte, Unterrichtseinheiten der Laborschule Bielefeld).

Malkova, Z.A. and Vulfson, B. (Eds) (1987). *Secondary Education in the World Today.* International Yearbook of Education, Vol. 34. Geneva: UNESCO, International Bureau of Education.

Mann, A.J., Harrell, A., and Hurt, M. Jr. (1977). *A Review of Head Start Research Since 1969 and An Annotated Bibliography.* DHEW Publication No. OHDS-78-31102. Washington, D.C.: United States Government Printing Office.

Mappa, S. (1987). Education for Immigrants' Children in the OECD Member Countries. In: *The Future of Migration.* Paris: OECD.

Maresceau, M. (Ed.) (1989). *The Political and Legal Framework of Trade Relations Between the European Community and Eastern Europe.* Dordrecht: Martinus Nijhoff.

Marklund, S. (1983). *The IEA Project: An Unfinished Audit.* Report No. 64. Institute of International Education, Stockholm University.

Marklund, S. (1980-1989). *Skolsverige 1950-1975,* 6 Vols. Stockholm: Liber.

Marklund, S. (1988). Education in Sweden: Assessment of Student Achievement and Selection for Higher Education. In: S.P. Heyneman and I. Fägerlind (Eds), *University Examinations and Standardized Testing.* Technical Paper No. 78. Washington, D.C.: The World Bank.

Martinson, H. (1935). *The Nettles Bloom.* Stockholm: Bonniers.

Marx, G. (Ed.) (1989). *Proceedings of International UNESCO Conference on Energy and Risk Education.* Veszprem: National Centre for Educational Technology.

Marx, G. (1990). *Educating for an Unknown Future.* Budapest: Department of Atomic Physics, Roland Eötvös University.

Marx, G. and Toth, E. (1981). Models in Science Education. *Impact of Science on Society*, Vol. 31, 26-34.
McDonald, M. (1986). *We are not French: Language, Culture and Identity in Brittany*, London, 1990.
McKnight, C.C., Crosswhite, F.J., Dossey, J.A., Kifer, E., Swafford, J.O., and Travers, K.J. (1987). *The Underachieving Curriculum: Assessing U.S. School Mathematics from an International Perspective*. Champaign, Ill.: Stipes Publishing Co.
McLaughlin, B. (1984-85). *Second-language Acquisition in Childhood, 2 Vols*. Hillsdale, N.J.: Lawrence Erlbaum Associates.
Meadows, D.H. (1972). *The Limits to Growth: A Report for the Club of Rome's Project on the Predicament of Mankind*, 2nd edn. New York: Universe Books.
Mehlhorn, H.G. and Urban, K.K. (Eds) (1989). Hochbegabtenförderung international. *Bildung und Erziehung,* Beiheft 6. Köln: Böhlau-Verlag.
Mesarovic, M. and Pestel, E. (1974). *Mankind at the Turning Point*. London: E.P. Dutton.
Mincer, J. (1962). On-the-job Training: Costs, Returns and Some Implications. *Journal of Political Economy*, Vol. 70 (5, Part 2), S50-S79.
Mincer, J. (1989). Human Capital and the Labour Market: A Review of Current Research. *Educational Researcher*, Vol. 18, (4), 27-34.
Mitter, W. (1983). *Kann die Schule Erziehen? Erfahrungen, Probleme und Tendenzen im europäischen Vergleich*. Köln: Böhlau-Verlag.
Mitter, W. (1984). *Education for All*. International Yearbook of Education, Vol. 36. Geneva: UNESCO, International Bureau of Education.
Monk, D.H. (1990). *Educational Finance*. New York: MacGraw-Hill.
Mulder, L.H., Rijn, J.G., and van der Goot, A. (1989). *De kleine talen van Europa*. Ljouwert, Groningen: Nederlandse Commissie van het Europees Bureau voor Kleine Talen.
Murray, M. (1988). *Utbildningsexpansion, jämlikhet, och avlänkning* (Educational expansion, equality, and diversion). Department of Education and Educational Research, University of Göteborg.
Neave, G. (1984). *The EEC and Education*. Stoke-on-Trent: Trentham Books.
Neave, G. (1987). Challenges Met: Trends in Teacher Education, 1975-1985. In: Council of Europe (1987). *New Challenges for Teachers and their Education: National Reports on Teacher Education* (introduction). Documents prepared for the Fifteenth Session of the Standing Conference of European Ministers of Education, Helsinki, 5-7 May 1987 (Document M ED-15-4). Strasburg: Council of Europe.
Neave, G. (1988). *The Status and Employment of the Teaching Profession in the European Community*. Brussels: Commission of the EC.
Neave, G. (1990). On Preparing for Markets: Trends in Higher Education in Western Europe, 1988-1990. *European Journal of Education,* Vol. 25 (2), 105-122.
Netherlands Advisory Council for Education. (1986). *Een boog van woorden tot woorden*. Utrecht: ARBO.

Netherlands Ministry of Education and Sciences (1989). *Richness of the Uncompleted. Challenges Facing Dutch Education.* (The Netherlands Report to the OECD). The Hague: Author.

Nicol, W. and Salmon, T.C. (1990). *Understanding the European Communities.* London: Philip Allan.

Nicolitsas, Ch. (1990). Notes on Surveys Concerning Functional Illiteracy. Paper for the seminar on Functional Illiteracy in Eastern and Western Europe, UNESCO Institute for Education, Hamburg, November 20-22, 1990.

Noah, H.J. and Eckstein, M.A. (1989). Tradeoffs in Examination Policies: An International Comparative Perspective. *Oxford Review of Education,* Vol. 13 (3).

Nordic Council of Ministers (1987). *Recommendations on Environmental Education: Ten Years of Development.* Copenhagen: Nordiska ministerrådet.

Norway, Ministry of Education and Research (1989). *Development of Education: 1988-1990.* National Report of Norway to the 42nd Session of the International Conference on Education in Geneva. Oslo: The Norwegian Ministry of Education and Research.

Opper, S. and Teichler, U. (1989). Educational Programs in the European Community. *International Encyclopedia of Education.* Suppl. Vol. 1, 342-347.

Oreja, M. (1987). *Social Cohesion and the Dangers Facing It.* Report of the Secretary General to the Parliamentary Assembly, May 6th. Strasburg: Council of Europe.

Organization for Economic Cooperation and Development (1962). *Policy Conference on Economic Growth and Investment in Education.* General report on the OECD conference held in Washington, D.C., 16-20 October 1961. Paris: OECD.

Organization for Economic Cooperation and Development (1963). *Planning Education for Social and Economic Development.* Paris: OECD.

Organization for Economic Cooperation and Development (1967). *Education, Human Resources and Development in Argentina,* 2 Vols. Paris: OECD.

Organization for Economic Cooperation and Development (1971). *Educational Policies for the 1970s.* General report of the OECD conference on policies for educational growth, June 3-5, 1970. Paris: OECD.

Organization for Economic Cooperation and Development (1973). *Recurrent Education: A Strategy for Lifelong Learning.* Paris: OECD.

Organization for Economic Cooperation and Development (1986). *Reviews of National Policies for Education: Spain.* Paris: OECD.

Organization for Economic Cooperation and Development (1987a). *Multicultural Education.* Paris: OECD.

Organization for Economic Cooperation and Development (1987b). *The Future of Migration.* Paris: OECD.

Organization for Economic Cooperation and Development (1987c). *Adolescents and Comprehensive Schooling.* Paris: OECD.

Organization for Economic Cooperation and Development (1989a). *Education and Structural Change. A Statement by the Education Committee.* Paris: OECD.

Organization for Economic Cooperation and Development (1989b). *Schools and Quality: An International Report*. Paris: OECD.
Organization for Economic Cooperation and Development (1989c). *Education in OECD Countries 1986/87: A Compendium of Statistical Information*. Paris: OECD.
Organization for Economic Cooperation and Development (1989d). *One School, Many Cultures*. Paris: OECD.
Organization for Economic Cooperation and Development (1989e). *Employment Outlook* (July). Paris: OECD.
Organization for Economic Cooperation and Development (1989f). *Main Economic Indicators*. Paris: OECD.
Organization for Economic Cooperation and Development (1990a). *Analytical Report: High Quality Education and Training for All*. Report to the Ministers of Education. Paris: OECD.
Organization for Economic Cooperation and Development (1990b). *Education in OECD Countries, 1990 Special Edition*. Paris: OECD.
Organization for Economic Cooperation and Development (1990c). *Reviews of National Policies for Education: The Netherlands*. Paris: OECD.
Organization for Economic Cooperation and Development (1990d). *Lone-parent Families: The Economic Challenge*. Social Policy Studies No. 8. Paris: OECD.
Organization for Economic Cooperation and Development (1990e). *Employment Outlook* (July). Paris: OECD.
Organization for Economic Cooperation and Development (1990f). *The Training of Teachers* (document CERI/AW/gk/600). Paris: OECD.
Organization for Economic Cooperation and Development (1990h). *SOPEMI: Continuous Reporting System on Migration*. Paris: OECD.
Oxenham, J. (1985). Policies for Educational Equality. *International Encyclopedia of Education*, Vol. 3, 1689-1698.
Pallas, A.M., Natriello, G., and McDill, E.L. (1989). The Changing Nature of the Disadvantaged Population: Current Dimensions and Future Trends. *Educational Researcher*, Vol. 18 (5), 16-22.
Parnes, H. (Ed.) (1962). *Forecasting Educational Needs for Economic and Social Development*. Paris: OECD.
Passow, A.H. (Ed.) (1979). *The Gifted and the Talented: Their Education and Development*. Chicago: University of Chicago Press.
Pelgrum, H. and Plomp, Tj. (1991). *The Use of Computers in Education Worldwide*. Oxford: Pergamon Press.
Peschar, J. (1991). Educational Opportunities in East-West Perspective: A Comparative Analysis of the Netherlands, Hungary, Poland and Czechoslovakia. *International Journal of Educational Research*, Vol. 15 (1), 107-121.
Peterson, A.D. (1972). *The International Baccalaureate: An Experiment in Education*. London: Harrap.
Picht, G. (1964). *Die deutsche Bildungskatastrophe*. Freiburg im Breisgau: Walter Verlag.

Plowden Report (1967). *Children and Their Primary Schools,* 2 Vols. A report of the Central Advisory Council for Education. London: Her Majesty's Stationery Office.

Poignant, R. (1973). *Education in the Industrialized Countries.* The Hague: Martinus Nijhoff.

Poland, Ministry of National Education (1990). *The Development of Education Within 1989-1990.* Warsaw: Ministry of National Education.

Postlethwaite, T.N. (1991). Achievement in Science Education in 23 Countries in 1984. In: T. Husén and J.P. Keeves (Eds), *Issues in Science Education* (pp. 35-59). Oxford: Pergamon Press.

Postlethwaite, T.N. and Wiley, D.E. (Eds) (1992). *Science Achievement in Twenty-three Countries.* Oxford: Pergamon Press.

Psacharopoulos, G. (1984). The Contribution of Education to Economic Growth. In: J.W. Kendrick (Ed.), *International Comparisons of Productivity and Causes of the Slowdown* (pp. 335-360). Cambridge: Ballinger.

Psacharopoulos, G. (1987). Earnings Functions. In G. Psacharopoulos (Ed), *Economics of Education. Research and Studies* (pp. 218-233). Oxford: Pergamon Press.

Psacharopoulos, G. (1990). Education and the Professions in Greece in the Light of 1992. *European Journal of Education,* Vol. 25 (1), 61-74.

Psacharopoulos, G. and Woodhall, M. (1985). *Education for Development. An Analysis of Investment Choices.* Oxford: Oxford University Press.

Purves, A.C. (Ed.) (1989). *International Comparisons and Educational Reform.* Alexandria, Virginia: Association for Supervision and Curriculum Development.

Rasmussen, H. (1986). *On Law and Policy in the European Court of Justice.* Dordrecht: Martinus Nijhoff.

Ravea, F.M. (1987). Ethnicity, Migrations and Societies. In: *Multicultural Education.* Paris: OECD.

Renaud, G. (1974). *The Experimental Period of the International Baccalaureate: Objectives and Results.* Geneva: UNESCO, International Bureau of Education.

Research voor Beleid B.V. (1987). *Final Report on the Conditions of Employment Within the EC.* Leiden: Stichting Research voor Beleid. [See also document ECSC-EEC-EAEC,1988. Luxemburg: Office for Official Publications of the EC.]

Rex, J. and Tomlinson, S. (1979). *Colonial Immigrants in A British City: A Class Analysis.* London: Allen and Unwin.

Rey, M. (1986). *Training Teachers in Intercultural Education.* Strasburg: Council of Europe.

Richards, J.C. and Rodgers, T.H. (1986). *Approaches and Methods in Language Teaching: A Description and Analysis.* Cambridge: Cambridge University Press.

Rist, R.C. (1973). The Urban School: A Factory for Failure. Cambridge, Mass.: MIT Press.

Rist, R.C. (1978). *Guestworkers in Germany: The Prospects for Pluralism.* London: Praeger.

Ritzen, J.M.M. (1987). Onderwijs en economie. In: J.A. van Kemenade (Ed.), *Onderwijs: bestel en beleid,* Vol. 2b. (pp. 11-147). Groningen: Wolters-Noordhoff.

Ritzen, J.M.M. (1990). De financiering van het onderwijs. In: *Onderwijseconomie, Vol. 1* (pp. 77-93). Heerlen: Open Universiteit.

Robitaille, D.F. and Garden, R.A. (Eds) (1989). *Second International Mathematics Study, Vol. 2: Contexts and Outcomes of School Mathematics.* Oxford: Pergamon Press.

Rockefeller Brothers Fund (1958). *The Pursuit of Excellence: Education and the Future of America* (Panel Report V of the Special Studies Project, Rockefeller Brothers Fund). New York: Doubleday & Co.

Roney, A. (1990). *The European Community Fact Book.* London: Kogan Page.

Rosen, S. (1977). Human Capital: A Survey of Empirical Research. In: R.C. Ehrenberg (Ed.), *Research in Labor Economics* (pp. 3-14). Greenwich, Conn.: JAI Press.

Runnymede Trust (1989). *Fortress Europe? The Meaning of 1992.* London: The Runnymede Trust.

Rust, V.D. (1989). *The Democratic Tradition and the Evolution of Schooling in Norway.* London: Greenwood Press.

Rutter, M, Maughan, B., Mortimore, P. Ouston, J. and Smith, A. (1979). *Fifteen Thousand Hours: Secondary Schools and Their Effects on Children.* London: Open Books.

Schmidtke, H.-P. (1986). Education and the Pluralistic Society (Report of the International Congress of the "Spanish Society for Education"). *Western European Education,* Vol. 18 (3), 25-30.

Schultz, H. and von der Groeben, A. (1985). *Laborschulabsolventen '80. Zwei Interpretationen.* Bielefeld: IMPULS 8 (Informationen, Materialen, Projekte, Unterrichtseinheiten der Laborschule Bielefeld).

Schultz, T.W. (1961). Investment in Human Capital. *American Economic Review,* Vol. 51, 1-17.

Schultz, T.W. (1980). Nobel Lecture. The Economics of Being Poor. *Journal of Political Economy,* Vol. 88 (4), 639-652.

Schwartze, J. (1987). Towards a European Foreign Policy: Legal Aspects. In: J.K. de Vree, P. Coffey and R.H. Lauwaars (Eds), *Towards a European Foreign Policy. Legal, Economic and Political Dimensions* (pp. 69-96). Dordrecht: Martinus Nijhoff.

Sharp, M. and Shearman, C. (1987). *European Technological Collaboration.* Chatham House Paper No. 36. London: Routledge.

Silverman, M. (1990). The Racialization of Immigration: Aspects of Discourse from 1968-1981. *French Cultural Studies,* Vol. 1, Part 2, (2).

Simon, B. (1989). *Bending the Rules: The Baker "Reform" of Education.* London: Lawrence and Wishart.

Skilbeck, M. (1990). *Curriculum Reform: An Overview of Trends.* Paris: OECD.

Skutnaab-Kangas, T. (1984). Bilingualism or Not? In: *The Education of Minorities.* Clevedon: Multilingual Matters Ltd.

Slavin, R.E. and Madden, N.A. (1989). What Works for Students At Risk: A Research Synthesis. *Educational Leadership,* Vol. 46 (5), 4-13.

Smith, A. (1980). From 'Europhoria' to Pragmatism: Towards a New Start for Higher Education Co-operation in Europe? *European Journal of Education*, 15 (1), 77-95.

Smith, F. (1986). *Insult to Intelligence*. New York: Arbor House.

Smith, M.S. and O'Day, J. (1988). Teaching Policy and Research on Teaching (Working paper, draft October 13, 1988). Stanford: School of Education, Stanford University.

Smolicz, J.J. (1990). The Mono-ethnic Tradition and the Education of Minority Youth in West Germany from an Australian Multicultural Perspective. *Comparative Education* (Oxford), Vol. 26 (1), 27-43.

Snow, C.P. (1959). *The Two Cultures and the Scientific Revolution*. New York: Cambridge University Press.

Spencer, D. (1990). Millions lose Out on Numeracy. *Times Education Supplement*, 12th of October, 1990.

Statistics Sweden (1989). *The Future Population of Sweden. Projection for the Years 1989-2025*. Stockholm: SCB.

Statistics Sweden (1990). *Statistical Yearbook 1990*. Stockholm: SCB.

Steinkamp, M. and Maehr, M.L. (Eds) (1984). *Women in Science*. Greenwich: JAI Press.

Svensson, N.-E. (1962). *Ability Grouping and Scholastic Achievement. Report on a Five-year Follow-up Study in Stockholm*. Stockholm: Almqvist and Wiksell.

Swann, D. (1983). *Competition and Industrial Policy in the European Community*. London: Methuen.

Swedish National Board of Education. (1989). Swedish Schools in an International Perspective. *School Research Newsletter*, Vol. 1989 (6). Stockholm: NBE.

Szabo, A. (1990). *Physics Teaching in Socialist Countries*. Moscow: Nauka.

Taft, R. (1976). Coping with Unfamiliar Structures. In: Warren, N. (Ed.), *Studies in Cross-cultural Psychology*, Vol. 1. London: Academic Press, pp. 121-153.

Taylor, P. (1989). The New Dynamics of EC Integration in the 1980s. In: J. Lodge (Ed.), *The European Community and the Challenge of the Future* (pp. 3-25). London: Pinter.

Teichler, U. (1988a). *Convergence or Growing Variety: The Changing Organization of Studies*. Strasburg: Council of Europe.

Teichler, U. (1988b). *Changing Patterns of the Higher Education System. The Experience of Three Decades*. London: Jessica Kingsley.

Teichler, U., Smith, A. and Steube, W. (1988). *Auslandsstudienprogramm im Vergleich: Erfahrungen, Probleme, Erfolge*. In: Schriftenreihe Studien zur Bilding und Wissenschaft No. 68. Bonn: Bundesministerium für Bildung und Wissenschaft.

Terman, L.M. and Oden, M.H. (1947). *The Gifted Child Grows Up. Twenty-five Years' Follow-up of a Superior Group: Genetic Studies of Genius, Vol. 4*. Stanford: Stanford University Press.

The Other Languages of England, Linguistic Minorities Project, Institute of Education, University of London, 1985.

Thorndike, R.L. (1973). *Reading Comprehension Education in Fifteen Countries*. Stockholm: Almqvist and Wiksell.

Tibi, C. (1989). The Financing of Education: Impacts of the Crisis and the Adjustment Process. In: Caillods, F. (Ed.), *The Prospects for Educational Planning* (pp. 100-133). Paris: UNESCO, International Institute for Educational Planning.

Tillmann, K.-J. (1988). Comprehensive Schools and Traditional Education in the Federal Republic of Germany. *International Journal of Educational Research*, Vol 12, 471-480.

Tobin, K. (1988). Differential Engagements of Males and Females in High School Science. *International Journal of Science Education*, Vol. 10, 239-252.

Tobin, K., Butler-Kahle, J., and Fraser, B.J. (Eds) (1990). *Windows into Science Classrooms: Problems Associated with Higher-level Cognitive Learning*. Basingstoke: The Falmer Press.

Tomiak, J.J. (1982). *The Soviet Union*. London, World Education Series.

Travers, K.J. and Westbury, I. (Eds) (1989). *Second International Mathematics Study, Vol. 1: International Analyses of Mathematics Curricula*. Oxford: Pergamon Press.

Troger, W. (1968). *Elitenbildung. Überlegungen zur Schulreform in der demokratischen Gesellschaft*. München: Ernst Reinhardt.

af Trolle, U. (1990). *Mot en internationellt konkurrenskraftig akademisk utbildning* [Towards an internationally competitive academic education]. Lund: Studentlitteratur.

Troyna, B. and Williams, J. (1986). *Racism, Education, and the State*. Beckenham: Croom Helm.

Tsakaloyannis, P. (1987). Political Constraints for an Effective Community Foreign Policy. In: J.K. De Vree, P. Coffey and R.H. Lauwaars (Eds), *Towards a European Foreign Policy. Legal, Economic and Political Dimensions* (pp. 143-156). Dordrecht: Martinus Nijhoff.

Tuijnman, A. (1989). *Recurrent Education, Earnings, and Well-being. A Fifty-Year Longitudinal Study of a Cohort of Swedish Men*. Acta Universitatis Stockholmiensis. Stockholm: Almqvist and Wiksell.

Tuijnman, A. (1990). Dilemmas of Open Admissions Policy: Quality and Efficiency in Swedish Higher Education. *Higher Education*, Vol. 20, 443-457.

Turkey, Ministry of National Education (1990). *Development of Education 1989-1990*. National Report of Turkey to the 42nd Session of the International Conference on Education in Geneva. Ankara: Ministry of Education.

United Kingdom, Department of Education and Science (1985). *Education for All* (The Swann Report). London: Her Majesty's Stationery Office.

United Kingdom, Department of Education and Science (1988). *The New Teacher in School*. London: Her Majesty's Stationery Office.

United Kingdom, Department of Education and Science (1989). *The National Curriculum: From Policy to Practice*. London: Her Majesty's Stationery Office.

United Nations Educational and Scientific Organization (1972). *Unesco Source Book: Teaching School Physics*. Harmondsworth, Middlesex: Penguin Books.

United Nations Educational and Scientific Organization (1986). *The Integration of General and Technical and Vocational Education*, 3 Vols. Paris: UNESCO.

United Nations Educational and Scientific Organization (1988a). *The Evolution of Wastage in Primary Education in the World*. Paris: UNESCO, Office of Statistics.

United Nations Educational and Scientific Organization (1988b). *Development of Education in Europe: A Statistical Review* (ED-88 MINEDEUOPE/Ref. 2). Paris: UNESCO, Office of Statistics.

United Nations Educational and Scientific Organization (1989a). *Statistical Yearbook 1989*. Paris: UNESCO.

United Nations Educational and Scientific Organization (1989b). *Trends and Projections of Enrolment by Level of Education and Age* (ST/1989/WS/13). Paris: UNESCO, Office of Statistics.

United Nations Educational and Scientific Organization (1990a). *Statistical Yearbook 1990*. Paris: UNESCO.

United Nations Educational and Scientific Organization (UNESCO) (1990b). *Compendium of Statistics on Illiteracy, 1990 Edition*. Paris: UNESCO, Office of Statistics.

United States, Office of Education (1989). *Educational Evaluation and Reform Strategies in the United States of America* (A Report to OECD). Washington, D.C.: U.S. Office of Education.

United States, Office of Education (1990). *The Progress of Education in the United States of America, 1983-89*. Washington, D.C.: U.S. Government Printing Office.

United States, National Commission on Excellence in Education (1983). *A Nation At Risk: The Imperative for Educational Reform*. Washington, D.C.: U.S. Government Printing Office.

United States, National Commission on Excellence in Teacher Education (1985). *A Call for Change in Teacher Education*. Washington, D.C.: American Association of Colleges for Teacher Education.

Van Leeuwen, S. and Dronkers, J. (1991). Effects of Continuing Education: A Study on Adult Education, Social Inequality, and Labour Market Position. In: A.C. Tuijnman (Ed.), *Effectiveness Research in Continuing Education*. Oxford: Pergamon Press.

Verma, G.K. and Ashworth, B. (1986). *Ethnicity and Education: Achievement in British Schools*. Basingstoke: The Falmer Press.

Verne, E. (1987). Multicultural Education Policies: An Appraisal. In: *Multicultural Education*. Paris: OECD.

Vernon, Ph.E., Adamson, G. and Vernon, D.F. (1977). *The Psychology and Education of Gifted Children*. London: Methuen.

Vickers, M. (1987). *Microcomputers and Secondary Teaching. Implications for Teacher Education*. Report on an International Seminar arranged by the Scottish Education Department in cooperation with the OECD, Glasgow, October 12-15, 1987.

Vogel, J., Andersson, L-G., Davidsson, U., and Häll, L. (1988). *Inequality in Sweden: Trends and Current Situation. Living Conditions 1975-1985*. Stockholm: Statistics Sweden.

Von Friedeburg, L. (1989), *Bildungsreform in Deutschland: Geschichte und geschichtlicher Widerspruch*. Frankfurt: Suhrkamp.

Von Hentig, H. (1988). Eight Characteristics of a Humane School. *Western European Education*, Vol. 20, 6-25.

Von Hentig, H. (1990). Bilanz der Bildungsreform in der Bundesrepublik Deutschland. *Neu Sammlung,* Vol. 30 (3), 366-384.

Von Recum, H. (1985). The Financing and Future of the Educational System of West Germany. In: W. Tellmann (Ed.), *Handbuch Schule und Unterricht,* Vol 7 (2). Dusseldorf: Schwann-Verlage.

Von Weizsäcker, E. (1989). Die Umweltkrise: eine Herausforderung für unser Bildungssystem. In: *Umweltbildung in der EG*. Dokumentation einer internationalen Fachtagung am 23./24 Februar 1988 in Dortmund (pp. 28-38). Bonn: Bundesminister für Bildung und Wissenschaft.

Vonk, J.H.C. (1989). The Education and Training of Teachers in a Changing Europe. In: J.T. Vorbach and L.G.M. Prick (Eds), *Teacher Education: Research and Developments on Teacher Education in the Netherlands,* Vol. 5 (pp. 189-199). The Hague: SVO.

Wagner, A. (1987). *Social and Economic Aspects of Teaching: The Attractiveness of the Profession.* CERI working paper. Paris: OECD.

Walberg, H.J. (1989). Productive Teaching and Instruction: Assessing the Knowledge Base. *Phi Delta Kappan,* Vol. 71 (6), 470-478.

Wallace, W. (1990). *The Transformation of Western Europe.* London: Royal Institute of International Affairs.

Weiss, M., von Recum, H. and Doring, P.O. (1990). *Prospective Trends in the Socio-economic Context of Education in European Market Economy Countries.* Working document for UNESCO International Congress on Planning and Management of Educational Development, Mexico, 26-30 March 1990, German Institute for International Educational Research, Frankfurt am Main.

White, J. (1990). Educational Reform in Britain: Beyond the National Curriculum. *International Review of Education,* Vol. 36 (2), 131-143.

White, M. (1987). *The Japanese Educational Challenge: A Commitment to Children.* London: Collier Macmillan Publishers.

Wiliam, D. (1990). National Curriculum Assessment Arrangements: The Secretary of State's Proposals. *British Journal of Curriculum and Assessment,* Vol. 1 (1), 11-12.

Wilkins, D.A. (1976). *Notational Syllabuses.* Oxford: Oxford University Press.

Woodhall, M. (1985). Economics of Education. *International Encyclopedia of Education,* Vol. 3, 1546-1553.

World Bank (1989a). *Improving the Quality of Primary Education in Developing Countries.* Washington, D.C.: The World Bank.

World Bank (1989b). *Sub-Saharan Africa: From Crisis to Sustainable Growth.* Washington, D.C.: The World Bank.

Wyatt, D. and Dashwood, A. (1987). *The Substantive Law of the EEC,* Second Edition. London: Sweet and Maxwell.

Yarmolinsky, A. (1960). *The Recognition of Excellence.* New York: Free Press.

Young, M. (1958). *The Rise of the Meritocracy.* London: Thames and Hudson.

Zigler, E. and Valentine, J. (Eds) (1979). *Project Head Start: A Legacy of the War on Poverty*. New York: Free Press.

Zupnick, E. (1989). Europe, the United States and 1992: A Prelude to Trade Wars? (pp. 47-58). In: Beliën, H.M. (Ed.), *The United States and the European Community: Convergence or Conflict?* The Hague: Nijgh and Van Ditmar.

Index

Access, problems of, 54-57
Adult education, 101-105
　advantages, 102-103
　career development, and, 104-105
　cycle of accumulation, and, 103
　disadvantages, 102-103
　equity, and, 104
　expenditure in EFTA countries, 313
　home background, and, 102
　inequality, and, 103-104
　longitudinal studies, 104
Adult training programmes
　financing of, 308
Apple, M.
　equality of opportunity, on, 53
Assessment of students, 87-89

Basic education
　supply and demand. *See* Supply and demand of basic education
Belgium
　languages, 167
Bernstein, Basil
　immigrant minority, on, 205
Braddock, J.H.
　equality of opportunity, on, 52
Bruner, Jerome, 35
Bullivant, B.
　multiculturalism, on, 206-207
Bureaucratization, 74

Capelle, J.
　"languages" of knowledge, and, 236
Cebrian, Juan-Lewis
　Europe, on, 163-164
Cha, Yun-Kyung
　language teaching, on, 169-171
Chevènement, Pierre
　vocationalism, and, 242
China
　science education, 139
Coherence of schooling, 96-101
Coleman, James S.
　"corporate actors", on, 111-112
Comparative studies in education, 330
Comprehensive schools, 45-47
　arguments against, 46
　arguments in favour, 46
　pedagogical argument in favour, 47
　principles, 45
Conant, James B., 35

Continuing education
　access to, 60-61
Cultural minorities, 185-220
Cultural parochialism
　increase in, 317
Culture, teaching of, 206-209
　"colour prejudice", and, 208
　cultural distance, and, 209
　curriculum "packages", 207
　"ethnocentrism", and, 209
　immigrant minority, and, 206-209
　intermarriage, and, 209
　multiculturalist policies, 206-207
　xenophobia, and, 208

Decentralization, 62-64
Decision-making
　opinions on, 15
Decrementalism, 73
Demographic factors
　school policy, and, 5-7
Developing countries
　school education, 327-328
Displaced persons, 190
Docimilogie, 224-225

Eastern Europe
　migration from, 191
Ecology of education, 318
Economic and financial aspects of
　　　education, 289-314
　adult training programmes, 308
　decentralization reforms, 312
　economic competitiveness, 311
　firms, financing by, 312
　increase in costs, 311
　management efficiency, and, 311-312
　new sources of finance, 310
　objectives of education, and, 309
　prospects, 310-314
　research studies, 309
　rise in public expenditure, and, 310
　sources of finance for job training, 314
　surveys, 309
　tendency for increase in costs, 310
Economic competitiveness
　education as essential element in, 311
Economic factors
　school policy, and, 5-7
Economic growth
　contribution of education to, 292

Economic integration
    knock-on effects, 22-27
Economism, 34
Economy
    globalization, 316
Educational expenditures, 296-303
    change in prospects, 302
    decrease in, 302
    "demand for education", and, 297
    forecasts of manpower needs, 296-297
    "human capital model", 297
    increase in, 297-298
    "integrated planning of human
        resources", and, 296
    limits, 296-303
    Netherlands, 299
    "quality of education", and, 302
Educational policy
    major themes 1960-1985, 32-33
EFTA countries
    expenditure on adult education and
        job training, 313
Elitism
    revival of, 57
Elvin, Lionel, 36
Emerging issues, 77-106
Environmental education, 148-150
Equality of opportunity, 38, 50-54
ERASMUS
    basis of, 23
Ethnic conflicts
    increase in, 317
Ethnicity, 185-186
European cultures, 4
European identity, 2-5
European school education, 303-304
    choices as to priorities of spending,
        305-306
    comparative study on expenditures, 306
    costs of, 303-304
    cost per pupil, 303
    cost variations, 303-304
    development of cost studies, 303-304
    financing of, 308-310
    new sources of finance, 308
    optimal size of classes, 305
    participation of families in financing,
        309
    rising costs of schooling, 304-308
    teacher/pupil ratio, 305
    teacher studies, 304
    unit costs, 306
    value for money, and, 308
European schooling
    quality of, 66-73
European school setting, 316-319

EURYDICE, 180-182
Examinations
    Sweden, 231-234
Eysenck, Hans
    school performance, on, 214

Family
    changing role and structure, 111-119
    role of, 318
Feminization of teaching, 269-270
Financial issues, 296
    limit of national expenditure, 296
    rising costs, 296
    sources other than public purse, 296
Formal education
    economy, and, 290
Fourastie, Jean
    economic growth, on, 290
France
    Diwan movement, 187
    immigration, 199-200
    Loi Savary, 235
    national assessment of student
        performance, 90
Full-time education
    participation rates, 20

German Democratic Republic
    gifted pupils, 59-60
Germany
    immigration, 198-199
    science education, 139
    Turkish children in schools, 215
Gifted pupils, 58-60
    acceleration, and, 58
    German Democratic Republic, 59-60
    Soviet Union, 58-59
Goal conflicts, 81-85
    economic and social needs, and, 83
    economic forces, and, 82
    equality of opportunity, 82
    functions of equality and quality, 83
    reassessment, need for, 84
    "revolution of rising expectations", and,
        84
Gorbachev, Mikhail
    "European house" proposal, 41
Governance and management of school
        education, 85-87
    centralized, 85
    national divergences, 85
    "school plan", 86
Grading of students, 87-89
    attitude of teachers, 88
    functions, 87-88

Halsey, A.H.
  social engineering, on, 51
Hartog, R.J. and Rhodes, E.C.
  examinations, on, 238
Helsinki Meeting, 249
Higher education, 221-247
  age-specific enrolment rates, 234
  coping with rising demand for, 240-241
  decline in links with secondary
    education, 243
  diversification, 240-241
  efficiency indicators, and, 241
  EC countries, in, 245
  flow of students, and, 240
  links with secondary school education,
    243-244
  national differences in admission
    procedures, 228-231
      assessment, 228-229
      baccalaureat, 230
      lottery system, 229
      positive discrimination, 229
      scholastic aptitude tests, 228
      state examination, 229
      uniform national examination, 228
      work experience, 228
  role of school in preparing for, 221-247
  selection of students, 240-241
  upsurge in technical and vocational
    education, and, 242-243
Holidays, 273-274
Home and school relationships, 107-136
  activities, 117
  adult households, 115
  agents competing with family, 115
  change from agricultural work, 114-115
  change in composition of labour force,
    126-127
  changing role and structure of family,
    111-119
  changing role of school, 119-121
  "client society", 115-116
  common activities, 123
  consequences for school education,
    123-124
  crucial problems, 133-134
  deprived families, 118
  disadvantaged families, 118
  discipline, maintenance of, 119
  discontinuities, 121-122
  discrepancies, 121-122
  divorce rate, 109
  "ecological" interventions, 134
  economic growth, and, 120-121
  effect of social status of home, 118-119
  family as "legal anachronism", 117
  "F-connections", 124-125
  fertility rate, 108-109
  "fragmentation and discontinuities", 134
  fragmentation of adult contacts, 116
  functional participation, 132
  indicators of household conditions,
    Sweden 1900-1985, 114
  industrialization, 112
  intervention programmes, 130-132
  "latch-key children", 123
  male life expectancy, 109
  marriage rate, 109
  measures of social capital, 126
  modal age of mother at birth of first
    child, 109
  "new underclass", and, 124
  "old" and "new" educational
    underclass, 126-130
  poverty, and, 134
  poverty incidence, 129
  poverty rates, 128
  "preparation" for adulthood, 132
  residential segregation, and, 122
  school and "social capital", 124-126
  school as substitute for family, 133-134
  school intervention, 130-132
  single-parent families, 110-111
  size of household, 110
  size of school, and, 123-134
  statistics on changing household
    structures, 108-111
  television, and, 135
  truancy, 123
  two-generation family, 114
  unionization, 124
  urbanization, 112
  withdrawal of parents, 123
  working women, 112-113
Household structures
  statistics, 108-111

IEA comparative studies, 67
Illiteracy, 96-98
  estimates and projections, 97
Immigrant minorities, 185-220
  Afro-Caribbeans, 211-212
  banning of books, 212
  culture gaps, 211-213
  culture, teaching of, 206-209
  developing countries, from, 190
  displaced persons, 190
  Eastern Europe, 191
  educational problems, 192-197
  France, 199-200
  Germany, 198-199
  language teaching, 201-206

Index

majority school system, and, 200
overall picture, 189-192
refugees, 190
religion, 210-211
repatriation, and, 197
school performance, and, 213-215
    social factors, 213
    teacher expectation, 214
    West Indians, 214
schooling and employment, 216-217
teacher education, and, 217-218
    ways of knitting with host community, 212
Immigration, 320
    bi- or multilingual teaching, and, 320
Incrementalism, 73
In-service training, 280-282
    complementary, 280
    computers, 280
    supplementary, 280
Inequalities in income, 57
Internationalization
    promotion of, 16
    trend towards, 316-317
Iron Curtain
    removal, 5
Islam, 210-211

Jenkins, Roy
    multiculturalism, on, 208
Job training
    expenditure in EFTA countries, 314
    sources of finance, 314
Johnson, President Lyndon, 130

Keeves, J.P.
    science education, on, 70
Kellaghan, Thomas
    Task Force, 131
Kindergarten, 54-55
    statistics, 55
Koran schools, 210
Kungälv Conference, 36

Labour market
    vocational qualifications, and, 292
Lambert, W.E.
    language, on, 205
Language policy
    harmonization, and, 24-25
Language teaching, 163-184, 201-206
    approaches, 176-177
    Belgium, 167
    commercial and industrial value, 179
    development, 168-172
    early teaching, 178

    English, 171
    EURYDICE, 180-182
    Germany, 167
    immigrant minority, and, 201
    bilingualism, 205-206
    bussing, 201
    cost, 205
    Kindergartens, 203
    language of origin, 203
    "linguistic deficit", 205
    Netherlands, 201-202
    proportion, 201
    typology for teaching majority language, 202-203
    international competition, and, 164-165
    language and literature as expressions of culture, 163-168
    Latin, 168-169
    LINGUA, 182-183
    linguistic minorities, 172-175
        numerical extent, 174
    methods, 176-177
    Netherlands, 168
    policy, 180-183
    practice, 180-183
    secondary school statistics, 170
    Single European Market, and, 166
    small countries, 175
    "small" languages, 172-175
    Spain, 166-167
    syllabuses, 177-180
    variety, 180
Lifelong education, 101-105
    strategy for, 291
LINGUA, 182-183
Linguistic minorities, 185-220
    alternative national language, 187
    Breton, 187-188
    Denmark, 188
    diversity, 186-187
    language as medium of instruction, 186-188
    "optional extra" to curriculum, 187
    Welsh, 187
Lomé Agreements, 328

Malta "Summit" 1989, 8
Mann, Horace
    elementary schools, on, 120
Mappa, Sophia
    immigrant minority, on, 194
Martinson, Henry, 116-117
Mathematics achievement
    terminal grade of upper secondary schooling, at, 68

Mathematics education, 152
Meritocratic tendencies, 50-54
Minorities, 185-220
    foreign school enrolment, 195
    general policies, 192-197
    integration, 196-197
    interculturalism, 196-197
    multiculturalism, 196
    national, 186-189
    percentage increase or decrease in number in school education, 195
    percentages, 192
    school enrolments, 194
Monnet, Jean
    Europe, on, 314
Monory, Rene
    vocationalism, and, 242
Multiculturalism, 196
    aim of, 207
Multilingualism, 327

National assessment programmes, 89-93
    aims, 89, 91
    criticisms, 93-95
    democracy, and, 94
    dilemmas, 93-95
    division between public and private schooling, 95
    European approach, 95
    France, 90
    "high tech" revolution, and, 92
    long term trend, 92
    planned evaluation, 91-92
    United States, 90
National school policy, 13-16
Nationalism, 3
    orientation of education, and, 317
"Nations at risk", 64-66
Netherlands
    compulsory school age, 321
    educational expenditure, 299
    literature, 168
    recommendations of Dutch Advisory Council for Education for education of minorities, 219-220
New Europe
    repercussions on school systems, 1-29
Nordic Countries
    costs and quality of child care provision, 307
North–south perspective, 327-329
    increasing gap in living standards, 328
    knowledge of development problems, 328-329
    public opinion, and, 328-329
Nursery schools, 321

Pacific Rim
    trade competition, and, 6-7
Pan-European co-operation, 7-10
Parental choice, 62-64
Picht, Georg, 254
Pieron, H.
    "Docimilogie", 224-225
Pluralism
    management of, 330-331
Politicization, 74
Poverty, 325
Preschool education, 54-56
Primary schools
    percentage with lower average scores in science than lowest scoring school in Japan, 69
Professionalism, 61-62
Provincialism
    increase in, 317

Quality of schooling, 96-101
    curricula, and, 99-100
    harmonization, and, 100
    illiteracy, 96-98
    meaning, 96
    school structures, and, 99
    science and technology teaching, 98-99
    teachers, and, 99

Reagan, President
    National Commission on Excellence, 65-66
Recruitment of teachers, 249-288, 325-326
    career of would-be recruit, 286
    competition from other occupations, 255-256
    demographic factors influencing, 252
    enlarging pool, 258-260
    enrolment statistics, 255
    factors influencing decision to become teacher, 267-275
        career prospects, 274-275
        feminization, 269-270
        holidays, 273-274
        job satisfaction, 275
        role of teacher, 271-272
        status, 267-269
        teacher–pupil ratio, 274
        teaching load, 272-273
    limitations of educational planning, 260-261
    "mature" graduates, 259
    mobility, and, 250
    oversupply, 256-258
    population estimates, and, 252
    quality, 261-267

salaries, and, 261-267
shortages, 251
supply and demand, problems of, 253-254
surpluses, 251
teacher:pupil ratio, 274
vocational education, and, 254
Recurrent education strategy, 101-102
Refugees, 190
Religion, 210-211
  equality of sexes, and, 210-211
  Islam, 210-211
  Koran schools, 210
Religious education, 210-211
  state-aided Moslem schools, and, 211
Rex and Tomlinson
  school performance, on, 214

School education
  "baby boom", and, 40
  basic concerns, 31-75
  challenges, 315-331
  criticisms, summary of, 41-43
  curriculum, 324
  decentralization, 321
  developing countries, 327-328
  divergence of models and ideas, 331
  elite, creation of, 324
  extension of autonomy, 321
  framework, 319-322
  Germany, 40-41
  goals, 322-323
    accountability, 323
    clarification, need for, 323
    increased opportunities, 323
    nation-state, and, 322
  "grassroot" participation, 325
  interaction of Eastern and Western traditions, need for, 329
  leaving age, 321-322
  limited burden of tasks, 324
  low achievers, 324-325
  means, 324-327
  opportunities, 315-331
  perceptions of role in 1960's, 33-36
  principles of pedagogy, and, 42
  questioning, 39-41
  structural reforms, 330
  technological competence, 320-321
  technology, 326-327
  trends, 31-75
School leaving examinations, 221-247
  efforts towards cross-national convergence, 244-247
    "forced", 246
    "natural", 246
  examining methods, 238-239
  insufficiencies of, 234-235
  maintenance of standards, and, 227
  oral examination appended to, 239
  pass rates as check for accountability, 227
  pedagogical problems, 238-239
  procedures, 238-239
  standardization in pass rates, 239
  syllabuses, and, 236-238
    expansion of knowledge 238
    "specialization in depth", 237
  upsurge in technical and vocational education, and, 242-243
  uses of, 224-227
  variable pass rates, 225-226
School marks, 88
School policy
  diversity, and, 16-22
  European dimension, 318
  harmonization, and, 16-22
School, role of, 78-81
  complementary, 79
  "European dimension", 80
  instrument of social policy, 79
  "moulder", 79
  protective, 79
  sub-society, as, 79
  traditional, 78-79
Schultz, Theodore, 35
Science achievement
  lower secondary school level, at, 70
  mean scale scores, 14 year olds, 72
Science education, 137-161
  associations, 144
  balance of sciences, 146-148
  beginning of, 138-139
  change post Second World War, 139-143
  China, 139
  computers, 153-154
  design, 155
  enrolment, 158-160
  environmental problems, 148-150
  examination system, 152
  experimental work, 142
  "gender problem", 146-148
  Germany, 139
  influence of American educators, 141
  information culture, 152-154
  information technology, 152-154
  "integrated science", 146
  international cooperation in development of, 143-144
  life sciences, 143
  mathematics, 152
  Nature, 145

Index

Nuffield Science Projects, 142
participation, 158
physics, 143
primary science, 144-145
search for understanding, 145
social implications, 150-151
society, and, 148-152
Soviet Union, 141-142
teaching methods, 155-157
technology, 155
Science teaching, 326
Secondary education
accountability, and, 44-45
Central and Eastern Europe, 49-50
common school, move towards, 49
constraints on, 43-45
different European systems, 47
economic recession in 1970s, and, 43-44
main trends in reform, 47-50
participation, and, 45
reforms in Britain, 48
reforms in France, 48
reforms in West Germany, 48-49
Swedish model, 47
"unit costs", and, 44
Secondary school education
links with higher education, 243-244
Sexism, 50-51
Single-parent families, 110-111
Single European Act, 10-13
implications, 10-13
national sovereignty, and, 11
objective, 12-13
syllabuses, and, 13
Social demand, 54-57
Social engineering, 51
Soviet Union
developments in education, 26-27
gifted pupils, 58-59
science education, 141-142
Snow, C.P., 140
Spain
languages, 166-167
urbanization, 319-320
"Special education", 56
Specialist schools, 58-60
Sputnik, 34
State
role of, 316
Status of teachers, 267-269
Supply and demand of basic education, 293-295
demographic development, and, 295
economic crisis 1975-85, and, 290-291, 294-295
expansion of school system, and, 294
matching of, 293
public opinion, and, 296
social demand, and, 294
teacher recruitment, and, 295
utility of notions of, 293-294
Sweden
examinations, 231-234
demand for higher education, and, 233
prolongation of vocationally oriented studies, and, 232-233
indicators of household conditions 1900-1985, 114
Syllabuses
school leaving examinations, and, 236-238

Taft
reciprocal role relationships, on, 196
Teacher–pupil ratio, 274
effect of reduction, 305
Teacher recruitment
pace of, 295
Teacher training
immigrant minority, and, 217-218
Teacher training programmes, 282-284
Teaching methods, 318
Teachers
comparative income position, 263
decline in professional status, 326
deprofessionalization, 326
recruitment. *See* Recruitment of teachers
training. *See* Training of teachers
Technical education
upsurge in, 242-243
Technological cooperation
extension, 25-26
intensification, 25-26
Television
home and school relationships, and, 135
Trade competition, 6-7
Training of teachers, 249-288, 325-326
after selection, 276
appointment, 276-278
elements of European model for, 276-278
in-service. *See* In-service training
programmes, 282-284
selection, 276
selection process, 278
specialization, 285-286
trainers, 279

Unemployment
    formal educational attainment, and, 292-293
United Kingdom
    compulsory school age, 321
United States
    "magnet" schools, 64
    Rockefeller Report, 28-29
Upper secondary schooling
    goals of, 222-224
Urbanization, 319-320

Vocational education
    recruitment of teachers, and, 254
    upsurge in, 242-243

Vocationalism, 61-62
Von Hentig, Hartmut, 251
Vonk, J.H.C.
    teacher training, on, 250

Wallace, William
    Europe, on, 164-166
Wastage, 325
Welfarism, 34
Williamsburg Conference 1967, 37
World Crisis in Education, 37-39
World economic crisis 1975-85, 290-291, 294-295

17

WP 0813233 X